PROCESS EQUIPMENT SERIES
VOLUME 3

Air Movement and Vacuum Devices

Edited by
Mahesh V. Bhatia, P.E.
Paul N. Cheremisinoff, P.E.

Contributors to Volume 3

M. P. Boyce
A. M. Dave
R. G. Mote
R. F. Neerken
K. S. Panesar
G. Szczepankiewicz

Technomic Publishing Company, Inc.
265 Post Road West, Westport, CT 06880, USA

PROCESS EQUIPMENT SERIES
VOLUME 3

Air Movement and Vacuum Devices

265 Post Road West, Westport, CT 06880

Printed in U. S.A.
ISBN 87762-291-4
LCC 7963114

FOREWORD

This is the third volume in the *Process Equipment Series* which presents useful information relating to equipment and devices used in the chemical and related process industries. The editorial material has been prepared and written by specialists in their respective fields.

The equipment described in this series is used by more than 10,000 chemical manufacturing and processing firms employing more than 1.1 million Americans, with sales of approximately 100-billion dollars. Chemical industry products represent 7.5 percent of the U.S. Gross National Product (GNP) and enjoy an annual growth rate of nearly twice that of the overall GNP.

Equipment, apparatus and devices employed by the process industries are so numerous and varied that the problem was to reduce such information to a moderate and practical size. The editors have given preference to equipment that has general application, and diverse uses in the process industries. The presentation has been divided among a series of volumes.

Material presented by authors is primarily written in general technical language and the text liberally supplemented with drawings and photographs. The intent has been to provide sufficient theoretical matter to the reader with a satisfactory understanding of the equipment or devices covered. Equipment discussed is of the type that can be purchased by contrast to in-house-designed or constructed apparatus.

The editors express their sincere thanks to the experts who contributed to these collected volumes.

Mahesh V. Bhatia
Paul N. Cheremisinoff

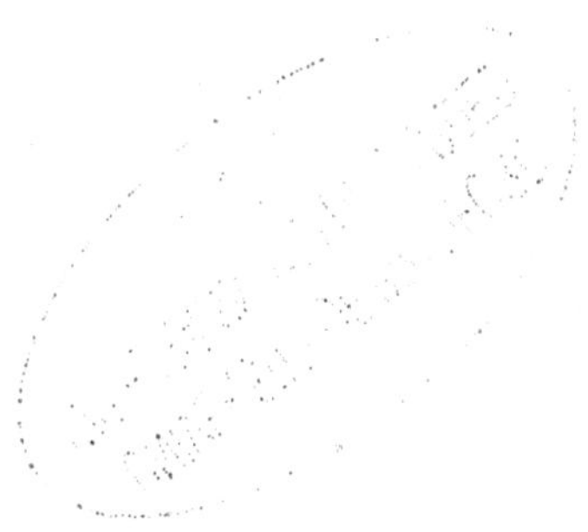

Mention of trade names or commercial products does not constitute endorsement or recommendation for use.

TABLE OF CONTENTS

PAGE

CHAPTER 1

FANS AND BLOWERS

K. S. PANESAR
Houston, TX

INTRODUCTION

Pumps, compressors, blowers and fans all belong to the same family of machines called TURBOMACHINERY or ROTATING EQUIPMENT. Pumps can handle only incompressible fluids viz. liquids while compressors, blowers and fans, on the other hand, can handle only compressible fluids like air and other gases. These machines can be damaged very easily if air or gases are pumped through centrifugal pumps or water etc. is pumped through blowers or fans. Most of this discussion will be limited to the centrifugal and axial machines, which are also called the constant pressure machines as compared with the constant volume machines which are the positive displacement machines. The performance characteristics of both are shown in Figure 1.1.

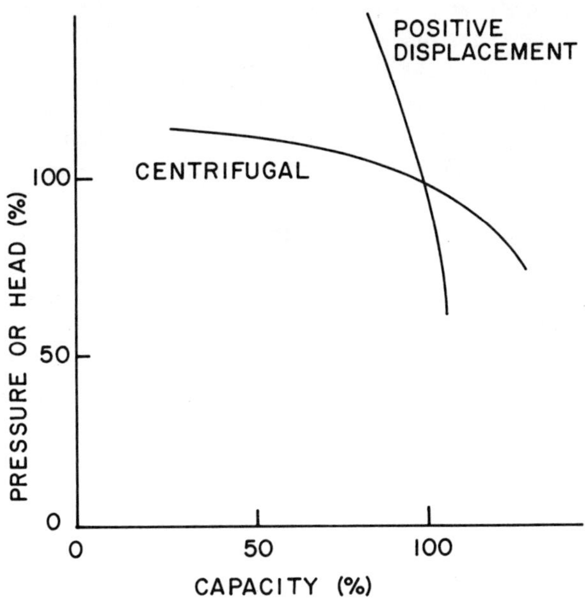

Figure 1.1. Pressure-volume diagram for the centrifugal and the positive displacement types of machines. Capacity is usually specified in terms of the inlet CFM (at the inlet conditions) and the head in feet or pressure in PSI for compressors and blowers. For fans, however, the pressure is usually specified in inches of water gage.

Most people, other than those involved in the design or application of the rotating equipment, get confused in the definitions of fan, blower or a compressor, and use these terms interchangeably. All these machines can handle fluid flows ranging from a few hundred cubic feet per minute to several million cubic feet per minute. The main distinction lies in the amount of pressure generated by each class of machine. Fans are supposed to have a pressure range of from a fraction of an inch of water gage to about 50 inches of water gage, which is approximately 2 Psi. Blowers, on the other hand, develop pressures from 2 Psi to about 15 Psi and pressures above 15 Psi fall in the category of compressors. Compressors can develop pressures up to several thousand pounds per square inch. Of course, extremely high pressures are developed in several stages (wheels) or sometimes in two or three cases, each case having several stages.

FUNDAMENTALS

Most centrifugal machines have a housing with an inlet and an outlet. Inside the housing is a wheel or an impeller which rotates and imparts kinetic energy to the fluid. The fluid comes in at a pressure, P_1, and leaves the housing at a higher pressure, P_2. These machines will, therefore, always generate a constant pressure differential when operating at the same flow.

Theoretically, the pressure-volume line is supposed to be a straight line, when the discharge angle on the blades is 90 degrees. The ideal pressure-volume lines for discharge angle less than and greater than 90 degrees are as shown in Figure 1.2. In actual practice, however, this is never the case. They have a curvature to them as already shown in Figure 1.1. The reason for this is that there are certain losses within the housing: – disk friction (wheel friction), blade inefficiency, circulation within the blades, etc.

The kinetic energy is generated by the rotary motion of the impeller and is imparted to the fluid moving through the machine. Each machine, in fact, each impeller is designed to produce a certain pressure rise or (sometimes called "Head") at a given capacity with a minimum loss or, in other words, with a maximum efficiency, as shown in Figure 1.3.

In order to understand how the head is generated by an impeller, let us look at Figure 1.4. Figure 1.4a shows a typical sectional elevation view of hydraulic path of an impeller, whereas Figure 1.4b shows the same impeller in plan view with inlet and discharge velocity triangles.

Let us assume that in time 'dt,' there is a mass 'dm' of very thin layer of fluid (air, gas or liquid) leaving the impeller and at the same time an equal amount of mass of fluid enters the impeller. This change in moment of momentum is equal to moment of all the external forces imposed on the fluid contained between the two blades. Let T denote the moment of external forces. Then T is given by the following equation:

$$T = dm/dt\,(r_2 c_2 \cos\alpha_2 - r_1 c_1 \cos\alpha 1) \qquad (1)$$

and when the term (dm/dt) is applied to all the fluid in all the blades, it becomes $Q\rho/g$; where

Q = the flow through the machine in cubic feet per second;
ρ = the specific weight in pounds per cubic feet;
r_2 = the outer radius of the impeller;
r_1 = the inner radius of the impeller;
c_2 = the absolute velocity of the fluid at the discharge;
c_1 = the absolute velocity of the fluid at the inlet;
$\alpha_{2,1}$ = the angle between the absolute and the peripheral velocities at the exit and the inlet, respectively;
u_2 = the peripheral velocity at the exit;
u_1 = the peripheral velocity at the inlet;

Now, substituting for (dm/dt), equation number 1 becomes:

$$T = Q(\rho/g)\ (r_2 c_2 \cos\alpha_2 - r_1 c_1 \cos\alpha_1) \tag{2}$$

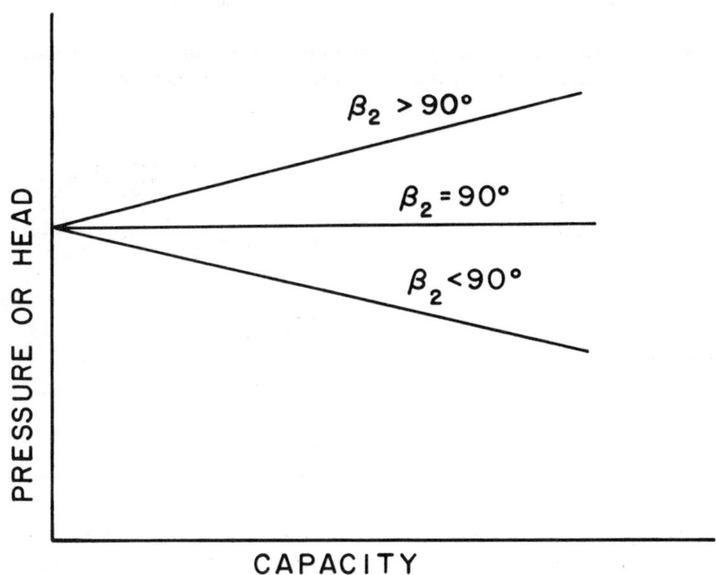

Figure 1.2. This figure shows the theoretical pressure-volume curves for the three different type of blade discharge angles.

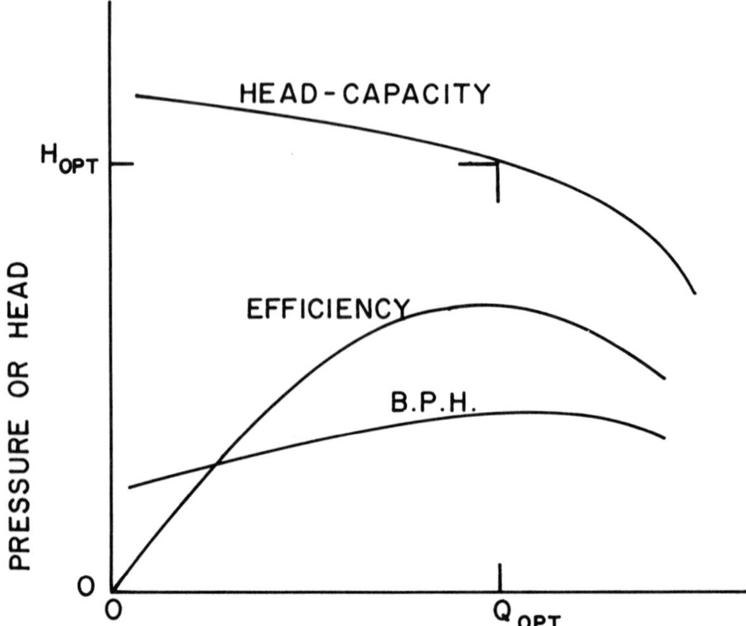

Figure 1.3. This figure shows a typical head-capacity curve for a centrifugal machine with maximum efficiency at an optimum flow.

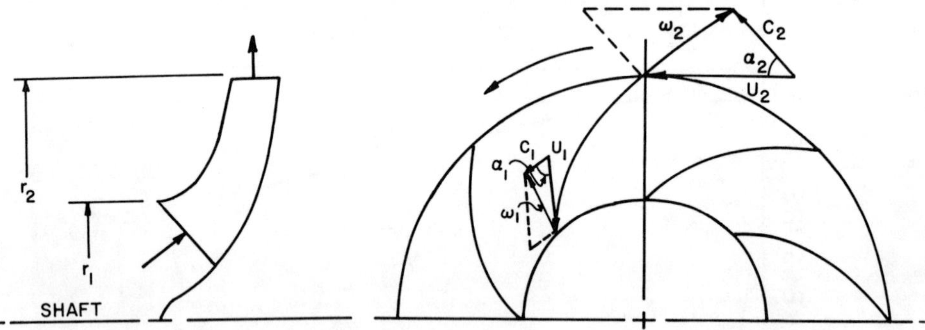

Figure 1.4. This figure shows a typical layout of the impeller. Fig. 1.4a shows the elevational view of the impeller and Fig. 1.4b is the plan view of the blades layout. The inlet and the outlet velocity triangles are shown.

Multiply both sides by w the angular velocity. Then, T•w equals power denoted by P, and substitute: $u_2 = w \cdot r_2$; $u_1 = w \cdot r_1$; and $c_{u2} = c_2 \cos\alpha_2$ $c_{u1} = c_1 \cos\alpha_1$ Then equation (2) becomes:

$$P = Q(\rho/g)\,(u_2\,c_{u2} - u_1\,c_{u1}) \tag{3}$$

Since $P = Q\,\rho\,H_i$ where H_i = the ideal Head, we get:

$$H_i = (u_2 c_{u2} - u_1 c_{u1})\,/\,g \tag{4}$$

No losses are considered in deriving the above equation and is called the Euler's Head or theoretical head equation.

The actual process that a gas or fluid undergoes in an impeller is approximated by the polytropic process, which stated mathematically is as follows:

$$P\,V^n = \text{constant} \tag{5}$$

Using thermodynamic equations (refer to any text book on thermodynamics), we arrive at the polytropic head equation given below:

$$H_p = \frac{Z_{av}\,R\,T_1 \cdot n}{(n-1)} \left[(P_2/P_1)^{\frac{n-1}{n}} - 1 \right] \tag{6}$$

where:

H_p = Polytropic head in ft-lbf/lbm
R = Gas constant in ft-lbf/lbm-°R
n = polytropic exponent,sdimensionless
Z_{av} = Average compressibility factor, $(Z_1 + Z_2)/2$
T_1 = Absolute inlet temperature, in °R
P_2 = Absolute discharge pressure in Psia
P_1 = Absolute suction pressure in Psia

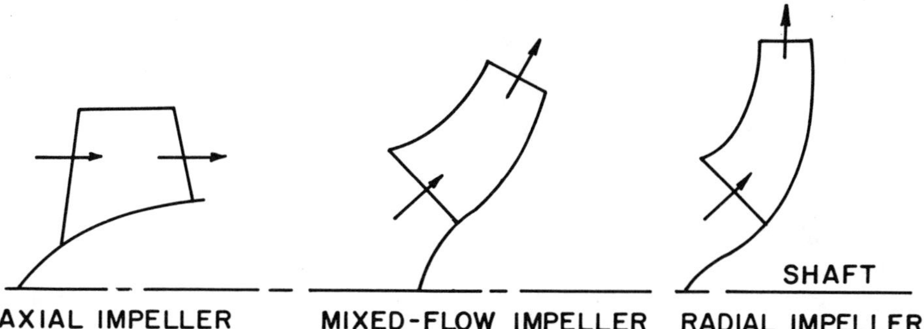

Figure 1.5. Showing the three different types of the impellers used in the pumps, compressors blowers and the fan industry.

Horsepower Calculations

Once you calculate the polytropic head, it is relatively easy to calculate the theoretical or actual brake horsepower as shown:

$$THP = w\,H_p/33000 \tag{7}$$

$$BHP = w\,H_p/33000 \times \eta_p \tag{8}$$

where:

THP and BHP are theoretical and brake horsepowers respectively.
w = flow in lbs per minute, and
η_p = polytropic efficiency in decimal.

There is another way of calculating the theoretical or actual horsepower when you do not know or do not have to know the polytropic head. The calculation of head equation is very useful in centrifugal and axial machinery because from this you can determine the number of stages or wheels required to get the desired pressure or head. For reciprocating compressors, on the other hand, you do not need to calculate the head, There compression is achieved in a cylinder, and if the pressure ratios are quite high you add additional cylinders until the desired pressure ratio is achieved. Thus to calculate the horsepower, without the head, you must know or calculate the inlet volume flow, called the ICFM (inlet cubic feet per minute) or also sometimes called the ACFM (actual cubic feet per minute). (They both mean the same). This is the flow in the machine at the inlet temperature and pressure. If the process flow is given only in terms of the pounds per hour, pounds per minute or moles per hour, use the following equations to get the ICFM and then finally the horsepower:

w = (Moles/hr) $\times$ (M.W.)/60 lbs per minute

SCFM = (Moles/hr) $\times$ (379/60)

(Note: Each gas occupies the 379 cubic feet volume at standard conditions. (9)

SCFM = (w/M.W.) $\times$ 379

ICFM = SCFM $\times$ $(14.7/P_1) \times (T_1/520)$

where M.W. is the molecular weight, and 14.7 and 520°R are the standard conditions. The blower and the fan manufacturers, however, use 68°F as the standard temperature, instead of the 60°F. Let's call V_1 as the inlet CFM or the ACFM, the horsepower is given as follows:

$$\text{THP} = \frac{P_1 V_1}{229.17} (n/n - 1) \left[(P_2/P_1)^{\frac{n-1}{n}} - 1 \right] \tag{10}$$

Just divide the THP by the polytropic efficiency to get the BHP. In fans, however, the horsepower is calculated in a little bit different form, which is shown below:

$$\text{BHP} = .000157 \times \text{ACFM} \times \text{S.P./S.E.} \tag{11}$$

or

$$\text{BHP} = \text{ACFM} \times \text{S.P.}/6356 \times \text{S.E.} \tag{12}$$

In fans the discharge temperature is hardly ever a problem because they are a very low pressure ratio machine. In blowers, and compressors the discharge temperature can be very high, and therefore intercooling (between stages) or aftercooling (i.e. cooling of the gas after it leaves the machine housing) is usually required. To calculate the discharge temperature, the following equations are used:

$$T_2 = T_1 (P_2/P_1)^{\frac{n-1}{n}} \tag{13}$$

As is quite evident from the above equation, the discharge temperature varies linearly with the inlet temperature (T_1), but more so with the pressure ratio (P_2/P_1). That's why intercoolers and aftercoolers are used in the multistage centrifugal compressors, or reciprocating compressors. There is another way of calcualting the discharge temperature:

$$T_2 = T_1 + H_p / [Z_{av} R (n/n - 1)] \tag{14}$$

Impeller Design

Basically there are three different designs used in the centrifugal machines, as shown in Figure 1.5. In the radial design, the fluid enters the impeller along the shaft, but leaves in a direction perpendicular to the shaft. In the mixed flow design, the fluid enters along the shaft as before, but leaves the impeller at some angle. In the axial design, however, the fluid enters and leaves parallel to the shaft. The radial blades are used in low specific speed range, mixed flow blades are used in the medium range and axial impellers or propeller design is used in high specific speed range.

Specific speed is a kind of a dimensionless number used in the industry to help engineers and designers get an idea of the geometry of the machine they have to design. Machines having same specific speed are geometrically similar and therefore you can predict the performance of the new machine. Specific speed, N_s, is defined as the speed which will produce one foot of head at a flow of one CFM, and mathematically stated it is:

$$N_s = N. (Q)^{.5}/H^{.75} \tag{15}$$

where:

N_s = specific speed
Q = flow in cubic feet per second for compressors and GPM for pumps.
N = machine rotative speed in RPM
H = feet of head

The above equation is applicable to pumps, compressors and blowers but for fans the equation may be rewritten and defined as follows:

$$N_s = N (Q)^{.5}/P_s^{.75} \tag{16}$$

where:

Q = flow in ICFM (inlet cubic feet per minute)
and P_s = Static Presure in inches of water gage.

For fans, the specific speed is defined as the speed which will deliver one cubic feet per minute of flow at one inch of water gage.

Specific speed, N_s, is related with another dimensionless number called the Specific Diameter, D_s, which is defined as:

$$D_s = D\ H^{1/4}/Q^{1/2} \tag{17}$$

where:

D = Diameter of the wheel in feet.

The above equation is applicable to pumps, compressors and blowers. The equation for fans, however, is a little different, as shown below:

$$D_s = D\ P_s^{1/4}/Q^{1/2} \tag{18}$$

where:

D = Diameter of the wheel in inches
P_s = Static pressure in inches of water gage
Q = Flow in Cubic feet per minute

The relationship between the specific speed and the specific diameter was developed and presented by Dr. O.E. Balje, and is shown in Figure 1.6. The higher the specific speed, the higher the flow and smaller the pressure or head and vice versa. Figure 1.6a covers a wide range of turbomachinery viz. centrifugal and positive displacement pumps and compresors. Figure 1.6b, however, covers just low

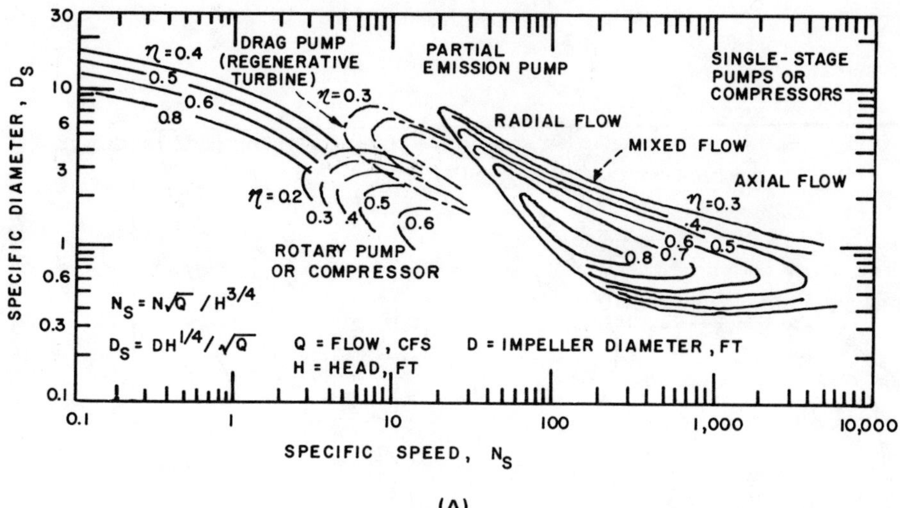

Figure 1.6a. $N_s - D_s$ *diagram for single stage pumps and compressors.*

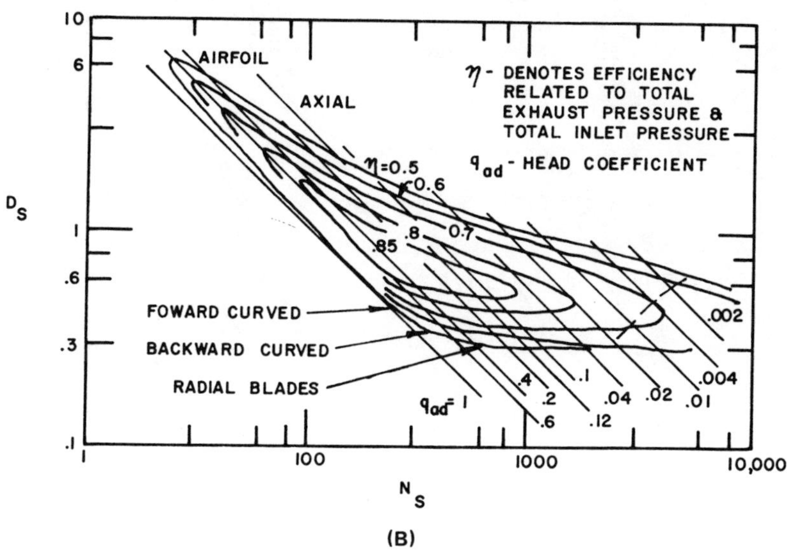

Figure 1.6b. Showing $N_s - D_s$ *curves for single stage pumps and low pressure blowers and fans.*

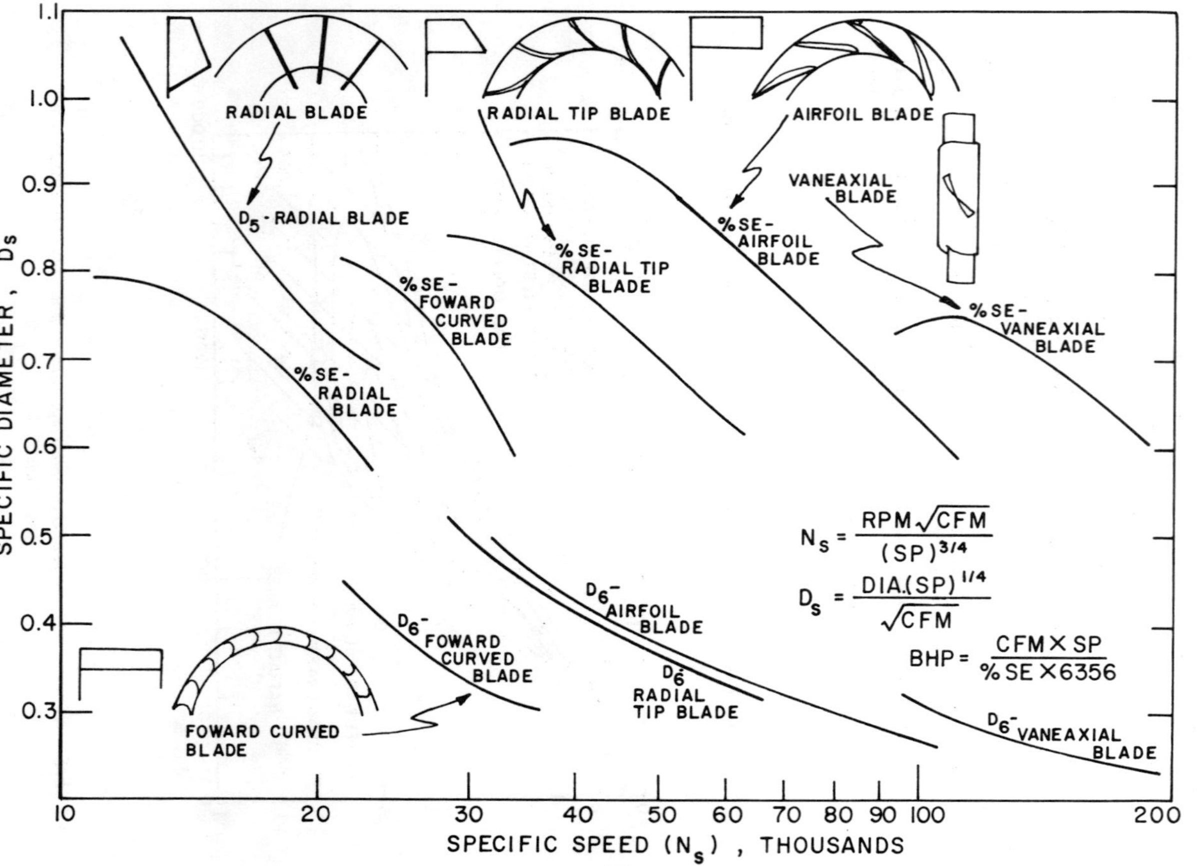

Figure 1.6c. Showing the $N_s - D_s$ diagram specially adapted for the fan industry. (Courtesy of American Standard)

pressure, single stage, pumps and blowers or fans. Figure 1.6c is a modified version of the Balje's N_s – D_s, diagram developed by the fan industry for their convenience.

The following examples will illustrate the use of Figures 1.6b and 1.6c for fans.

EXAMPLE: It is requirea to have 54,000 inlet CFM at 30 inches of WG. Find the optimum wheel diameter for the best efficiency for an airfoil type blade. Also find the optimum speed.

SOLUTION: A (Using curves in Figure 1.6b)

$$Q = 54{,}000 \text{ CFM}$$
$$= 900 \quad Ft^3/Sec$$
$$H = 30 \cdot (69.3)^*$$
$$= 2079 \quad Ft.$$

For airfoil blade, for $N_s = 200$, the efficiency is the highest you can get. This gives $D_s = .82$ and using equation (17), we get:

$$D = D_s \cdot \sqrt{Q} / H^{1/4}$$
$$= 0.82 \times 30 / 6.75$$
$$= 3.644 \text{ Ft. or } 43.7 \text{ inches.}$$

To find the speed, use equation (15) which gives,

$$N = N_s \cdot H^{.75} / \sqrt{Q}$$
$$= 200 \times 307.9 / 30$$
$$= 2052.6 \text{ RPM}$$

SOLUTION: B (Using Figure 1.6c)

$$N_s = N \cdot \sqrt{Q} / P_s^{.75}$$
$$= 2052.6 \times 232.38 / 12.82$$
$$= 37206$$

Now using Figure 1.6c,sfor airfoil blades, we get a value of 0.44 for D_s and the wheel diameter can be found as follows:

$$D = D_s \cdot \sqrt{Q} / P_s^{1/4}$$
$$= 0.442 \times 232.8 / 2.35$$
$$= 43.6 \text{ inches}$$

*At sea level under normal temperature and pressure, the atmospheric pressure of about 14.7 Psia will hold a water column of approximately 34 feet or a mercury column of about 29.92 inches. In other words, the column height is inversely proportional to the densities of the fluids. Expressed mathematically,

$$h_1 \, d_1 = h_2 \, d_2$$

where:

h_1 = column height of air in 'Feet.'

d_1 = density of air in pounds per cubic feet.

h_2 = column height of water in 'Inches.'

d_2 = density of water in pounds per cubic feet.

Substituting the actual values of the densities of each fluid at the same temperature, we get:

$$\begin{aligned} h_1 &= h_2 \quad (d_2 \,/\, d_1) \\ &= h_2 \quad (62.4 \,/\, .075 \times 12) \\ &= 69.3 \end{aligned}$$

Thus in order to convert inches of water to feet of Head, you multiply by 69.3.

FANS AND BLOWERS

We have discussed the basic theory of the centrifugal machines and their design. Now the discussion will be limited mostly to fans, blowers and exhausters. Fans are used for a wide variety of application, in the petrochemical and chemical industry, steel mills, paper mills, heating-ventilating and air-conditioning industry. For fan components see Figures 1.21a and 1.21b.

Basically there are two types of fans used in the industry viz, the centrifugal fans which are low to medium flows and relatively higher pressures than the axial flow fans which are mostly large volume and low pressure fans.

Centrifugal fans are further sub-classified as: (see Figure 1.7) (Depending on blade geometry).

a. Radial blade	Fig. 1.7a
b. Forward-Curved blade	Fig. 1.7b
c. Backward-Curved blade	Fig. 1.7c
d. Airfoil	Fig. 1.7c

Axial fans are sub-divided into the following classes: (see Fig. 1.8)

a. Vane-Axial Fans	Fig. 1.8a
b. Tube-Axial Fans	Fig. 1.8b

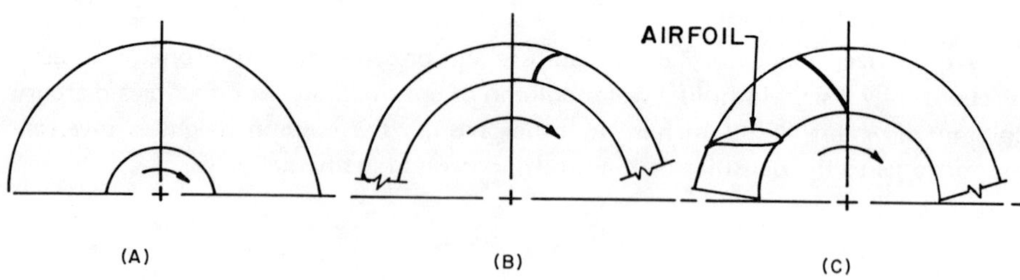

Figure 1.7

Figure 1.7a. Showing the radial blade design.

Figure 1.7b. Showing the forward curved blade.

Figure 1.7c. Showing the backwardly inclined blade and the airfoil design.

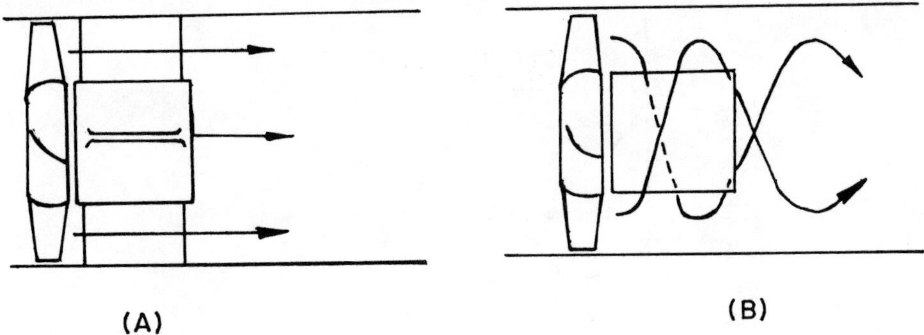

Figure 1.8.
Figure 1.8a. Vane-axial fan with the discharge guide-vanes showing smooth air flow.
Figure 1.8b. Tube-axial fan showing spiral air flow (no guide-vanes).

Radial Blades

This is a simple and a popular design which has many applications in systems requiring high resistance and low flows. The manufacturing is relatively simple. The performance of this type blade is also simple, as shown in Figure 1.9a.

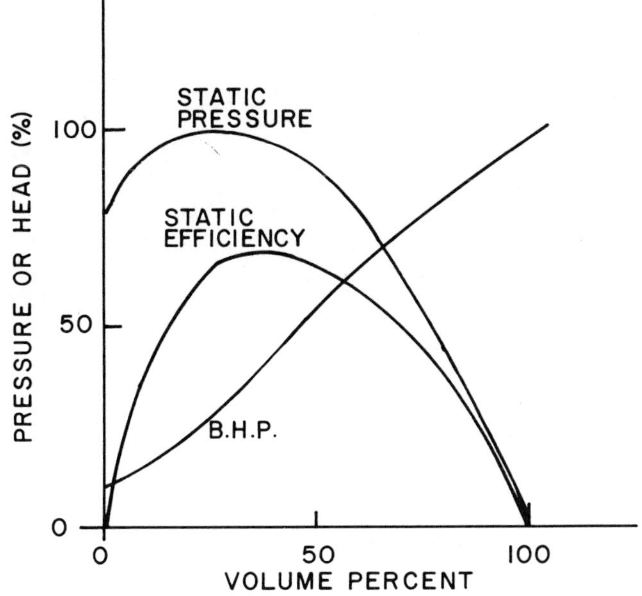

Figure 1.9
Figure 1.9a. Typical performance curve of a radial blade fan.

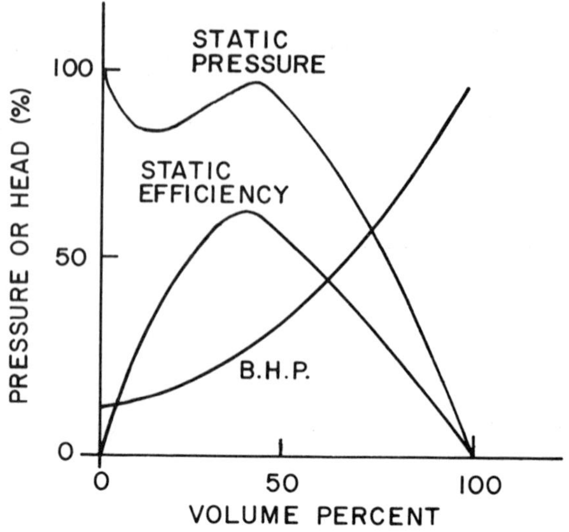

Figure 1.9 (Continued)
Figure 1.9b. Characteristics performance curve of a forward curved blade fan.

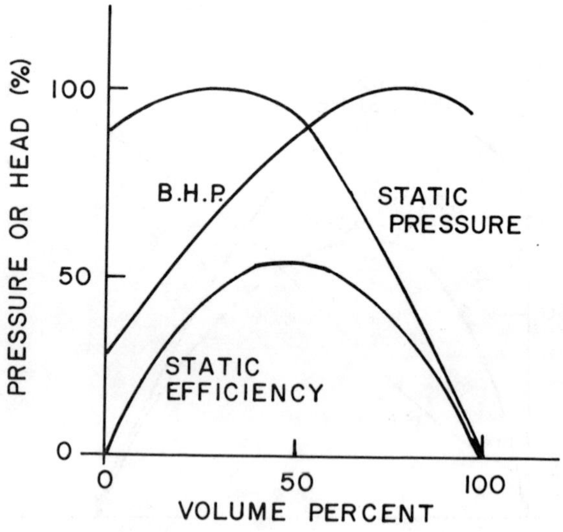

Figure 1.9c. A typical performance curve of a backwardly inclined blade fan.

Forward Curved Blades

The fluid leaving the blade has a greater velocity than when it leaves the backward curved blade, assuming the two have the same rotary speed and wheel diameter. The performance of this type blades is shown in Figure 1.9b. These machines should not be operated below about 40 to 45 percent of the flow, as is evident from Figure 1.9b.

Backward Curved Blades

These blades are tilted backwards as opposed to the forward tilt. The performance for these type of blades is shown in Figure 1.9c.

Airfoil Blades

The performance is similar to any blade but the only difference is in the shape of the blade which looks like an airfoil and is more efficient than the other types. These types are more stable and less turbulent.

Axial Fans

Axial fans have a much narrower range of stable operation than the centrifugal fans. Centrifugal fans are easier to control. That is why there should be no obstructions before and after the axial fan. A good rule of thumb is to keep two to four diameters of ducting clear of any obstructions and bends etc. The performance of a typical axial fan is shown in Figure 1.10. The axial flow machines, including fans have a hump under 50 to 75 percent of their optimum design capacity. This hump is due to a stall point or a reversal of flow within the blade. The horsepower at the shut-off point (zero capacity) is always higher for axial machines than for centrifugal machines, as is evident from Figure 1.10.

Vane-Axial Design

In this design the propellers are contained in the cylindrical housing and have discharge guide-vanes to smooth out the flow, as shown in Figure 1.8a. Vane-axials can develop pressures up to 25 inches of water gage or a little more. The motor can be direct-connected, thus taking less space and eliminating the need for the belts etc. These fans may also have inlet guide-vanes as well.

Tube-Axial Design

This design is similar to that of the vane-axials except that these fans do not have any guide-vanes to smooth out the flow. (see Figure 1.8b).

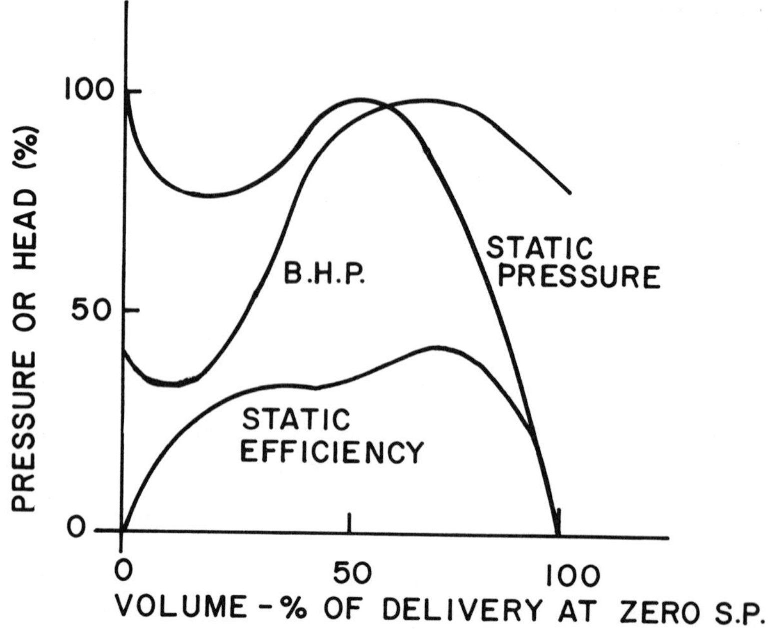

Figure 1.10. Performance curve of an axial flow fan. Note the hump in the pressure curve especially.

Fan Laws

Fans, like any other rotating equipment obey certain laws, called fan laws. The fan laws are useful in predicting performance of a new fan or an existing one due to change in the inlet conditions etc. These laws hold good for geometrically similar fans i.e. those that have same specific speed and the same blade geometry as well.

There are basically six fan laws out of which the second one is applicable to pumps as well and is called the AFFINITY LAW in the pump industry.

1. DIAMETER CHANGE: (Tip speed and density constant)
   a. Flow varies as the square of the Diameter $Q_2/Q_1 = (D_2/D_1)^2$
   b. Speed varies inversely with the Diameter $N_2/N_1 = D_2/D_1$
   c. Horsepower varies as the square of Diameter $HP_2/HP_1 = (D_2/D_1)^2$
2. SPEED CHANGE: (density and diameter constant)
   a. Flow is proportional to speed $Q_2/Q_1 = N_2/N_1$
   b. Pressure is proportional to square of speed $P_2/P_1 = (N_2/N_1)^2$
   c. Horsepower is proportional to cube of speed $HP_2/HP_1 = (N_2/N_1)^3$

3. DIAMETER VARIES: (RPM and density constant)
   a. Volume varies as the cube of the diameter $Q_2/Q_1 = (D_2/D_1)^3$
   b. Pressure varies as square of diameter $P_2/P_1 = (D_2/D_1)^2$
   c. Tip speed varies as the diameter $TS_2/TS_1 = D_2/D_1$
   d. Horsepower varies as fifth power of diameter $HP_2/HP_1 = (D_2/D_1)^5$
4. DENSITY VARIES: (volume, diameter, RPM constant)
   a. Pressure varies as density $P_2/P_1 = d_2/d_1$
   b. Horsepower varies as density $HP_2/HP_1 = d_2/d_1$
5. DENSITY CHANGES: (Pressure diameter)
   a. Volume varies inversely as the square-root of density $Q_2/Q_1 = (d_1/d_2)^{.5}$
   b. RPM varies inversely as square-root of density $N_2/N_1 = (d_1/d_2)^{.5}$
   c. Horsepower varies inversely as the square root of density $HP_2/HP_1 = (d_1/d_2)^{.5}$
6. DENSITY CHANGES: (air weight, diameter, tip speed constant)
   a. Volume varies inversely as density $Q_2/Q_1 = d_1/d_2$
   b. Pressure varies inversely as density $P_2/P_1 = d_1/d_2$
   c. RPM varies inversely as density $N_2/N_1 = d_1/d_2$
   d. Horsepower varies inversey as the square of density $HP_2/HP_1 = (d_1/d_2)^2$
7. NOISE:
   a. Noise is usually proportional to the tip speed.
   b. Noise is proportional to the air velocity leaving the impeller.
   c. Noise is proportional to the pressure developed.

   NOTE:
   Since the backward curved fans have discharge velocities a little less than those leaving the forward curved fans, the backward curved fans tend to be less noisy.

Fan Dynamics and Vibrations

All rotating equipment impellers have two major forces acting on them. The axial force, due to the unbalance of pressures acting on the two sides of the impeller, and the centrifugal force. The double inlet fans do not have any unbalance, and therefore do not have much of an axial thrust. For single inlet centrifugal fans, the axial thrust can be calculated as follows: (see Figure 1.11)

$$T = (A_1 - A_s)(P_2 - P_1) \qquad (19)$$

where:

T = axial thrust in pounds.
A_1 = Inlet area in square inches.
A_s = Shaft sleeve area in square inches.
P_2 = Pressure in the back shroud in psi.
P_1 = Pressure in the inlet in psi.

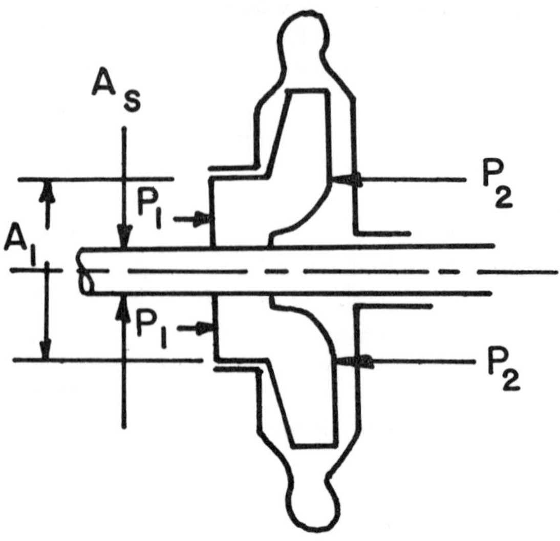

Figure 1.11. Showing axial thrust on a single inlet, impeller.

In pump industry there are generally two ways to reduce this thrust. In the first method, pumping rings or ribs are provided to reduce the pressure (P_2), but this method uses a little more horsepower. In the second method, holes are drilled in the back shroud, which reduces the thrust but also reduces efficiency. This method is seldom used in the fan industry.

In the axial fans, the thrust is approximated by the following equation:

$$T = 4.078 \cdot P_t \cdot D_{tip}^2 \tag{20}$$

In centrifugal fans, the thrust may be approximated as follows:

$$T = 4.078 \cdot P_s \cdot D_i^2 \cdot C \tag{21}$$

where:

P_t = fan total pressure in inches of water gage.
P_s = fan static pressure in inches of water gage.
D_{tip} = Tip diameter in feet
D_i = Inlet diameter in feet
C = Constant. If the pressure in the fan housing is positive, then $C = 1.0$, but if the pressure in the fan housing is negative (for example, exhausters), then $C = 2.0$

The amount of the centrifugal force acting on a blade can be determined by:

$$F = m\,v^2\,/\,g\,r \tag{22}$$

where:

F	=	Centrifugal for in lbf.
m	=	Mass of the wheel in lbm.
v	=	Tip velocity in feet per second.
r	=	Radius of the wheel in feet.
g	=	Gravitational constant in ft-lbm/lbf-sec^2.

The above equation can be simplified by replacing 'v' by 2π r N/60 then equation (22) becomes:

$$F = m\,r\,N^2\,/\,2934 \tag{23}$$

The magnitude of the centrifugal force helps in the mechanical design of the blades and their attachment to the fan hub. There is another force that acts on the housing and that is the net radial force or radial thrust, due to the volute design. There are basically two types of volutes viz. single volute and double volute. In a single volute casing, as shown in Figure 1.12a, the radial force varies along the periphery of the impeller at off-design condition, and the net resultant radial force will act in a direction shown with an arrow. The radial forces are pretty much balanced when the machine is operting at the design point. In a double-volute design, as shown in Figure 1.12b, the radial forces are balanced even at off-design conditions. Actually there are two net resultants acting opposite to each other and are almost equal in magnitude. The resultants F_1 and F_2 are shown in Figure 1.12b. They tend to balance out even though they are not eliminated. Any unbalanced radial force or load is taken up by the radial bearings which can be either anti-friction (ball bearings) or journal sleeve bearings.

If there is any unbalance in the center of gravity and the center of rotation of the wheel, it would cause vibrations which are passed on to the bearings and further down to the foundations. That is why some manufacturers recommend that their equipment be mounted on vibration isolating pads. No matter how carefully the dynamic balancing of the rotor is done, it is impossible to have a perfect balance. The industry practice is to have the operating speeds at least 20% above or below the critical speed. The shafts that operate below the critical speed are called the "Stiff shaft design" and the ones that operate above the critical speed are called the "Flexible shaft." In flexible shaft design the shaft has to go through the critical speed during the starting and the shutdown of the machinery. The critical speed of a shaft is calculated by:

$$N_{cr} = \frac{60}{2\pi} \sqrt{9/y}$$

$$= 187.7 / \sqrt{y} \qquad (24)$$

where:

y = Static deflection of the shaft in inches.

There are various other ways of calculating the critical speed of shaft, using Rayleigh's or Dunkerley's method. Reader is advised to refer to any text on vibrations for these and other techniques of calculating the critical speeds. Simple cases of 'Overhung shaft' and 'Simply supported shaft' are, however, discussed below. For a single bearing overhung design, the shaft deflection can be calculated as: (see Figure 1.13a).

$$y = -\frac{W}{6E\ I} (x^3 - 3L^2 + 2L^3) \qquad (25)$$

For a two bearing design, the shaft deflection is given by: (see Figure 1.13b).

$$y = -\frac{W}{48E\ I} (3L^2 x - 4x^3) \qquad (26)$$

$$= -\frac{W L^3}{48\ E\ I} \qquad (26a)$$

Equation (26a) is for the case when the load acts in the center, between the two bearings i.e. when $x = L/2$.

Sometimes there are other ways when minor or secondary vibrations are excited. They can be caused by oil-whip, sleeve bearings, couplings etc. separately or in combination with each other. Let us discuss each case separately.

a. Oil Whip

Most high speed and high load machines usually use sleeve bearings which are normally oil-lubricated. The oil in these type of bearings sometimes forms what is called oil wedge, which travels around the shaft at about half the speed of the shaft. If this wedge speed happens to coincide with the shaft's first critical speed, the shaft will start to vibrate. Once these vibrations get started, they will not disappear even if the shaft speed is increased to about twice the critical speed. These vibrations are very difficult to eliminate, but the following procedures may help reduce them.

(1) Switch to anti-friction bearings.

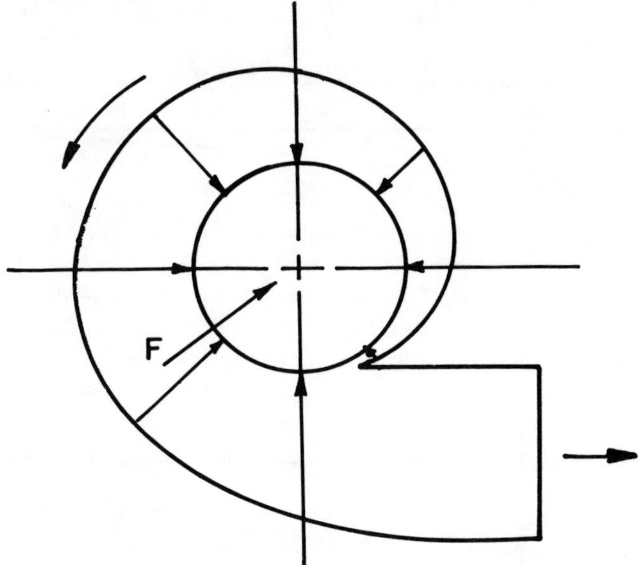

Figure 1.12a. A typical single volute or a scroll showing the radial forces acting at off-design point. Force F is the net resultant acting on the casing.

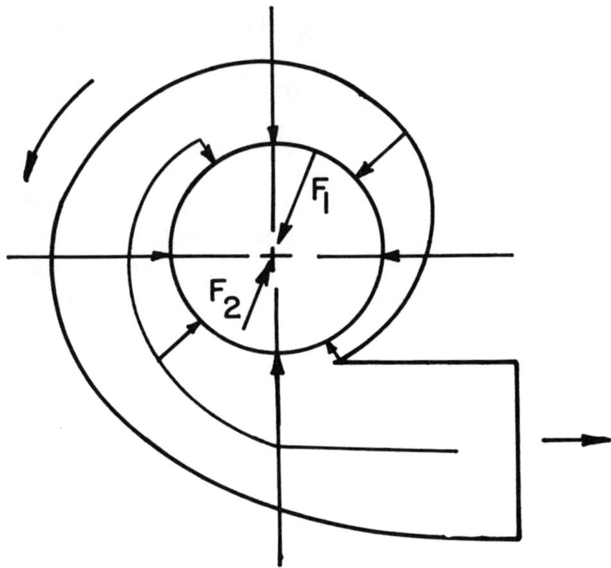

Figure 1.12b. This shows the double volute design and the radial forces which are balanced. Forces F_1 and F_2 act opposite to each other.

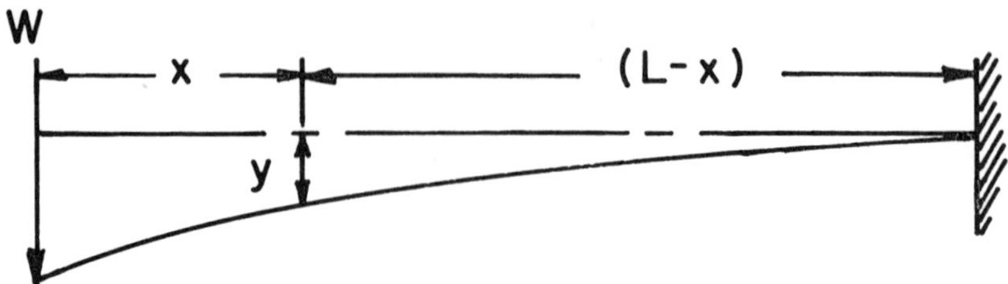

Figure 1.13a.

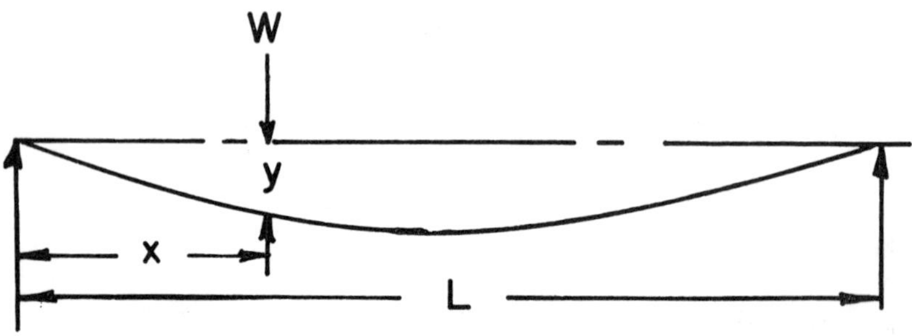

Figure 1.13b.

(2) Decrease the bearing length to increase the load on the oil wedge and thus reduce the oil-whip induced vibrations.
(3) Do not run the machinery at twice the critical speed.

b. Sleeve-Bearings

With sleeve-bearings small disturbances are transmitted from the coupling, driver or any other minor fluctuations will easily produce shaft vibrations, since the shaft does not have a continuous contact with the bearings. Sometimes these vibrations disappear if you reduce the clearances between the shaft and the sleeve bearing.

c. Coupling

These vibrations are usually caused by either misalignment or are transmitted by the driver. Even with the elastic couplings, if the shaft runs above the first critical speed, a slight misalignment will produce periodic disturbances.

Fan Controls

Capacity control is achieved in several ways and some of the most commonly used methods of control are described below.

1. Inlet Guide Vanes

These are just before the air or gas enters the impeller. They are designed such that the fluid leaving these vanes would match the impeller blade angle with a minimum shock. This is an efficient way of saving energy, especially when the fan has to operate at reduced capacities. The variable inlet guide vanes are very efficient and desirable when the fan operates at different capacities varying quite a bit.

2. Outlet Dampers

This is one of the most inefficient methods of capacity control. The inlet capacity (ICFM) stays the same and the discharge capacity is varied by letting the extra capacity to atmosphere if it is safe to do so. Toxic gases are either recycled or disposed of in a safe manner.

3. Speed Variation

Variable speed motor drivers are sometimes used to handle reduced operation. Sometimes steam turbines are also used, if the initial and the operation costs are justifiable.

Fan Construction

After the air or gas leaves the impeller tip, it is collected in a scroll or volute which converts kinetic energy imparted by the blades, into pressure energy. A typical volute or scroll is shown in Figure 1.12. This conversion into pressure energy is achieved by increasing the area continuously around the impeller. The inlet box is usually specified with the fan. As with any other air handling machine, inlet conditions are critical, a proper selection of an inlet box will assure proper performance. Even though the insulation is installed in the field, it should be specified so that the manufacturer can design the insulation attachment method on the fan housing.

Materials

Housing is usually made of fabricated steel, which is used up to a maximum of 900°F in a normal non-corrosive atmosphere. Above 900°F, stainless steels or high nickel alloys are employed, especially when the atmosphere is corrosive or you are pumping corrosive gases. In low temperature but corrosive atmosphere or corrosive gases, the wheels are made of FRP (fiber reinforced plastics).

Bearings

In small sizes, single row ball bearings are used very commonly and they are grease lubricated. Grease lubriication is satisfactory up to about 200°F. Oil lubrication is recommended above this temperature. Bearing cooling should be considered above 300°F and also some sort of barrier and or cooling device between the fan housing and the bearing housing should be used. These air cooling/circulating devices are called slingers or cooling disks. Usually above 600°F, the bearing housing or sub-assembly should be isolated from the fan housing. Fans and the inlet and the outlet ducting should be insulated, not only to cut down on the heat transfer, but also to protect the personnel from injury.

In small fans with higher pressures double row ball bearings are used. In large fans, say 1000 horsepower and above, some manufacturers or users may require sleeve bearing which are oil-lubricated. The American Petroleum Institute Specifications No. 610 (API-610) for centrifugal pumps for refineries has clearly stated as to when to use sleeve bearings. The same guide lines can be used for fans. The guide lines are: (for use of sleeve bearings)

a. When the anti-friction bearings fail to meet the B-10 life.
b. Where D.N factors are greater than 300,000, where D = bore in millimeters, and N = speed in rpm.
c. Whenever pump rated speed times the pump rated horsepower equals or exceeds 2.7 million. Assuming the pump or fan is going to be operating at about 3550 rpm, sleeve bearings should be specified around 750 horsepower.

Some users are successfully using "oil-mist" lubrication system for the ball bearings for their machines. It has a central mist generating system, which supplies tiny oil molecules carried by air under pressure, for the bearings. They claim to have increased bearing life or reduced maintenance costs on their equipment with the use of the "oil-mist."

The bearing life is usually expressed in terms of "B-10" life. This simply means that only 10% of all the bearings operating under identical conditions would have failed.

Special Considerations

High temperature is very critical for the rotating equipment's life and alignment. For high temperature application, therefore, it is recommended that the fan housing be center-line mounted as opposed to base mounted or foot-mounted, for ease of alignment. In case of large equipment and also in the case of high temperature or both, use of turning gear is recommended. This would prevent the large shaft from getting a permanent set or bow.

If you are pumping gases with some abrasive materials like sand, flyash, or cement dust etc. the use of wear plates or hard-facing is recommended. With abrasive particles in the gases, select fans at lower speeds. This will cut down on the erosion.

Couplings

There are basically two types of couplings that have been used for years, viz.

gear type and the diaphragm type. Gear couplings for low speed applications are usually oil filled or grease packed and can operate from six months to a year before the oil or grease is changed. However, if the environment is dirty, the oil or grease may need changing sooner. The diaphragm type, on the other hand, has the advantage of no lubricant requirement and hence very little maintenance. They only need a periodic check to make sure that the discs, which are very thin, have not developed any fatigue cracks. There is another type of coupling that should be mentioned here, and that is the LIMITED-END-FLOAT type. As the name implies, this type just limits the motor rotor float. NEMA (National Electrical Manufacturers Association) MG1-14.38, specifies that 250-HP and larger AC motors, when fitted with sleeve bearings, have ½-inch minimum total rotor end float in the bearings (or ¼-inch end float in each direction). It also specifies that the coupling should limit that end float 3/16-inch maximum (or 3/32-inch in each direction. Limited end float couplings restrict float by locating the rotor axially using the fixed thrust bearing on the driven machine and prevent damage to the motor bearings.

Forces and Moments

Fan housings are usually not as rugged as those of large pumps and compressors, and the manufacturers normally do not allow any loads on the fan housing. Sometimes you cannot help avoid loads imposed by the piping. Especially in high temperature cases, the loads due to thermal growth cannot be eliminated. In such cases, you must consult the fan manufacturer and he may advise you to use flexible connections (also called expansion joints) at the discharge end or at both the discharge and the inlet connection as well. In the later case, you must, then, specify that the fan be supplied with a flanged inlet, so that a proper connection with the expansion joint can be made in the field. For a picture of a flexible connector, see Figure 1.14.

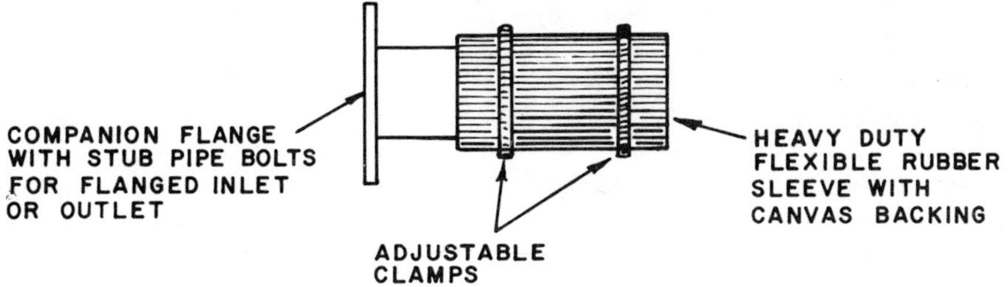

Figure 1.14. Showing flexible connector for blowers and fans.

The Air Moving and Conditioning Association (AMCA) Standards

The Air Moving and Conditioning Association has published quite a few standards for the air handling equipment, but here only a few important ones will be discussed.

1. Spark Resistant Construction

The AMCA Standard AS-401, lists the following types:

a. Type "A" Construction – This states that all parts of the fan that come in contact with the gas should be made of non-ferrous materials. Bronze and aluminum are commonly used for this type.

b. Type "B" Construction – This construction requires that only wheels and sleeves (or rings) around the shaft openings be made non-ferrous materials.

c. Type "C" Construction – This specifies that the design should be such that it should prevent the ferrous parts from rubbing together.

2. Construction by Class

The class designation just limits the pressures viz.

a. Class I – This is the lightest frame construction and limits applications to a pressure of about 4 inches of water gage.

b. Class II – This is a medium construction and limits pressures up to about 7 inches of water gage.

c. Class III – This is a heavier frame and the pressures are limited up to about 12 to 13 inches of water gage.

3. Motor Positions

The AMCA also specifies the motor position with respect to fan as shown below. The motor position is specified by the letters W, X, Y, or Z. (see Figure 1.15).

4. Fan and Motor Arrangements

The AMCA Standard No. 2404-66, has for the convenience of everybody come up with the following arrangements (see Figure 1.16).

5. Inlet Box Positions

The AMCA Standard No. 2405-66, has the following positions standardized for the inlet box (see Figure 1.17).

6. Rotation and Discharge Orientation

The AMCA Standard No. 2406-66, is to specify the rotation and orientation of the discharge of the centrifugal fans. The fan is viewed from the drive side.

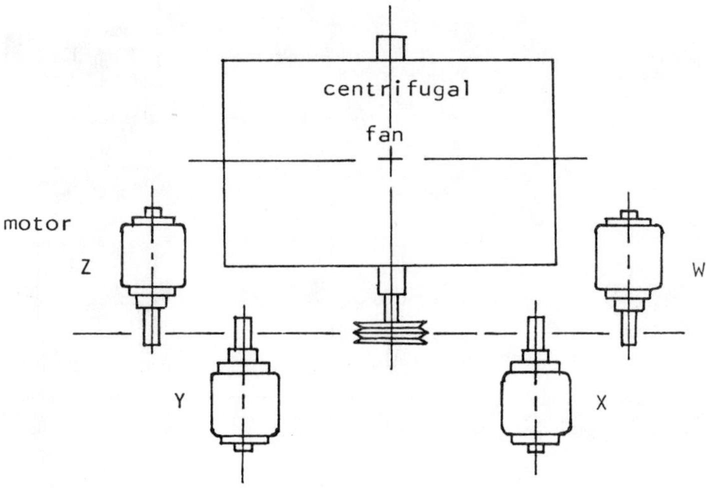

Figure 1.15. AMCA Standard 2407-66, showing the different motor positions with respect to the fan.

Fans Operation

Once you understand the fan fundamentals and have some knowledge of the system in which the fan is going to be used, the selection of a proper fan is relatively easier based on the capital cost, operation cost and the maintenance cost. Fans, like any other rotating equipment, should be selected as close to its maximum or best efficiency point as possible. If a fan is selected to the right of its best efficiency point (bep), it means either the fan is a little too small or is running at a higher speed. If a fan is selected to the left too far from its bep, it means that the selected fan is too large or is running too slow.

Fans, like other centrifugal machinery, operate only where the system curve intersects the fan curve. At this point the system pressure matches the fan pressure. System resistance is very simple to calculate for a pipeline or duct and it is proportional to the square of the capacity. If there are other obstructions or equipment in the duct system or pipeline, the pressure drop across each item should be added to the line resistance and a combined system resistance be drawn on the same graph which has the fan curve. If the system resistance consists mostly of the line loss, the curve will be a parabolic one as shown in Figure 1.19a and Figure 1.19b shows the system curve with some static pressure in addition to line resistance. Normally most fans are suitable to operate in a system shown in Figure 1.19a. The fan selection for systems shown in Figure 1.19b is, however, a little critical. The fan operation where the fan curve is flat or drooping will be unstable and should be avoided, as much as possible. Normally a single fan is satisfactory for a system's operation, but some-

SW – Single Width DW – Double Width
SI – Single Inlet DI – Double Inlet

Arrangements 1, 3, 7 and 8 are also available with bearings mounted on pedestals or base set independent of the fan housing.

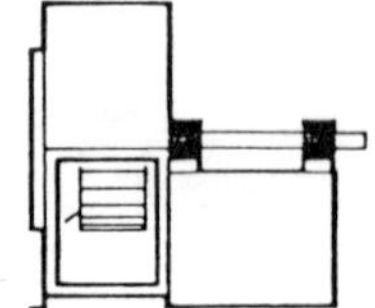

ARR. 1 SWSI For belt drive or direct connection. Impeller overhung. Two bearings on base.

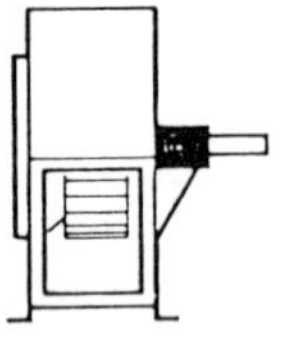

ARR. 2 SWSI For belt drive or direct connection. Impeller overhung. Bearings in bracket supported by fan housing.

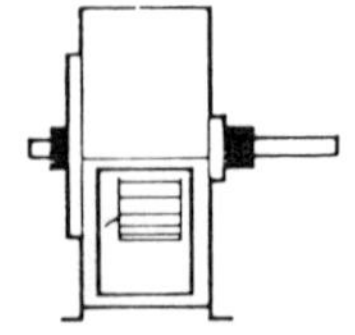

ARR. 3 SWSI For belt drive or direct connection. One bearing on each side and supported by fan housing.

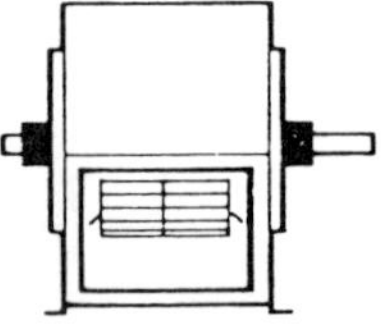

ARR. 3 DWDI For belt drive or direct connection. One bearing on each side and supported by fan housing.

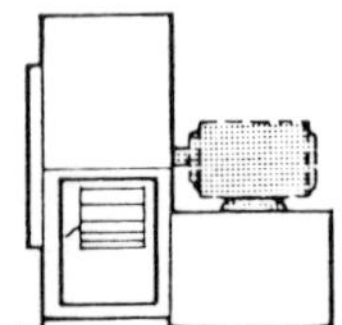

ARR. 4 SWSI For direct drive. Impeller overhung on prime mover shaft. No bearings on fan. Prime mover base mounted or integrally directly connected.

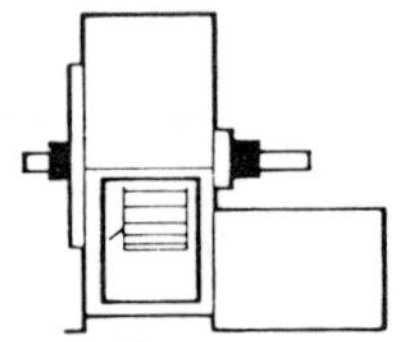

ARR. 7 SWSI For belt drive or direct connection. Arrangement 3 plus base for prime mover.

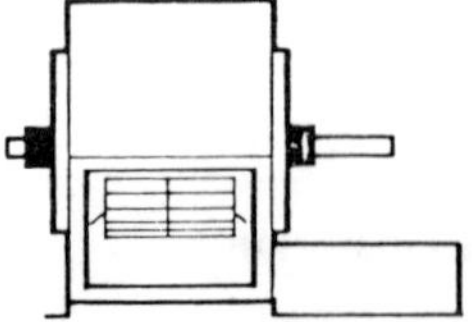

ARR. 7 DWDI For belt drive or direct connection. Arrangement 3 plus base for prime mover.

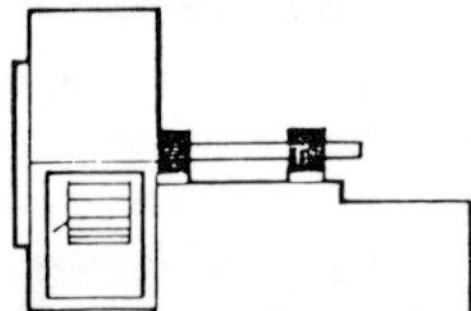

ARR. 8 SWSI For belt drive or direct connection. Arrangement 1 plus extended base for prime mover.

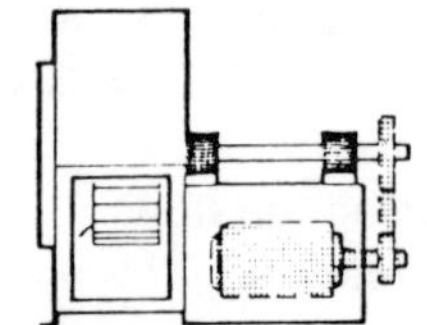

ARR. 9 SWSI For belt drive. Impeller overhung, two bearings, with prime mover outside base.

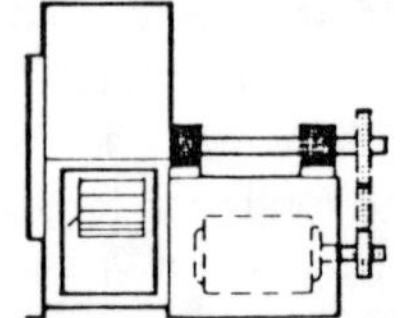

ARR. 10 SWSI For belt drive. Impeller overhung, two bearings, with prime mover inside base.

Figure 1.16. Showing different arrangements.

times you may require two or more fans to meet the demand which may vary from time to time. Sometimes two fans may be required in series to meet the higher pressures for a certain operation whereas normally a single fan will do. Similarly two fans in parallel may be used to meet the higher capacity demand at certain times. The two systems are discussed below.

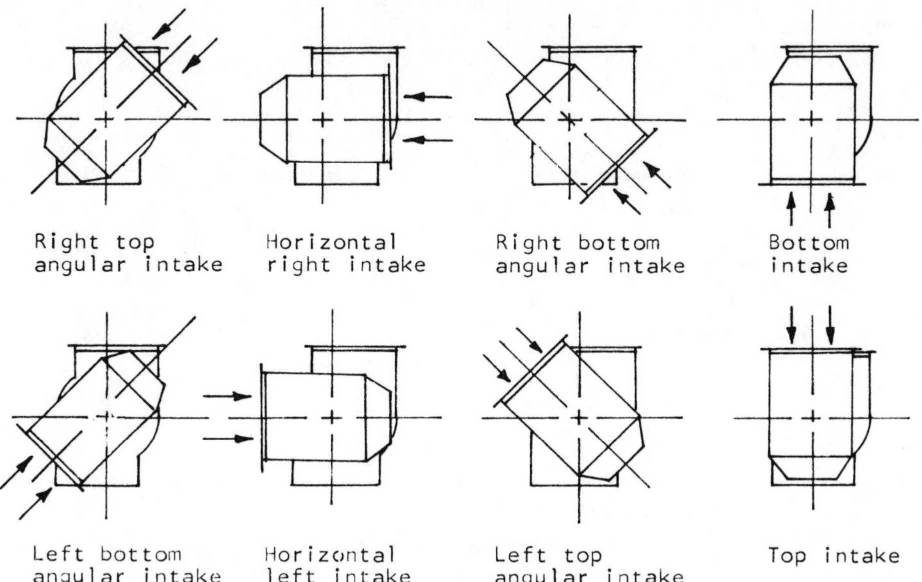

Figure 1.17. Showing the different positions for the inlet box. The angular positions shown are at 45-degree angle.

Series Operation

For two fans operating in series, the combined total pressure will be the sum of total pressure of each machine at the same inlet capacity. The static pressure for the two should not be added to get the combined static pressure. The weight flow-rate handled by the fans operating in series is the same unless there is a loss or gain of flow in between. The performance of two fans in series is shown in Figure 1.20a.

Parallel Operation

Any two fans similar or a little bit dissimilar can be operated in series without much trouble, but for parallel operation, it is difficult to do the same without sophisticated controls. For parallel operation, therefore, the two fans selected should be similar in performance if not identical. It is important they share the total load equally, otherwise one fan will be overloaded and the other would be underloaded or could even operate near shut-off conditions. This type of loading could even shift back and forth between the two fans and can damage the fans as well as the drivers. So you need controls to make sure the load is shared equally and no driver is overloaded. The performance of two fans in parallel is obtained by adding inlet capacities at the same total pressure. To find the point of operation for the two fans in parallel, you have again to draw a combined system curve (see

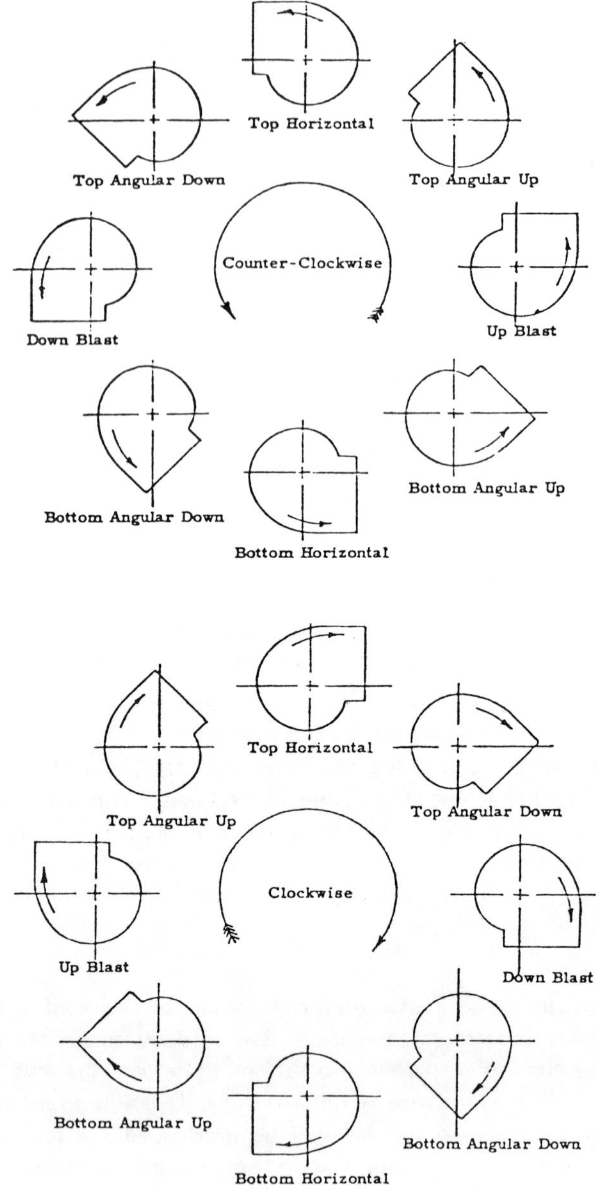

Figure 1.18. Showing different discharge positions and the direction of rotation.

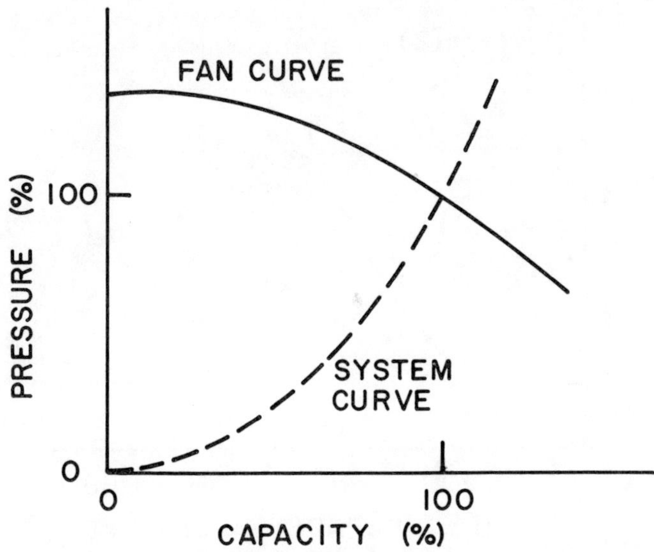

Figure 1.19a. In this figure all the system resistance consists of line losses.

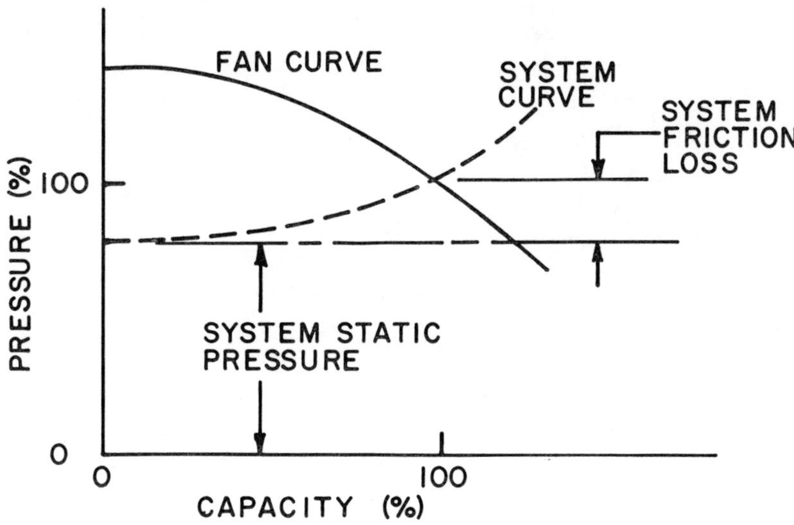

Figure 1.19b. In this system there is a lot more static pressure than the friction. The operation is satisfactory until near the flat part of the fan curve, which should be avoided.

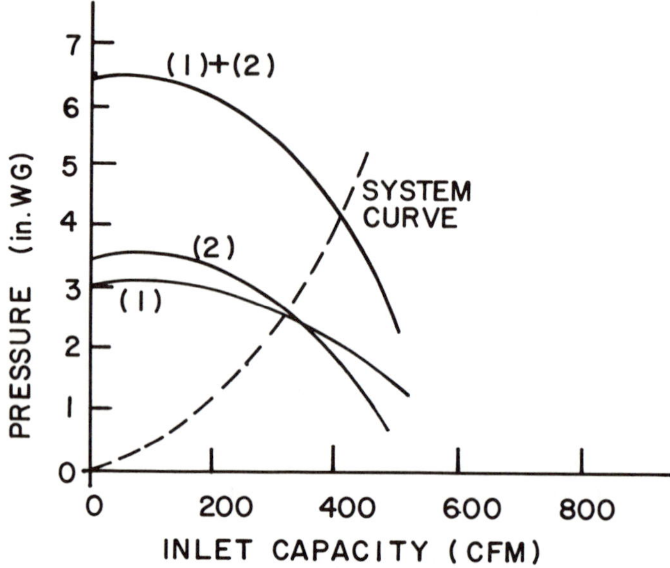

Figure 1.20a. The point of operation is where the system curve intersects the combined fan curve (1) + (2). The diagram shown is for two fans in series.

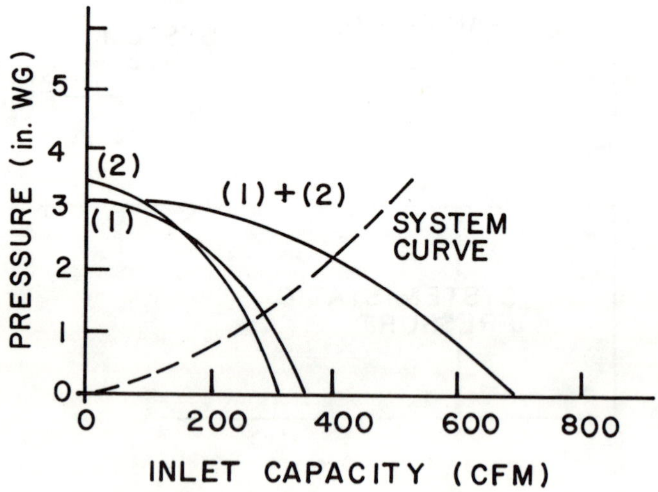

Figure 1.20b. Shows two similar fans but not identical, in parallel. The curves are not identical but are continuously rising, which makes the operation satisfactory.

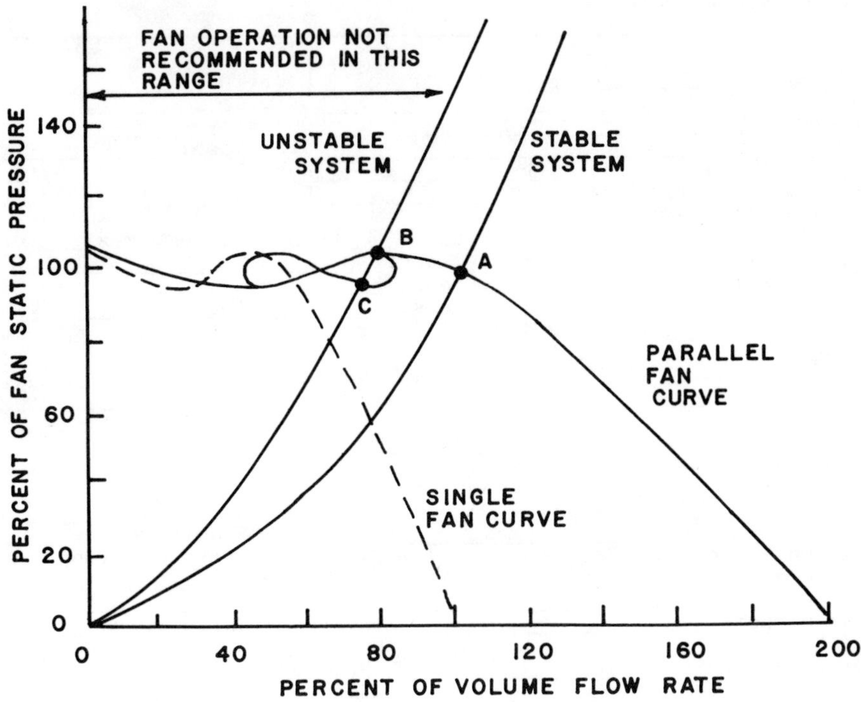

Figure 1.20 (Continued)

Figure 1.20c. The two fans have identical curve with a hump which when combined with another forms a loop. Operation in the loop area should be avoided.

Figure 1.20b for the parallel operation) and the point of intersection is the point where the fans would operate. The parallel operation is simple if the fan curve is continuously rising. The situation can become very delicate if the two fans, even though identical, have a hump at a reduced capacity. When you obtain the combined performance curve, it would have a loop as shown in Figure 1.20c. The loop is obtained by adding all the possible capacities at a given pressure. If the system curve intersects the fan curve at point 'A,' then there will be no problem in the satisfactory operation of the fans, provided the system does not fluctuate. But if the system curve happens to intersect in the loop area, it will have two points of intersection and thus two points of operation viz. 'B' and 'C.' This is not desirable, because one fan may operate at 'B' and the other fan may operate at 'C' and they may switch back and forth. This switching of loads back and forth is also called 'Hunting' and can damage the fans as well as the drivers. Controls in the fan outlet areas or inlet and outlet dampers may be required to rectify the problem.

Table 1.1. Specimen Data Sheet

Customer. ______ Location. ______
Unit . ______ Service . ______
Item No. ______ Quantity. ______
Mfr. . ______ Size, Model ______
Item No. (Motor Driven) ______ Furn By ______ Mtd By ______
Item No. (Turbine/other) ______ Furn By ______ Mtd By ______

OPERATING CONDITIONS:

Elevation ______ Atm. Press. ______ Psia: Amb. Temp. Max. ______ Min. ______
Capacity lb/hr ______ Density lb/cft ______ Acfm ______
Inlet Temp F ______ Inlet Press. ______ Psia ______ in.WG.
Disch Temp F ______ Disch Press. ______ Psia ______ in.WG.
Design ACFM ______ Design Diff. Press. ______ in.WG/Psi

PERFORMANCE:

Specific Speed. ______ Fan Speed Min. ______ Max ______
Static Eff. ______ Total Eff. ______
BHP ______ Max. H.P. ______
Design outlet velocity fpm ______ Sound Level dbA ______

FAN CONSTRUCTION:

Inlet Size ______ Discharge Size ______
Fan Type (Centrifugal/ axial) ______ Blade Type ______
Inlet Box Config. ______ Inlet Type (Single/Double) ______
Rotation, Discharge Position ______ Motor Position ______
Arrangement No. ______ Direct/Belt Drive ______
Housing Liners Type ______ Thickness ______
Inlet Dampers ______ Inlet Var. Guide Vanes ______
Outlet Dampers ______
Expansion Joints Inlet ______ Discharge ______
Insulation Material ______ Thickness ______
Spark Resist. Const. Type ______ Material ______
Shaft Material ______ Impeller Mat'l ______
Bearings Type ______ Size ______ Mfr. ______
Bearing Span in. ______ Bearing Lube ______
Heat Slinger/Cooling Disk Reqd? ______ Type (if Reqd) ______
Cooling Water Reqm'nt. Bearing ______ Pedestal ______
Rotating Element WR^2 ______ $Lb\text{-}ft^2$
First Critical Speed ______ Second Critical ______
Shaft Diameter inches ______
Inlet Screen Mesh ______ Drain Conn. ______

DRIVER DATA:

Motor Mfr. ______ Motor H.P. ______ RPM ______
Enclosure ______ Bearings ______ Lube ______
Motor Frame ______ Type ______ Insulation ______
Temp. Rise ______ ^{o}C. Volt/Phase/Cycles ______
Area Classfication: Class ______ Group ______ Division ______
Locked Rotor Amps ______ Full Load Amps ______

STEAM TURBINE DATA:

Mfr. ______ Type, Size ______ Model ______
H.P. ______ RPM ______ Steam Rate (lb/hp-hr) ______
Inlet Steam Psig, Min ______ Norm ______ Max. Temp. Min ______ Norm ______ Max ______
Exhaust Steam Psig, Min ______ Norm ______ Max ______ Exh. Temp. ______
Steam Flow lb/hr ______ Max. ______ . Governor Nema Type ______
Turbine Max. Cont. Speed ______ Rpm: Trip Speed ______
Inlet Size ______ Rating ______ Position ______
Exhaust Size ______ Rating ______ Position ______

Table 1.1 (Part 2) Specimen Data Sheet

Gas Analysis

It is a good practice to enclose the complete gas analysis data sheet with the inquiry package for vendor's information. Remember, the manufacture has the best experience and expertise in the design and material selection if he knows the complete gas analysis. The following is a smple of "Natural Gas" composition.

Composition	Symbol	% Mole	M.W.
Methane	CH_4	60.00	16.04
Ethane	C_2H_6	16.60	30.07
Propane	C_3H_8	8.50	44.10
i-Butane	C_4H_{10}	1.55	58.12
n-Butane	C_4H_{10}	2.49	58.12
Pentane	C_5H_{12}	1.21	72.15
Nitrogen	N_2	0.86	28.01
Carbondioxide	CO_2	6.10	44.01
Hydrogensulfide	H_2S	2.06	34.08
Water	H_2O	0.63	18.02
Total		100.00	

Note: From the above gas analysis, the average molecular weight of the mixture, and the ratio of specific heats should be calculated.

Table 2. Bid Tabulation (Sample)

Item No. ____________ For ____________

Location ____________ Service ____________

Description	Vendor A	Vendor B	Vendor C	Vendor D
Model, type & size				
Specific Speed/RPM				
Static Eff./Tot. Eff.				
Fan BHP/Fan Max H.P.				
Wheel Dia./Outlet Vel.				
Wheel Type/Blade Type				
Size Inlet/Outlet				
Shaft Dia./Shaft Mt'l				
Liners typ./Insul.				
Bearing type/Brg. Spar				
Miscell. ------------				

PRICES:				
Fan --------------				
Driver -------------				
Accessories ---------				
Total Price -------$				
Power Evaluation ---				
@$ X/hp-yr for 3 yrs.				
Total Evaluated Cost $				
Delivery ------ weeks				
Recommend Vendor.				

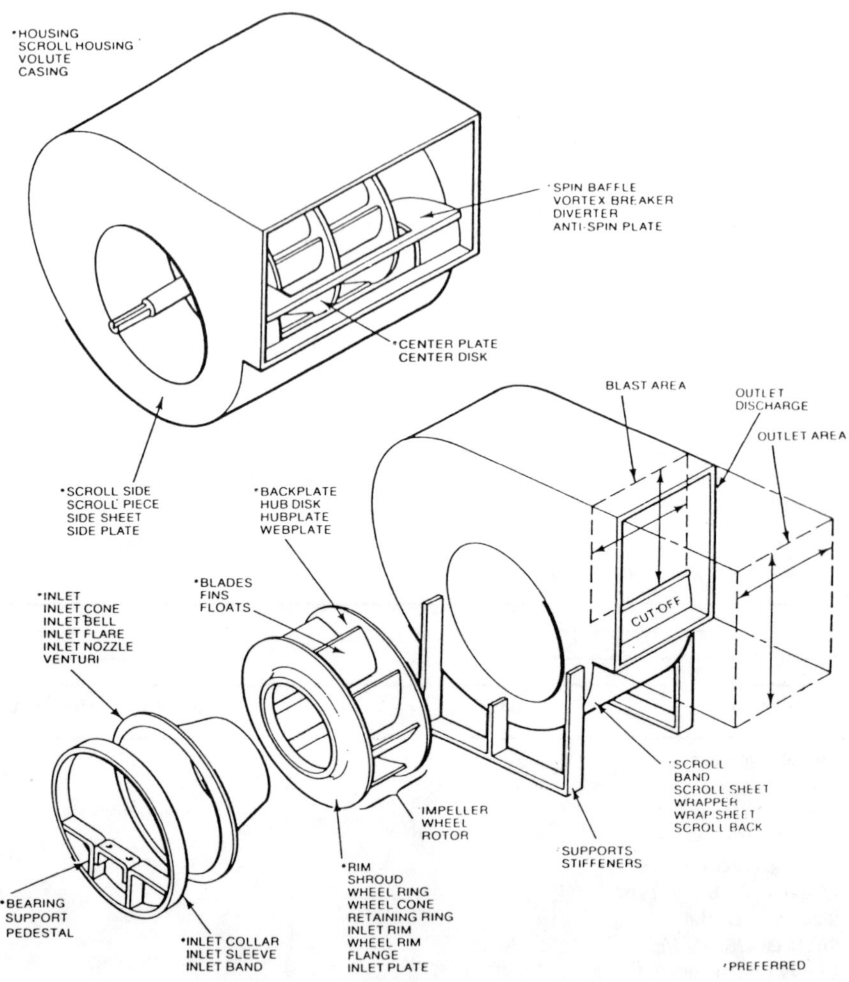

Figure 1.21a. Common terminology for centrifugal fan components. (Courtesy AMCA)

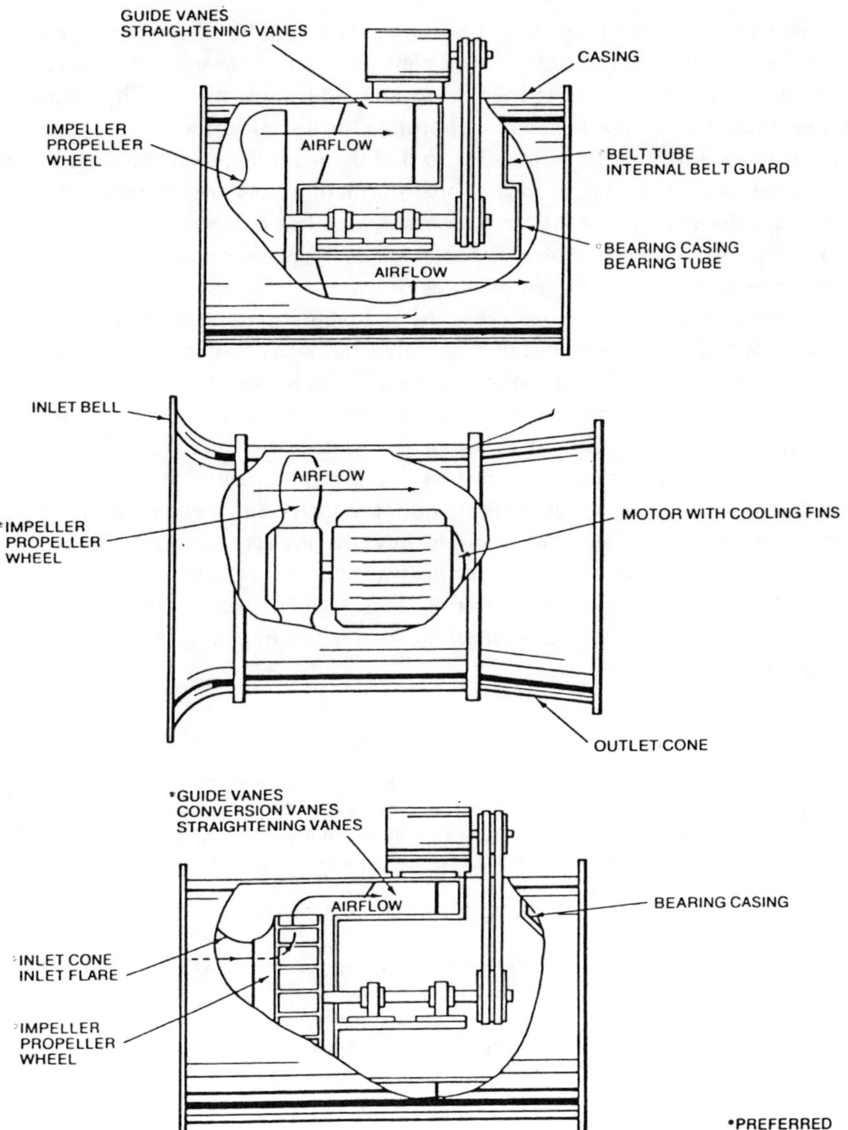

Figure 1.21b. Common terminology for axial and tubular centrifugal fans. (Courtesy AMCA)

Fan Selection and Evaluation

Fan selection is fairly easy once you understand the basic principles and the requirements. To get the right fan for the right job, you must send out inquiry with a set of specifications and a data-sheet to several fan manufacturers. The data-sheet should be filled out by the engineer responsible for the selection of fan and should contain as much information as possible so that the manufacturer has a good idea of what is required. It is better to have more information than not enough, otherwise you may get the wrong type of fan which may not last long enough or may have some kind of maintenance problem. Table 1.1 outlines the minimum information required to be conveyed to the fan manufacturers.

Next step in the selection process is the bid-tabulation and evaluation of each bid. A sample bid-tablulation is shown in Table 1.2. What may look like the cheapest (or least expensive) offering, might turn out to be higher than most of the bids. Power evaluation must be done for at least a period of three years, if not more. This cost should be added to the initial equipment cost, and then a proper evaluation should be made. In addition other factors like, bearings, couplings, seals (if any), material of construction, speed (RPM), outlet velocity (in feet per minute) etc. should be considered. It does not pay to buy the lowest evaluated fan when it would be down quite a bit more time for maintenance and repairs.

A thorough evaluation of each offering, therefore, is recommended and would pay out many times more in the long run. As a guide Figure 1.21 gives common terminology for various fans.

REFERENCES

1. Stepanoff, A. J., Pumps and Blowers, John Wiley & Sons, Inc., New York, 1966.
2. Stepanoff, A. J., Centrifugal & Axial Pumps, John Wiley & Sons, Inc., New York, 1957.
3. Jorgensen, Robert, Fan Engineering, Published by Buffalo Forge Co., Buffalo, N.Y., 1970.
4. Jennings, B. H. and Lewis, S. R., Air Conditioning and Refrigeration, International Textbook Co., Scranton, Pa., 1963.
5. AMCA (Air Moving and Conditioning Association), Fans & Systems, Publication 201, Arlington Heights, Ill., 1975.
6. Balje, O.E., A Study on Design and Matching of Turbomachines; Part B – Compressor and Pump Performance and Matching of Turbocomponents.
7. API – 610, Centrifugal Pumps for General Refinery Services, Washington, D.C. 1973.
8. American Standard, Industrial Division, Detroit, Michigan. Fan Composite Curve, Specific Speed/Specific Diameter, 1967.

CHAPTER 2

COMPRESSOR APPLICATION AND SELECTION

RICHARD F. NEERKEN
The Ralph M. Parsons Co.
Pasadena, CA

INTRODUCTION

Any chemical, petrochemical or petroleum process which involves the pressure rise of air or gas will require compression equipment. Compressor is the term applied to the rotating machinery which produces such a pressure rise, getting its name from the fact that the volume of gas is compressed as it flows through the machine. In lower pressure ranges the term blower may be used to mean compressor; there is no uniform agreement on where or when the term blower should be used instead of compressor. In this chapter our objective will be to present basic principles of air or gas compression, an overall description of all types of compressors and blowers, and certain sizing and performance methods which may be used by the process designer or others who are concerned with compressor application, selection, and analysis. Inasmuch as compressors are often rather large, complex, expensive, long delivery machines, it is readily apparent that considerable early effort must be made in any process design involving this machinery.

TYPES OF COMPRESSORS

Compressors can be broadly divided into two basic types: dynamic and positive displacement (Figure 2.1). Dynamic types produce a pressure rise by imparting velocity to the gas through one or more rotating impellers. The most well-known is the centrifugal type machine, although axial flow compressors are also correctly categorized as dynamic types. Positive displacement types have a reciprocating piston or plunger within a cylinder, or a rotating mechanism such as mating lobes, screws or vanes within a pressure containing casing. Any of these types displace a positive volume with each revolution of the drive shaft, which after allowances for inefficiences within the machine, will produce a nearly constant output at a given speed. Historically the most widely used positive displacement type has been the reciprocating compressor; currently there is increasing usage of the rotary positive displacement types in various designs as shown in Figure 2.1.

COMPRESSOR OPERATING CONDITIONS

Whether the reader's purpose is to select or specify a certain type of machine, to make a preliminary size estimate of a compressor, or to examine designs offered by

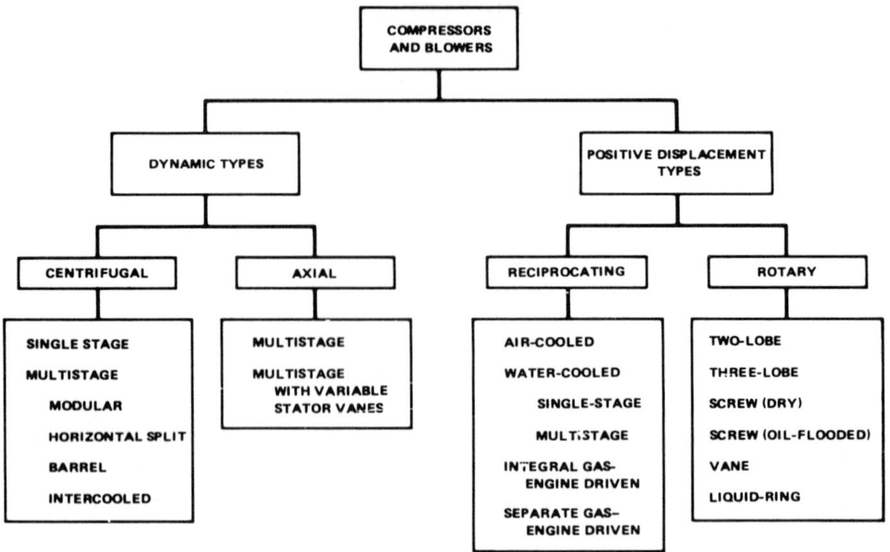

Figure 2.1. Compressor types.

manufacturers for given duty requirements, it is always necessary to determine the basic compressor operating conditions and obtain or develop certain information concerning the gas to be compressed.

Gas Analysis

If the air or gas to be compressed is pure, data is available in the form of pressure-enthalpy diagrams or tables of gas properties. If the gas is a mixture, often unique to the given process design, a gas composition, given either as a molal analysis or a volume percent analysis, will make possible the determination of properties of the gas mixture. For air compressors, it is usually necessary to know the relative humdity at the inlet conditions, or otherwise obtain data concerning the amount of water vapor in the entering air.

Molecular Weight and Ratio of Specific Heats

Essential to selection and sizing of any compressor are values of molecular weight, and ratio of specific heats ($k = C_p/C_v$) either at inlet conditions, or for more accurate calculations, at the average temperature during the compression cycle or at the entrance to each succeeding stage. These properties may be obtained from published data or readily calculated from a gas analysis.

Compressibility Factors

Compressibility factors show how the actual gas will deviate from an ideal gas. They are given or calculated for gas mixtures, usually at suction and discharge conditions, or at inlet and outlet of each stage of compression. For air or a pure gas, charts or tables are available; for gas mixtures these factors may be determined from generalized compressibility charts (reference 1, 2, 3), or from specific data which is available for a mixture such as 75% hydrogen—25% nitrogen in ammonia synthesis. From the gas analysis and known properties of each component, the critical pressure (P_c) and critical temperature (T_c) of the gas mixture can be calculated. Values of reduced pressure and temperature, defined as:

$$P_r = \frac{P}{P_c} \quad \text{and} \quad T_r = \frac{T}{T_c} \tag{1}$$

will enable the use of the generalized charts. Note that pressures and temperatures are expressed in absolute units. Figure 2.2 illustrates a typical gas analysis calculation for a compressor. Note that in this example, the value for molal specific heat, MC_p, is taken at 150 F, a common approximation for the average temperature during the compression cycle. For more accurate calculations, values of MC_p should be used which more accurately represent the actual values during a specific cycle.

Gas Analysis (mol %)		MW	MW x %	MC_p at 150°F	MC_p x %	P_c (psia)	P_c x %	T_c (°R	T_c x %
Methane	65%	16.04	10.43	8.95	5.82	673	437	344	224
Ethane	14%	30.07	4.21	13.77	1.93	708	99	550	77
Propane	6%	44.09	2.65	19.53	1.17	617	37	666	40
i-Butane	3%	58.12	1.74	25.75	0.77	529	16	735	22
n-Butane	2%	58.12	1.16	25.81	0.52	551	11	766	15
Carbon Dioxide	8%	44.01	3.52	9.37	0.75	1073	86	548	44
Water Vapor	2%	18.02	0.36	8.94	0.18	3187	64	1165	23
Total	100%		MW = 24.07		MC_p = 11.14		P_c = 750		T_c = 445

Computation:

$$\frac{C_p}{C_v} = k = \frac{MC_p}{MC_p - 1.986} = \frac{11.14}{11.14 - 1.986} = 1.217$$

Figure 2.2. Typical gas analysis calculation

Given: *Hydrocarbon gas mixture containing 65% methane, 14% ethane, 6% propane, 3% iso-butane, 2% normal butane, 8% carbon dioxide, 2% water vapor.*

Find: *Molecular weight (MW), ratio of specific heats (k), critical presure (P_c), and critical temperature (T_c).*

Pressure and Temperature

Pressures and temperatures are required for selection of a compressor. Pressure and temperature should be given at suction or inlet conditions and pressure at discharge conditions. Discharge temperature is calculated, based on the type of compression cycle expected, which affects the heat rise during compression. Pressures are normally expressed in pounds per square inch gage (psig) or absolute (psia), where psia = psig + barometric pressure (2).

Capacity

Capacity can be stated in numerous ways, based either on weight flow or volume flow requirements. Most common expressions for capacity by the process engineer are:

- Weight flow, pounds per hour or pounds per minute.
- Volume flow in "standard conditions," defined in the process industries as 14.7 psia and 60 F (or 32 F). Standard units are SCFM (standard cubic feet per minute), SCFH (standard cubic feet per hour), or MMSCFD (million standard cubic feet per day).
- Volume flow may also be stated at inlet flow conditions, usually ICFM (inlet cubic feet per minute) or ACFM (actual cubic feet per minute).

Unless capacity is given at inlet flow conditions, it is necessary to convert to these conditions. Any or all of the following relations may be used, as applicable:

$$\frac{P_1 V_1}{T_1 Z_1} = \frac{P_2 V_2}{T_2 Z_2} \quad \text{or} \quad V_2 = V_1 \cdot \frac{P_1}{P_2} \cdot \frac{T_2}{T_1} \cdot \frac{Z_2}{Z_1} \tag{3}$$

$$\text{SCFM} = \frac{\text{mols/hour} \times 379.46}{60} \tag{4}$$

$$\text{Pounds per hour} = \text{mols per hour} \times \text{MW} \tag{5}$$

$$\text{Specific volume, } v_s = Z_s \cdot \frac{1545}{\text{MW}} \cdot \frac{T_s}{144 \times P_s} \quad \text{ft}^3/\text{lb} \tag{6}$$

$$\text{Density } \rho = \frac{1}{v_s} \quad \text{lb/ft}^3 \tag{7}$$

$$\text{ICFM} = \frac{\text{MMSCFD} \times 10^6}{1440} \cdot \frac{14.7}{P_s} \cdot \frac{T_s}{520} \cdot \frac{Z_s}{1.0} \tag{8}$$

$$\text{ICFM} = \text{lb/min} \times v_s \quad \text{or lb/min}/\rho \tag{9}$$

PERFORMANCE CALCULATIONS

Recalling now the basic relationship of a gas:

$$PV = wZRT \tag{10}$$

it can be shown that the adiabatic work performed on a gas in raising its pressure from P_1 to P_2 is:

$$wZRT \left(\frac{k}{k-1}\right)\left(\frac{P_2}{P_1}^{\frac{k-1}{k}} - 1\right) \tag{11}$$

If this calculation is made in English units, with w equal to the weight flow in pounds per minute, the gas horsepower will be:

$$GHP = \frac{wZRT \left(\frac{k}{k-1}\right)\left(\frac{P_2}{P_1}^{\frac{k-1}{k}} - 1\right)}{33{,}000} \tag{12}$$

and the actual gas horsepower based on a known or assumed value of adiabatic efficiency will be this number divided by that efficiency, expressed as a decimal.

The expression

$$ZRT \left(\frac{k}{k-1}\right)\left(\frac{P_2}{P_1}^{\frac{k-1}{k}} - 1\right) \tag{12a}$$

is the adibatic head which the compressor must produce to meet the required pressure. It is developed from the adiabatic relation

$$P_1 V_1{}^k = P_2 V_2{}^k = \text{constant} \tag{13}$$

Certain types of compressors closely perform according to adiabatic cycles; other types, notably the uncooled multistage centrifugal compressor, do not. Deviation from adiabatic compression results in a polytropic cycle, where the value of the polytropic exponent, n, is substituted for the adiabatic exponent, k. For a known or assumed value of polytropic efficiency, η_p, the relation between the two exponents can be expressed by:

$$\frac{\dfrac{k-1}{k}}{\eta_{poly}} = \frac{n-1}{n} \tag{14}$$

Summarizing then, either of these may be used for preliminary sizing of compressor horsepower:

$$\text{adiabatic gas HP} = \frac{\text{weight flow, lb/min} \times \text{head, adiabatic}}{33{,}000 \times \text{adiabatic efficiency}} \tag{15}$$

$$\text{polytropic gas HP} = \frac{\text{weight flow, lb/min} \times \text{head, polytropic}}{33{,}000 \times \text{polytropic efficiency}} \tag{16}$$

It is important to remember not to mix adiabatic and polytropic values in a given calcuation; either use adiabatic exponent k and adiabatic efficiency, or polytropic exponent n and polytropic efficiency. Adiabatic efficiency may be defined as:

$$\eta_{adia} = \frac{\left(r_c^{\frac{k-1}{k}} - 1\right)}{\left(r_c^{\frac{n-1}{n}} - 1\right)} \tag{17}$$

Compressor calculations may also be performed using enthalpy differences as a means of determining adiabatic head. From charts or tables for the given gas or gas mixture, find enthalpy at inlet, h_1 and at discharge, h_2, then:

$$\text{adiabatic head} = (h_1 - h_2)\,(778) \quad \text{or} \quad (\Delta h)\,(778) \tag{18}$$

In polytropic compression the actual discharge enthalpy $h_2{}'$ will differ from the adiabatic value; such discharge point must be determined by including the compression efficiency:

$$h_1 - h_2{}' \quad \text{or} \quad \Delta h_{poly} = \frac{\Delta h_{adia}}{\eta_{adia}} \tag{19}$$

Temperature rise during a compression cycle may likewise be found either by calculation or from pressure-enthalpy charts or tables.

$$T_{d_{poly}} = T_s \cdot r_c^{\frac{n-1}{n}} \tag{20}$$

and

$$T_{d_{adia}} = T_s \cdot r_c^{\frac{k-1}{k}} \qquad (21)$$

CENTRIFUGAL COMPRESSORS

Although first applied in process use over fifty years ago, the centrifugal compressor has come into greatest popularity in the past thirty years and is today the preferred type for most process applications. Current U.S. manufacturing programs from major compressor builders include single and multistage compressors in sizes from somewhat less than 500 cfm inlet flow to well over 100,000 cfm. High pressure centrifugals have been installed on many gas pipelines, process plants and gas injection plants, ranging up to the highest pressure installation in a North Sea gas plant at approximately 10,000 psi.

Fundamental Principles

The centrifugal compressor consists of one or more impellers attached to a shaft, rotating inside of a pressure-casing which has stationary diffuser passages with or without vanes, or a scroll-type volute shaped much like a centrifugal pump casing. The rotating shaft and impeller impart energy into the air or gas due to the velocity of the impeller, then the high velocity gas moves through stationary diffusers or volutes in which the velocity is reduced before exit from the stage or the compressor casing.

The basic function of the compressor impeller can be seen from examination of Figure 2.3, simplified velocity diagrams for a typical impeller with backward-leaning blades. Velocity triangles are shown for the entrance and the exit of the impeller. The velocity components c (absolute fluid velocity), u (peripheral velocity of rotor), and ω (fluid velocity relative to rotor) are resolved into velocity triangles at inlet and outlet. The ideal head generated by an impeller can be expressed in the form known as Euler's equation:

$$\text{Head (ideal)} = \frac{U_2\, c_{u_2} - U_1\, c_{u_1}}{g} \qquad (22)$$

The peripheral velocity is easily calculated from the rotor dimensions:

$$U = \frac{\pi D N}{720} \qquad (23)$$

Thus these triangles are used only for theoretical determinations of basic

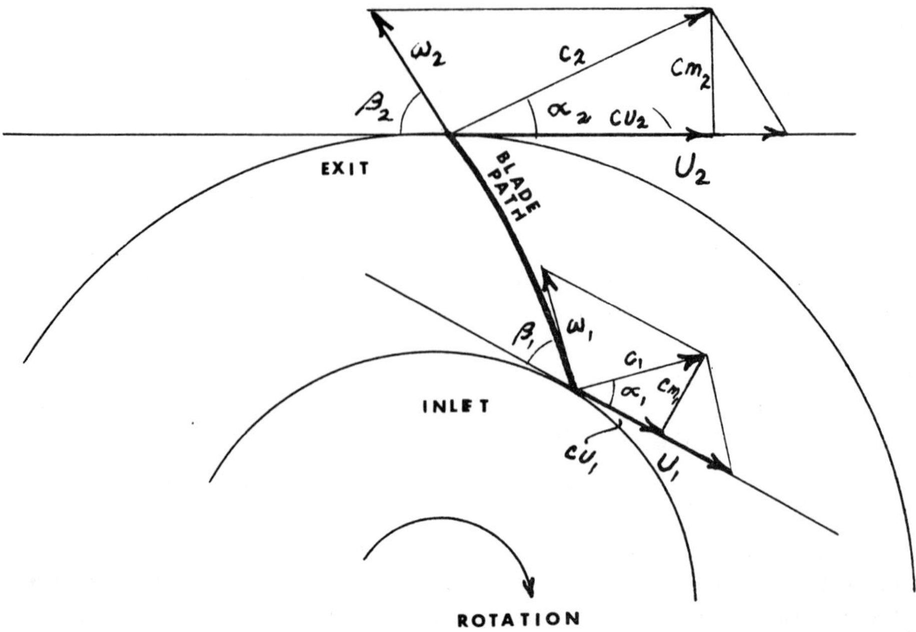

Figure 2.3. Impeller velocity diagram.

relationships of blade angles and prerotation at inlet. Assuming an impeller with no inlet swirl, $c_{u_1} = 0$ and the basic equation for head becomes:

$$H_{ideal} = \frac{U_2 \, c_{u_2}}{g}$$

Since $c_{u_2} = U_2 - \omega_{u_2}$ or $U_2 - \dfrac{c_{m_2}}{\tan\beta_2}$

then

$$H_{ideal} = \frac{U_2{}^2 - \dfrac{U_2 \, c_{m_2}}{\tan\beta_2}}{g} \qquad (24)$$

In a radial-bladed impeller, where $\beta_2 = 90°$, the expression becomes:

$$H_{ideal} = \frac{U_2{}^2}{g} \qquad (25)$$

and can be shown to decrease as the blade angle decreases in backward-leaning impeller designs.

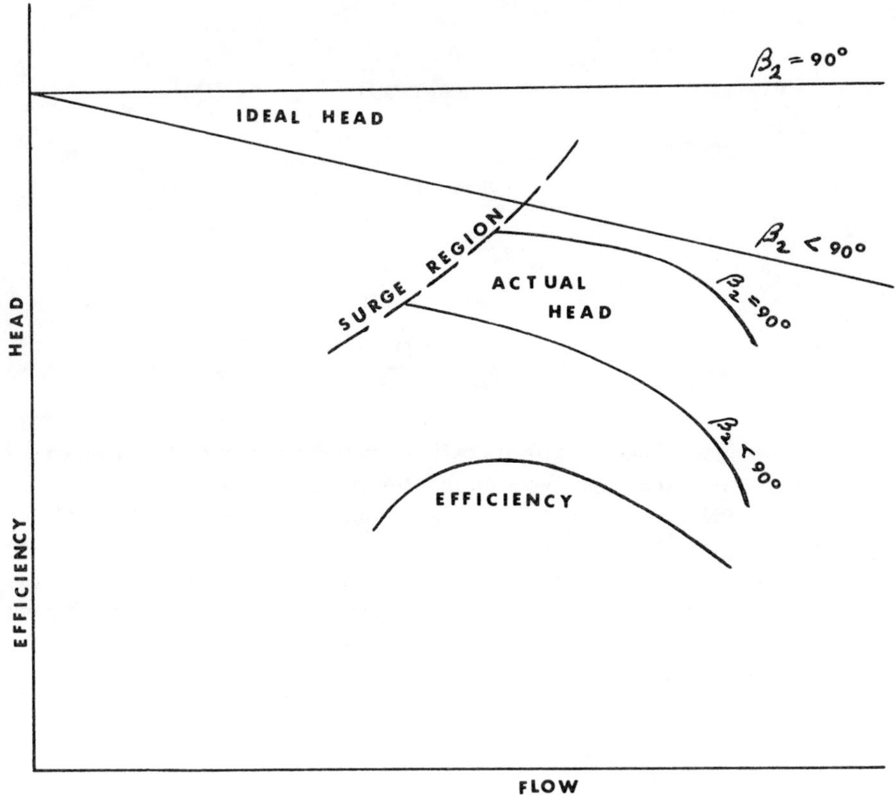

Figure 2.4. Ideal and actual heads.

The above relationships are theoretical; actual impellers produce significantly less head because of inefficiencies due to disc friction, slip, incidence losses (at off-design conditions), and other factors. Figure 2.4 shows theoretical heads and actual heads for radial-bladed and backward-curved impellers. It illustrates the fact that the radial bladed impeller produces higher head but has a curve shape which tends to be flat, whereas the backward-curved impeller curve shape has greater slope and a wider range of stable operation. Both impeller types are in wide use in commercial machines today. Although it is helpful to the process engineer to have an understanding of the basic fundamentals, selection of each impeller for a given application must be left to the compressor designer.

Head Coefficient and Flow Coefficient

Convenient terms for compresor design and analysis are the head coefficient (μ) and the flow coefficient (ϕ), defined as:

$$\mu = \frac{\text{Head (actual)}}{\text{Head (ideal)}}$$

or

$$\mu = \frac{\text{Head (actual)} \cdot g}{U_2{}^2} \tag{26}$$

and

$$\phi = \frac{700\ Q}{N\ D^3} \tag{27}$$

Figure 2.5 lists typical values of head coefficients and nominal diameter range for impellers for current U.S. manufactured centrifugal compressors. Figure 2.6 shows approximate overall polytropic efficiency as a function of inlet flow. Flow coefficient using ft^3/min for Q and impeller diameter, D, in inches, will range from 0.01 to 0.14 for radial flow impeller types. It is important to note that these values are based on stage performance; in a multistage compressor the overall head coefficient becomes:

$$\mu = \frac{\text{Head per stage (actual)} \cdot g}{U_2^2} \tag{28}$$

and the flow coefficient may be determined at the inlet or exit of each impeller. At other than the design rated condition, similar values of μ and ϕ can be found, resulting in the display of a given stage or an overall machine performance curve in several manners (Figure 2.7), showing flow versus head or flow versus pressure, both for one single speed, also flow coefficient versus head coefficient (dimensionless) which is useful in generating performance curves at varying speeds or on alternate gas conditions in the same compressor. These curves also show efficiency, recognizing from earlier discussion that inlet and exit flow angles for off-design conditions will be less than optimum, blade incidence losses higher, resulting in varying overall efficiency throughout the stable operating range.

For backward-leaning closed impellers which are used in conventional multistage compressors, impeller peripheral speed, U, will range from approximately 750 to 900 ft/sec unless higher molecular weight gas requires lower speed. Single stage and radial type wheels may utilize higher values of U. A quick check of the relation of

Inlet Flow Range ICFM or Q_s	Average Head Coefficient*	Nominal Impeller Diameter inches
500-2,500	0.48	13-16
1,500-7,500	0.49 to 0.50	17-19
4,000-12,000	0.50 to 0.52	21-22
6,000-18,000	0.51 to 0.52	24-26
8,000-35,000	0.52-to 0.53	30-32
20,000-60,000	0.53 to 0.54	36-42
60,000-100,000	0.54 to 0.55	48-60

*Based on impellers with backward-curved blades; higher values (0.60 to 0.63) are found in impellers with radial blades.

Figure 2.5. Preliminary selection data for centrifugal compressors.

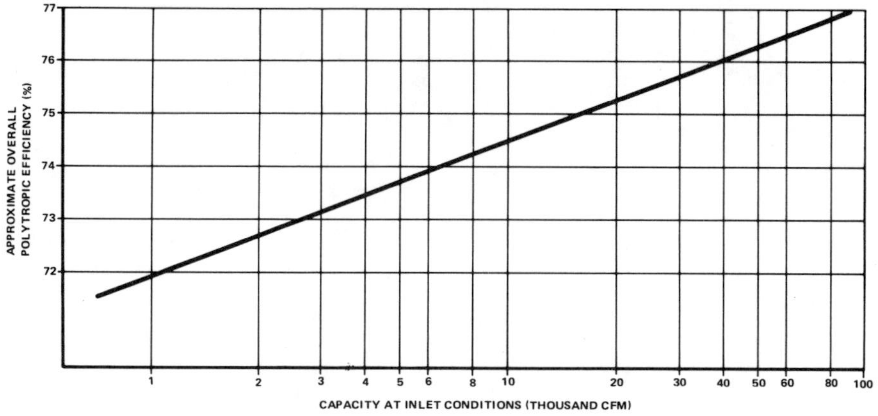

Figure 2.6. Overall polytropic efficiency.

the actual gas velocity through the compressor impeller versus the acoustic velocity (v_a) of the gas should often be made, especially on gases with higher molecular weight. Acoustic velocity, $v_a = \sqrt{k\ g\ R\ T\ Z}$. Although the primary interest in checking the ratio of impeller peripheral speed to acoustic velocity is at impeller inlet, it is seldom possible to obtain sufficient data about impeller inlet dimensions for use in preliminary calculations. Experience has shown that by keeping the ratio of impeller tip speed (U_2) to acoustic velocity (v_a) below approixmately 0.95 for conventional impeller designs, an acceptable, stable compressor design may be achieved.

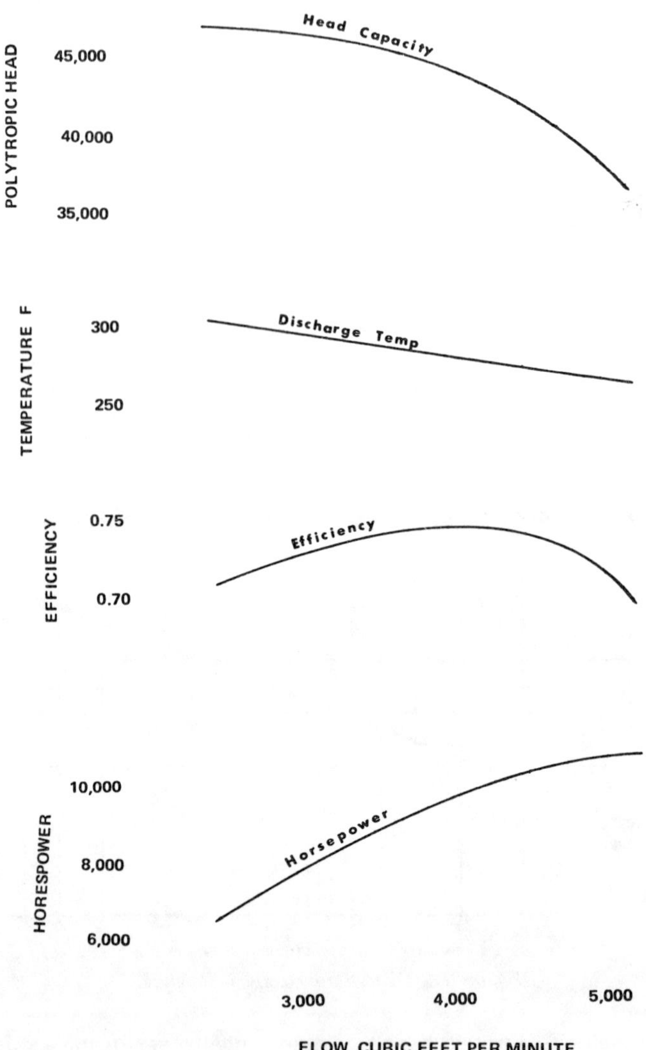

Figure 2.7a. Performance curve for calculation in Figure 9.

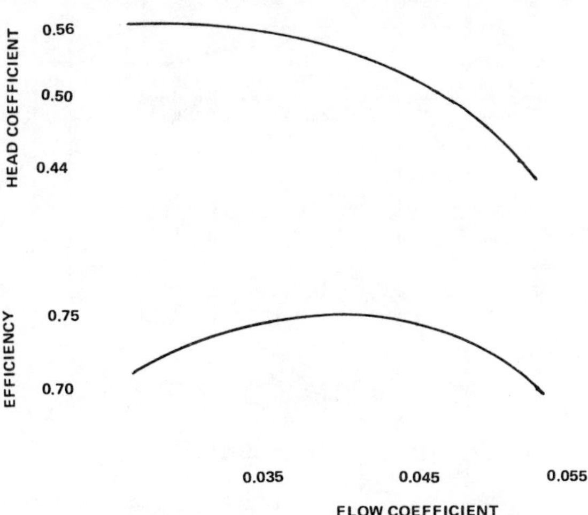

Figure 2.7b. Performance curve — Dimensionless.

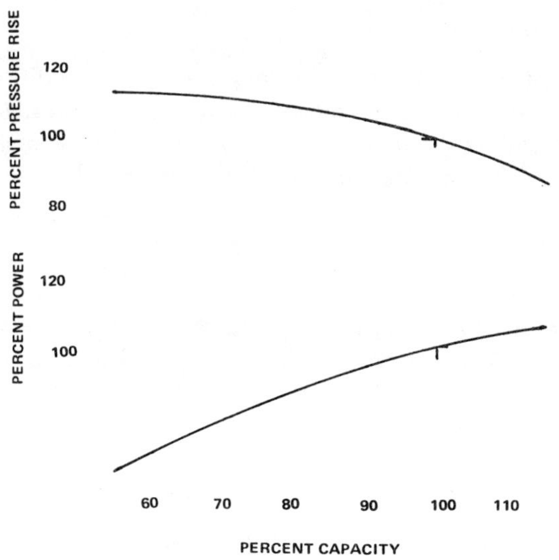

Figure 2.7c. Performance curve — Flow versus pressure rise.

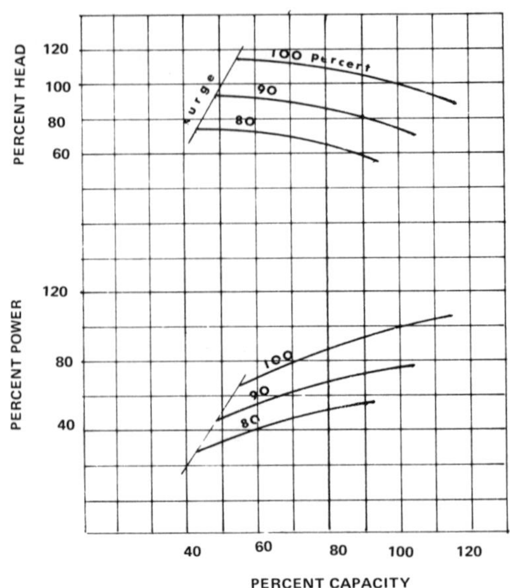

Figure 2.7d. Performance curve variable speed.

Bearing and Seal Losses

In any centrifugal compressor calculation, a small power addition must be made to cover losses due to friction and leakage in bearings and seals. While such losses will obviously vary with the exact size and design of bearings or seals, values such as those shown in Figure 2.8 may be used for preliminary calculations.

Sample Calculations

Figure 2.9 shows a typical preliminary calculation for a multistage compressor using the overall polytropic head mehod. The gas is the same as the example in Figure 2.2. Figure 2.10 shows how a calculation is made on a pure gas such as carbon dioxide, utilizing a pressure-enthalpy chart to determine values of head, specific volume, and temperature.

TYPES OF CENTRIFUGAL COMPRESSORS

Centrifugal compressors are built in one-stage and multistage designs. The single-stage compressor has one impeller, usually overhung from a bearing assembly and enclosed in a volute diffuser assembly similar to a large centrifugal pump (Figure 2.11). Such types are commercially available today in size ranges from about 1000 ICFM to 150,000 ICFM. The enclosed, backward leaning, two-dimensional impeller (Figure 2.12a) produces adiabatic heads up to about 12,000 ft lb/lb. The semi-open

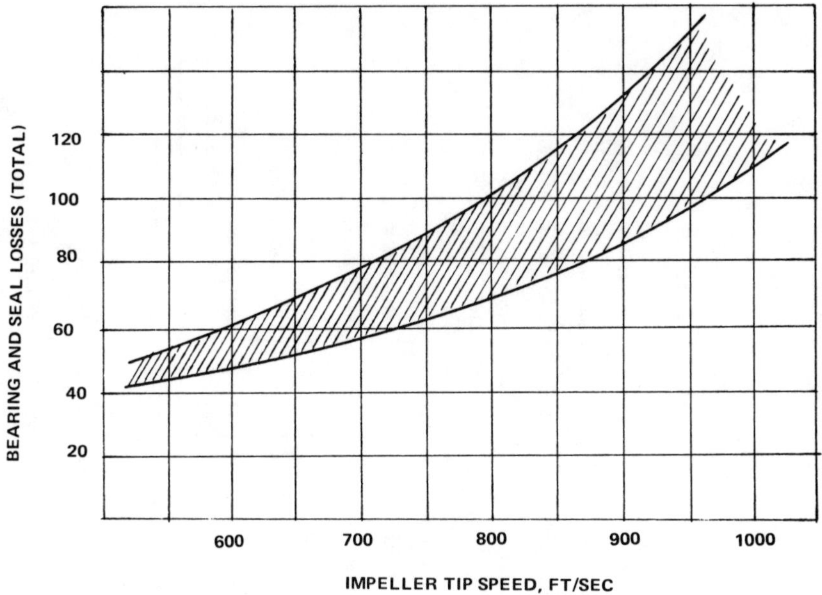

Figure 2.8. Bearing and seal losses.

radial bladed design, usually with prerotation or inducer vanes making it three-dimensional (Figure 2.12b) produces heads to 20,000 ft lb/lb. Similar designs use slightly curved blading and inducer blading (Figure 2.12c). Higher strength materials such as titanium may be employed in small machines, allowing higher impeller tip speeds and resulting in higher heads. Small industrial high speed designs are also available today, making this type in one form or another a very versatile machine.

Many process applications require higher heads or pressures than a single stage machine can produce. Multistage compressors thus have become the most widely used throughout industry. Many different design arrangements are available to suit widely varying service conditions. The most common type is the uncooled, "straight through" variety (Figure 2.13) where as many as eight or ten impellers are arranged on one shaft, running at the same speed, and each producing approximately the same head. Head per stage will vary to a certain extent as the volume flow through a given machine is reduced and succeeding stages or impellers are designed for lower volume flow and hence exhibit somewhat lower head coefficients and efficiencies. Figure 2.13 shows the casing split horizontally, parallel to the shaft, with a flat joint held tight by numerous casing studs. Figure 2.14 shows a similar type machine where the outer casing is a cast or forged barrel, having no horizontal joint. The rotating element is inserted from the outer end and the vertical casing joint is made to withstand higher pressures as well as to provide a

PROCESS REQUIREMENTS (GIVEN):		
Gas	Hydrocarbon Mix (Figure 2)	
Molecular Weight	24.07	
Specific Heat Ratio	1.217	
Flow, MMSCFD	132	
Suction pressure, psia temperature, °F	300 100°	
Discharge pressure, psia	900	
CALCULATION		**EXPLANATION**
Compressibility factors: At suction Z_s At discharge Z_d	 0.93 0.93	 Reference 2 "
Flow, cfm at inlet	4500	Equation 8
Specific volume, ft^3/lb	0.774	Equation 6
Weight flow, lb/min	5815	$ICFM/v_s$
Adiabatic exponent $\frac{k-1}{k}$	0.178	Calculate
Impeller diameter D	19"	Figure 5
AT RATED CONDITION		
Head coefficient, μ	0.50	Figure 5
Polytropic efficiency η_p	0.735	Figure 6
Polytropic exponent $\frac{n-1}{n}$	0.242	Equation 14
Ratio of compression	3.0	p_d/p_s
Polytropic head	42070	Equation 12_a using polytropic exponent

CALCULATION		
Gas horsepower	10086	Equation 16
Friction horsepower, bearings and seals	90	Figure 8
Total BHP	10176	
Discharge temperature °R °F	731° 271°	Equation 21 T_d - 460
Number of stages	4	Average 10000 ft/stg.
Impeller tip speed, U	823	Equation 28
Rotating Speed, rpm	9919	Equation 23
Flow coefficient, ϕ	0.046	Equation 27
Acoustic velocity, v_a	1145	$\sqrt{k\ g\ R\ T\ Z_s}$
"Pseudo"-Mach Number	0.72	U/v_a
AT SURGE CONDITION		**EXPLANATION**
Approximate Flow, ICFM	2475	55% of rated
Head coefficient, μ	0.56	Mfr's data
Efficiency, η_p	0.705	" "
Polytropic head	47118	Equation 12a
AT OVERLOAD CONDITION		
Approximate flow, ICFM	5175	115% of rated
Head coefficient, μ	0.43	Mfr's data
Efficiency, η_p	0.69	" "
Polytropic head	36180	Equation 12a

Figure 2.9. Centrifugal compressor calculation — polytropic head method.

PROCESS REQUIREMENTS (GIVEN):	
Gas	Carbon Dioxide
Molecular Weight	44.01
Specific Heat Ratio	1.30 @ 40°; 1.29 @ 100°
Flow, Pounds/hour	120,000
Suction pressure, psia temperature, °F	20 40°
Discharge pressure, psia	290

CALCULATION	1st Sect.	2nd Sect.	SOURCE
At Inlet:			
Pressure psia	20	95	Given;Note 1
Temperature °F	40	100°	" "
Specific volume ft^3/lb	6.0	1.38	P-H Diagram
Weight flow lb/min	2000	2000	Given
Volume flow ICFM	12000	2760	$w \times v_s$
Enthalpy h_s	-3760	-3750	P-H Diagram
Entropy S_1	1.132	1.082	" "
At discharge:			
Pressure psia	100	290	Given;Note 1
Enthalpy, adia. h_d	-3716	-3717	P-H Diagram
Entropy S_2	1.132	1.082	" "
Δh (adia.) $(h_d - h_s)$	44	33	
Efficiency, adia. η_{ad}	69.6%	69.4%	Eq. 17
Head, adia. H_{ad}	34232	25674	$\Delta h \times 778$

CALCULATION (CONT'D)	1st Sect.	2nd Sect.	SOURCE
Δh (poly.) $(h_d - h_s)$	63	47.5	$\Delta H_{ad}/\eta_{ad}$
Enthalpy, poly h_d	-3697	-3702.5	P-H Diagram
Entropy S_2	1.157	1.104	" "
Efficiency, poly η_p	74.5%	73%	Figure 6
Head, polytropic H_p	36642	27006	$H_{ad} \times \eta_p/\eta_{ad}$
Specific volume, ft^3/lb	1.9	0.64	P-H Diagram
Flow, cfm at discharge	3800	1280	$w \times v_s$
Temperature °F	335°	322°	P-H Diagram
Gas horsepower	2981	2242	Eq. 15 or 16
Friction horsepower	75		Figure 8
Total BHP	5298		
Frame Size	No. 4		Figure 5
Impeller diameter, D	24		" "
Head coefficient, μ	0.52	0.51	" "
Number of stages	4	3	Assume
Impeller tip Speed U	753ft/sec		Eq. 28
Rotating speed N	7186 rpm		Eq. 23
Acoustic velocity v_a	857	904	$\sqrt{kg\ RTZ}$
"Pseudo" Mach No.	0.88	0.83	U/v_a

Note 1: Second section suction conditions are determined by choosing approximately equal temperature rise across each section.

Note 2: Data from P-H diagram copyright 1972 Gulf Publishing Company.

Figure 2.10. Centrifugal compressor calculation — pressure-enthalpy method.

Figure 2.11. Single stage centrifugal compressor.

more positive seal to prevent gas leakage. The American Petroleum Institute Standard 617 for Centrifugal Compressors (Ref. 4) specifies that the vertically split or barrel type casing must be used for pressures above 200–250 psig if the hydrogen content of the gas mixture is 70% or greater. It is common practice to use the barrel type for any gas or gas mixture if the discharge pressure will be higher than about 500 psig.

Multistage compressors can be divided into sections, with intercoolers for gas cooling between each impeller or each group of two or three impellers. Overall compression efficiency is increased by this design as each stage or section performs more closely to adiabatic. Disadvantages include higher cost of casing designs with additional inlet and outlet nozzles, plus the costs of supply and installation of intercoolers and piping. In certain applications however, the advantages far outweigh the cost differential.

Another variation to the multistage type is the use of a double-flow impeller, usually on the first stage, with the flow divided into two streams, each of which flows through one side of a double-suction type impeller. Succeeding stages, usually single flow, are arranged with impellers facing in opposite directions on the same shaft, thus equalizing to a large degree the axial thrust produced by each impeller and thus providing better axial balance.

Air compressors for industrial or process applications have been developed in recent years which utilize fully the above principle of intercooling between stages. In this type (Figure 2.15), an intercooler is located between every impeller, making possible highest adiabatic efficiency. The individual impellers are mounted on pinion shafts driven from one main gear (Figure 2.16), thus providing different operating speeds for each impeller or pair of impellers, based on one driver input shaft speed. The hydraulic advantage of this arrangement is obvious when we recall the relationships of head coefficient (μ) and flow coefficient (ϕ) and their optimum selection with respect to efficiency (Figure 2.7). This type machine is available only for use on air or inert gas such as nitrogen, due primarily to mechanical arrangements of seals adjacent to the bearings and drive gear. Special oil film seals for flammable, toxic or dirty gases have not been applied to this design.

The modular type multistage machine (Figure 2.17) is also available for flows up to about 30,000 ICFM and pressures to about 15 psig. It is widely used as aeration

Figure 2.12a. Impeller – enclosed, backward-leaning, 2-dimensional type.

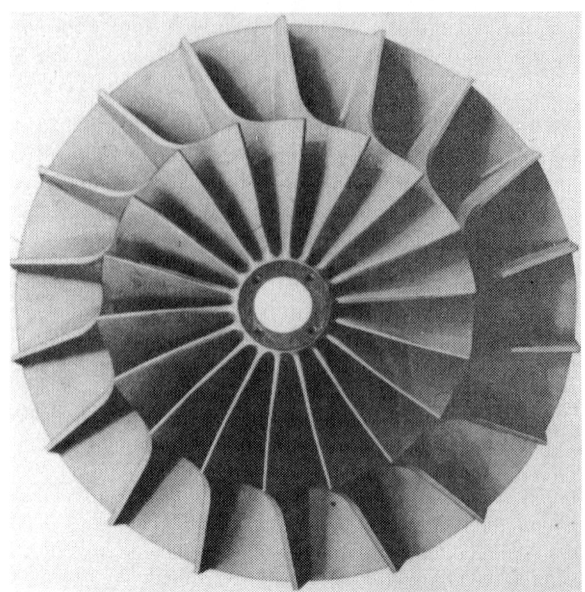

Figure 2.12b. Impeller — semi-open, radial bladed type.

Figure 2.12c. Impeller — semi-open, advanced inducer blading.

Figure 2.13. Multistage horizontally split compressor.

Figure 2.14. Multistage vertically split "barrel" compressor.

Figure 2.15. Multistage, intercooled type air compressor.

blower in sewage treatment plants or in certain chemical or petrochemical processes. It has the advantage of being low in first cost, yet in current designs does not sacrifice much in overall efficiency compared with more expensive designs. It is normally limited to speeds not exceeding 4000 rpm and service only on air or inert gases, due to availability of seal designs (labyrinth or carbon ring types only) casing materials, and casing joint assembly method. It should be noted that this style uses much lower impeller tip speeds and accordingly will require three or four stages to produce as much head rise as the previously described single-stage blower.

Multistage compressors may also be arranged in series, using two or three casings connected to the same drive shaft and driver (Figure 2.18). In this manner it is possible to provide a much higher pressure rise and yet stay within mechanical design limits of eight to ten impellers on a single shaft.

Temperature rise during compression must also be carefully checked before designing or specifying a machine. Usual limits for preliminary guidance are to stay below 350 F for flammable, toxic or dirty gases, or for air, and below 400 F for gases such as carbon dioxide or nitrogen, where higher discharge temperatures will

Figure 2.16. Bull gear and pinions from intercooled type compressor.

not present a hazard. By inserting intercoolers between stages or between casings, it is possible to achieve the higher pressures required without exceeding maximum safe discharge temperature limits. It should be noted that the discharge temperature varies over the head-capacity characteristic curve of a given compressor (Figure 2.7) and thus the engineer must check temperature rise throughout the entire proposed operating range of a given machine.

Bearings

Hydrodynamic, sleeve type bearings are used on all centrifugal compressors with a few exceptions in smaller single stage overhung types and modular multistage blowers, where antifriction bearings are used successfully. Higher speeds and larger power requirements needed for most process compressors, demand hydrodynamic bearings. Both journal type and tilting pad type bearings are used. Kingsbury type thrust bearings, or an equivalent design, will be found in every American or foreign design. Bearings such as these require forced-feed lubrication, which in turn requires a complete, self-contained lubricating oil system for each compressor or group of compressors. This system may be quite complex and must be given sufficient atten-

Figure 2.17. Modular type low pressure air blower.

tion by the specifying engineer or the designer to ensure a completely adequate system. American Petroleum Institute Standard 614 (Reference 5) has been written to guide users in the requirements of complicated lubricating and seal oil systems for special applications. Numerous diagrams appear in the appendix of API-614 which can aid the engineer in considering many optional arrangements.

Shaft Seals

Equally important in the design of the centrifugal compressor is the type of shaft seals to be used to prevent excessive leakage of the compressed gas to atmosphere. Seals may be classified into four basic types:

- Labyrinth type
- Restrictive-ring (carbon-ring) type
- Oil-film and pumping-ring type
- Mechanical-contact type

Labyrinth or restrictive ring types may be used only when some leakage of air or gas can be tolerated. Oil-film and mechanical contact types are normally used for

Figure 2.18. Compressors arranged in series.

any process gas or gas mixture. When even trace quantities of gas cannot be allowed to atmosphere, (such as in gas mixtures containing hydrogen sulfide), buffer gas is required, at a pressure somewhat higher than the sealing pressure. Buffer gas is injected into the seal and forms a barrier between the compressed gas and the atmosphere. Source of such buffer gas may be a problem, even requiring a small additional compressor in some applications, to supply a gas such as nitrogen at the required sealing pressure. Increasing restrictions on gas leakage to atmosphere because of pollution restrictions has made the use of buffered-gas seals more widespread. The specifying engineer as well as the designer of the compressor must consider this subject carefully. Figure 2.19 and reference 4 show some views of typical compressor shaft seal types.

For the oil-film and the mechanical-contact seal, oil is also required to form the oil film between seal faces or rotating and stationary members. Again, it is necessary to have a complete, self-contained oil circulating system similar to the lubricating oil system. The seal oil must be at a pressure slightly above seal operating pressure and frequently requires the use of high-pressure seal oil pumps, filters and coolers. In many designs, the seal oil pressure is kept at a fixed static head differential above the seal by the use of overhead seal oil tanks, thus automatically regulating sealing oil pressure even in event of changing gas pressures. Contaminated seal

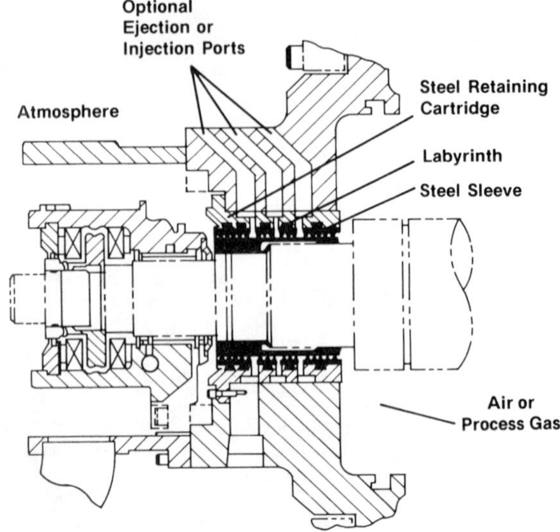

Figure 2.19a. Compressor shaft seal — labyrinth type.

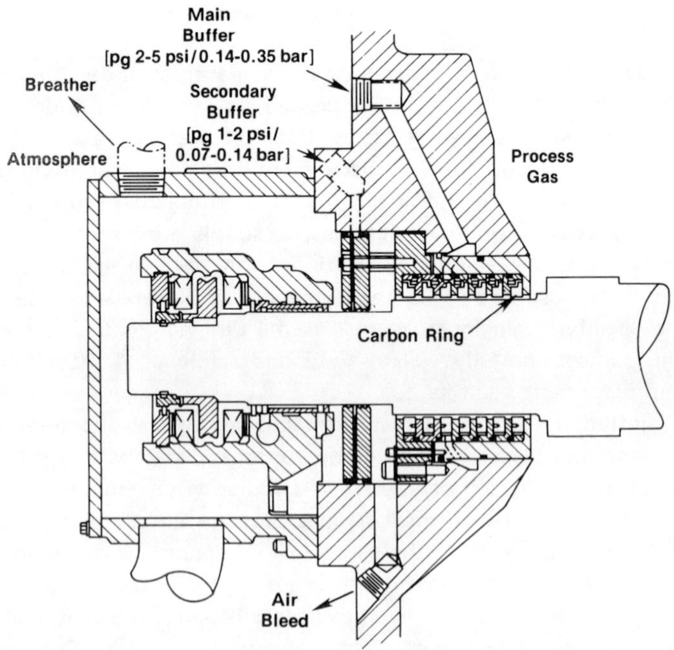

Figure 2.19b. Compressor shaft seal — carbon ring type.

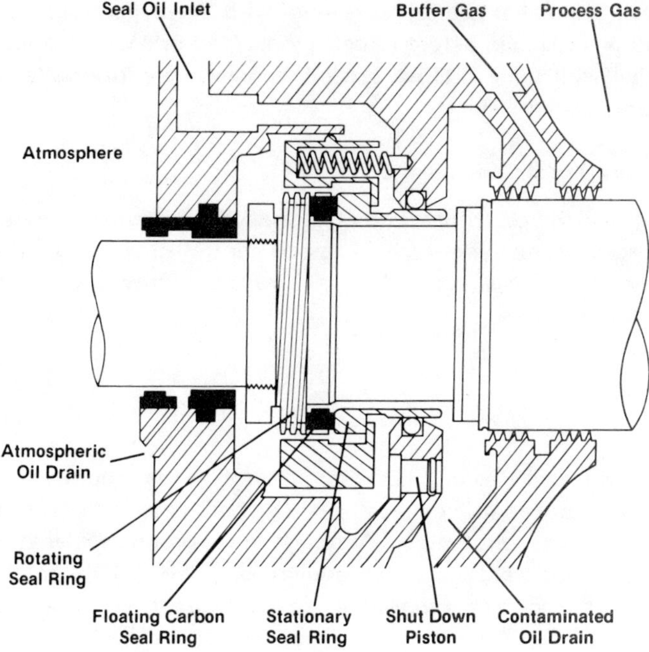

Figure 2.19c. Compressor shaft seal — mechanical contact type.

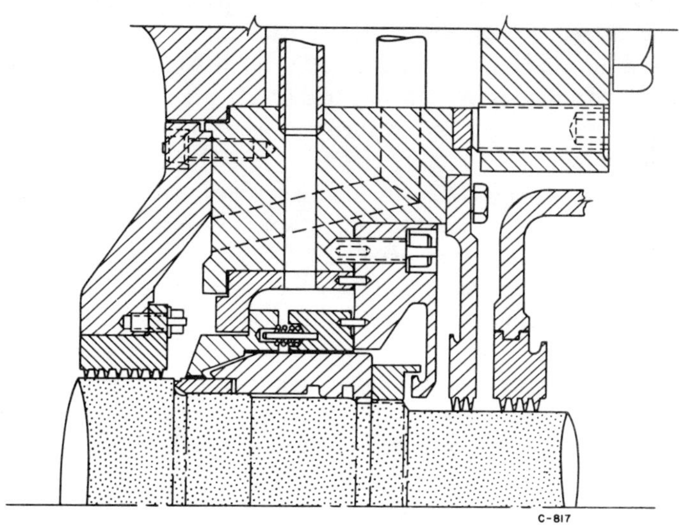

Figure 2.19c. Compressor shaft seal — oil film, pumping ring type.

oil, that is, oil which has come into contact with the compressed gas, may need to be discarded or reclaimed before reuse. Contamined or sour oil drain traps may be used, together with some external seal oil reconditioning facility, all a part of the complex compressor installation.

Control of Centrifugal Compressors

Recalling that the centrifugal compressor develops head, not pressure, it is easy to understand how this type machine can be controlled either by throttling or by varying speed. Centrifugal compressors follow the affinity laws, similar to fans or centrifugal pumps:

$$\frac{N_1}{N_2} = \frac{Q_1}{Q_2} = \sqrt{\frac{H_1}{H_2}} \tag{29}$$

The most effective way to match the compressor to the required output is to vary the speed. Many centrifugal compressors are driven by steam or gas turbines making variable speed an inherent feature. Although in the past most electric motor driven machines have been constant speed applications, the variable frequency electric motor driver has made a new appearance in the industry, and appears to have much application for the future.

Looking again at Figure 2.7, showing typical curves for a centrifugal compressor, it is readily seen what the effect will be if the speed is varied and how the use of head coefficient and flow coefficient will make it possible to find required speeds for different gases in the same compressor.

If fixed-speed operation is desired, the compressor may be controlled by throttling the discharge (least efficient), throttling the suction (more efficient), or use of variable inlet guide vanes (Figure 2.20). The latter method uses stationary vanes ahead of the first stage impeller to provide pre-rotation of the gas stream flowing into the compressor. This in turn will cause a variation in the shape of the head-capacity characteristic as the vane angle is increased, and will result in considerable power savings (Figure 2.21). Inlet guide vanes are most effective on single-stage machines, but may have limited effect on multistage machines depending on the number of sets of guide vanes installed.

Another aspect of compressor control which must be considered is that of surge control (or anti-surge control). Every centrifugal compressor of whatever size or design will have a minimum flow limit, known generally as its surge capacity. Below this capacity the operation is unstable and will cause rapid deteriation, increase in vibration and probable failure of the machine in a short time. Operation at or near the surge region is unstable and must be avoided, except as experienced during startup or shutdown cycles where the compressor may be allowed to pass through the surge region as it comes up to speed.

Figure 2.20. Variable inlet guide vanes for compressor.

The most simple solution to the surge problem is use of a bypass valve which will blow off the excess capacity to atmosphere in the case of an air blower, or air compressor application, or return the excess gas to the suction source or a suction vessel with or without cooling on the way. Since the temperature of the gas is increasing during compression, if much of this gas is bypassed back to suction without cooling, the gas suction temperature will rise to an unacceptable value in a very short time. Surge control systems may be designed or purchased from firms who specialize in such designs (Reference 6, 7). The added cost of such special systems can often be justified in a very short time because of the savings in operating costs and the increased dependability of such a custom designed system.

AXIAL FLOW COMPRESSORS

Dynamic type compressors also include axial flow machines, in which the flow of air or gas is parallel to the shaft axis. These types utilize two-dimensional flow analysis rather than three-dimensional analysis as required by radial flow centrifugal wheels. For industrial or process use, the axial flow compressor is generally applied only for very large flow volumes (at least 75,000 ICFM or higher) and is most often used on air, although it has seen limited application on certain gases. Efficiency of

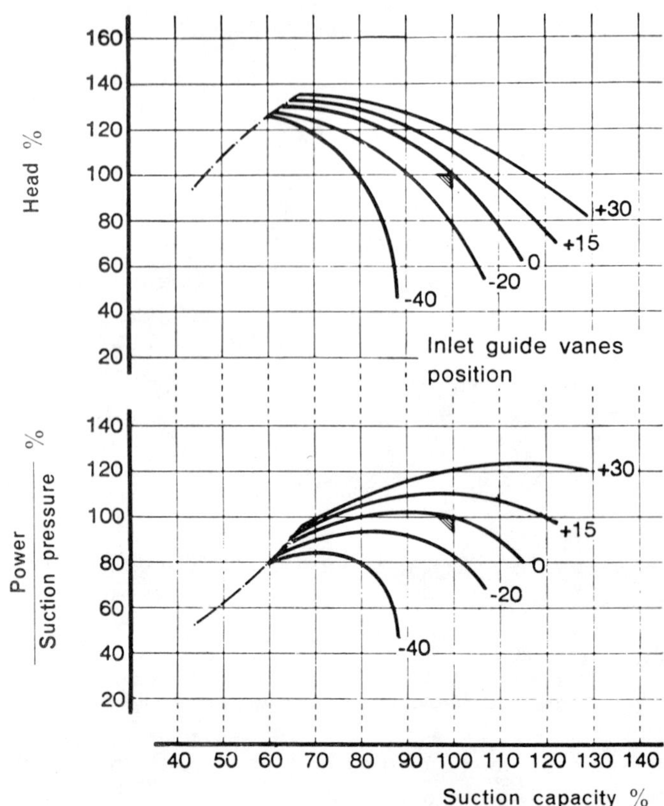

Figure 2.21. Effect of guide vanes on compressor performance.

the axial flow is generally higher than the large centrifugal of comparable size. The cost is usually also higher but can generally be justified in view of the power saving.

The axial compressor stage follows the same basic relationships of head, temperature rise, capacity and power determinations as previously given for centrifugals. Head per stage in an axial machine is usually not more than one-half the amount for a centrifugal. It is common to find axials applied with ten or fifteen stages on one shaft (Figure 2.22) when the required pressure rise demands that many stages. Probably the largest application for the axial flow compressor is in the combustion gas turbine, either for industrial use or as an aircraft engine. Several major processes in the hydrocarbon processing industry, notably fluid catalytic cracking, make wide use of the axial flow type. The process or specifying engineer should look carefully at the axial flow compressor if the volume of air or gas to be compressed is at or near 100,000 ICFM or above.

Axial compressors are readily controlled by varying the stationary (or stator)

Figure 2.22. Axial flow compressor with top half removed.

Stability Range of Axial Compressor

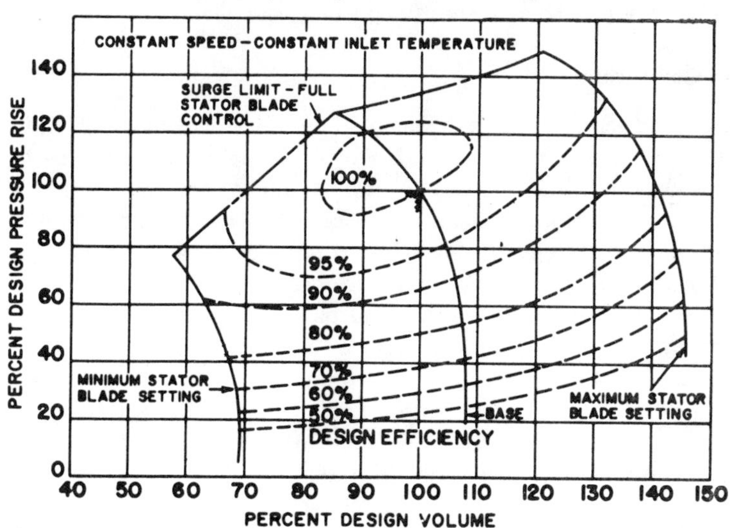

Figure 2.23. Axial flow performance curve.

blade settings, analogous to the inlet guide vanes on centrifugals. Due to the two-dimensional flow, it is easier to use adjustable stator blades on many stages if desired. It is common to find five stages of a ten-stage axial each with its own set of adjustable stator blades, thus giving a fairly wide capacity control range to a machine type which is somewhat narrow in its basic design. Figure 2.23 shows a typical axial-flow performance curve.

DRIVERS FOR CENTRIFUGAL AND AXIAL COMPRESSORS

Three types of drivers are commonly used for dynamic type compressors. Most common is the electric motor at fixed speed. Steam turbines and gas turbines are the other two types.

Electric Motor Drivers

Since the centrifugal compressor operating speed is almost always 3600 rpm or above, the two-pole induction motor may be used for direct connection to a 3600 rpm machine, or in conjunction with a gear speed increaser for higher speeds. Four-pole or six-pole motors are generally used when gear speed increasers are required, as they are somewhat less expensive and more readily available in larger sizes above about 1000 HP. Speed increaser gears are usually parallel shaft, single-reduction type, either double or single helical configuration. Epicyclic gears are also used, notably by foreign manufacturers, and this type will often show a cost saving over parallel shaft type. It is not the purpose of this chapter to cover the subject of drivers and gears in detail. The reader is referred to available standards (references 8, 9, 10, 11) for aid in sizing or specifying a motor or gear.

Steam and Gas Turbines

As mentioned earlier, turbine drivers offer the advantage of variable speed operation, plus sometimes a more dependable or available power source for remote installations or where electric supply may be subject to interruption. Steam turbines to drive centrifugal or axial compressors are usually multi-stage turbines due to size and speed requirements. Small centrifugals, especially the modular type air blower, may utilize standard single stage turbines.

Part of any specification for a steam turbine driven compressor must include a clear statement of steam conditions, both at inlet and exhaust. It is surprising how many process engineers ignore this until pressed for an answer. Just as the compressor cannot be sized or built without knowing both inlet and outlet conditions, neither can the turbine. Consideration should be given to use of a direct-connected turbine or a turbine with speed increaser gear, usually only desirable on smaller units as a means of saving on first cost. Lubrication requirements of turbines should be considered when specifying or engineering the total installation; it is quite common to combine lubrication requirements for compressor, turbine and gear (if used) into one lubrication system. Pneumatically or hydraulically controlled turbine governors provide the most dependable method for utilizing the variable speed feature which every turbine offers.

Gas turbines used to drive compressors are of two types: expansion gas turbines and combustion gas turbines. They are really both closely related except for the addition of the combustion chamber and fuel system in the latter type, which produces the hot gas at a pressure high enough for it to expand through a gas expansion turbine to produce power. The combustion gas turbine is usually a large machine and has excellent operating records driving pipeline compressors and gas compressors in gas treating, gas injection and gas processing facilities.

Expansion turbines may be the hot-gas type, as used in a combustion turbine, or may be cold-gas, radial inflow turbines, usually applied to smaller loads and often when the obtaining of a low temperature from the turbine discharge will serve a useful purpose in the overall chemical process. Again, the reader is referred for further details to numerous references for gas turbines. (references 12, 13, 14).

POSITIVE DISPLACEMENT COMPRESSORS

Reciprocating Compressors

Reciprocating compressors cover a range of capacity from the smallest flow required, such as a utility air compressor in a service station or a very small plant, up through sizes about 5000 ICFM. In large sizes for new installations, rotary or centrifugal types will be utilized today to avoid what some consider to be excessive maintenance on reciprocating machines due to piston rings, valves and packings. Many other users have found the reciprocating type to be their standard for dependable service and continue to install this type on new applications.

Small sizes may have single-acting cylinders and be air-cooled, resulting in mixing of oil vapors from the crankcase with the air or gas being compressed. This type is not recommended in process service or for instrument air service where oil in the compressed air would not be permissible. Several variations of the air-cooled design have appeared, where a double acting crosshead is inserted between the crankcase and the air cylinder, making the design "oil-free."

Small single-cylinder process types (15 to 200 HP) having water-cooled cylinder and arranged either horizontally or vertically (Figure 2.24) may be used for instrument air or process gas. The cylinder is double-acting type, with a separate packing box and distance piece, thus making it possible to have oil-free air or to conduct leaking gases away from the packing under slight pressure.

Larger reciprocating compressors for most process applications will usually have horizontal, water-cooled cylinders. More than one cylinder is generally required, with machines having four, five or six cylinders being quite common in process use (Figure 2.25). Cylinders are placed on opposite sides of a frame and crankshaft, giving a close approach to a balanced design and comparatively low forces and

Figure 2.24. Small reciprocating compressor unit.

moments to be absorbed by proper foundation design. Figure 2.26 shows current typical frame size ratings for medium to large process-type reciprocating compressors.

The reciprocating machine has the feature of producing considerably more pressure per stage than a centrifugal. On certain low molecular weight gases, the head required for a given compression cycle is far beyond the capability of a multistage centrifugal (refer to equation 11). Thus the reciprocating compressor has a distinct area of application on low molecular weight gases. In addition, this type can be built for very small flow requirements, considerably below the lower limits for centrifugal types.

The number of stages of compression is basically a function of frame rating and temperature rise across the stage. Temperature rise should be limited to approximately 250 F across one stage to avoid dangerous outlet gas temperatures.

To make an approximate selection of the number of stages required, the overall total ratio of compression is determined; if too high for one stage (about 3 to 3.5 maximum), the square root will give the ratio per stage for two stages, the cube root for three stages, and so on. Interstage pressure losses due to intercoolers, piping and pulsation devices will cause the actual ratio per stage to be somewhat higher.

Figure 2.25. Reciprocating compressor – process type.

A quick estimate of horsepower can be made either by use of the expression previously given (equation 12 or 15), or by the use of horsepower per million charts as given in standard references. (Reference 1) Actual quoted values for reciprocating compressor horsepower as given by U.S. manufacturers show allowances for gas specific gravity variations; thus the same basic compressor properly equipped with valves for low molecular weight gas appears to show higher compression efficiency than with standard valves for air service. Similarly, as the specific gravity increases beyond that of air, the efficiency appears to become lower. No widespread agreement on reciprocating compressor efficiency is available; Figure 2.27 is included to show a sample of adiabatic efficiencies for gases of varying molecular weights, as a function of compression ratio. Figure 2.28 shows a typical preliminary performance calculation for a process gas type machine.

Cylinder Sizing

When the pressures and temperatures for each stage have been established, the capacity at the inlet of each cylinder (ICFM) can be calculated. Because the piston

		Typical Stroke, inches	Typical Speed rpm	Horse-power Range	Typical Rod Load Pounds
Horizontal, Opposed Cylinders (2 or more)	slow speed	9 9½	600-514	200-800	25,000
		10, 10½	450	400-1,200	35,000
		11, 12	450-400	800-2,000	45,000
		14	327	1,000-2,500	60,000
		15, 15½, 16	327-300	1,500-4,000	100,000
		17, 18, 19, 20	277-257	3,000-10,000	150,000
	medium speed	5	1,000	150-400	25,000
		6, 8	720-900	500-4,000	35,000
		9	600	4,000-8,000	125,000
Single Crank		5, 7	675-514	15-40	5,000
		7, 9	514-450	60-100	10,000
		9, 11	450-400	100-150	15,000
		11, 13	400-327	125-200	20,000

Typical Frame Ratings
Reciprocating Compressors

Figure 2.26. Typical frame ratings – reciprocating compressors.

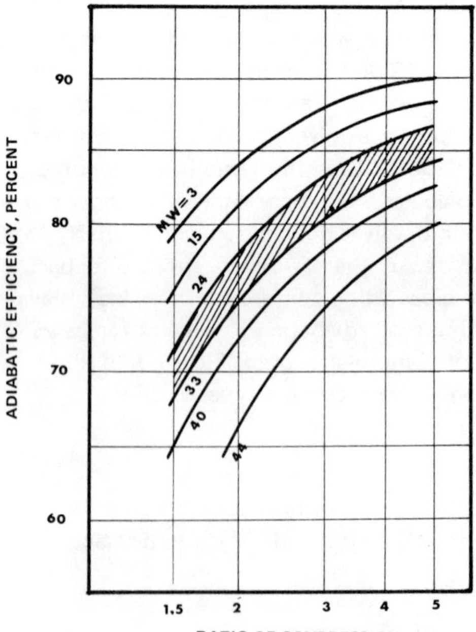

Figure 2.27. Reciprocating compressor adiabatic efficiency,

PROCESS REQUIREMENTS (Given):	
Gas	Hydrogen + Hydrocarbon Mixture
Molecular Weight, MW	6.0
Specific Heat Ratio, k	1.38
Flow, MMSCFD	15.0
Suction pressure, psia	300
Temperature, °F	100°
Discharge pressure, psia	800

HORSEPOWER CALCULATION		EXPLANATION
Compressibility factors: At suction z_s At discharge z_d	1.01 1.026	Compressibility charts (references 1,2,3)
Suction pressure at compressor inlet, p_s	297	Given - 1% pressure drop
Discharge pressure at compressor outlet, p_d	808	Given + 1% pressure drop
Ratio of compression, r_c	2.72	p_d/p_s
Flow, cfm at inlet	561	Equation 8
Spec.vol.at inlet ft^3/#	3.405	Equation 6
Weight flow, lb/min	164.8	$icfm/v_s$
Adiabatic head, H_{ad}	169157	Equation 12a
Adiabatic efficiency, %	86%	Figure 27
Brake horsepower req'd	982	$\frac{\text{Head X weight flow}}{\text{33,000 x efficiency}}$
Discharge temperature, °F	278°	Equation 21; $t_d = T_d - 460$

CYLINDER AND FRAME SIZING		EXPLANATION
Frame piston stroke, in.	14	Figure 26
Crankshaft speed, rpm	327	"
Number of stages	1	Assume
Number of cylinders	2	"
Cylinder clearance, %	14	"
Volumetric efficiency, %	81.9	Equation 32
Displacement req'd., cfm	685	icfm/V.E.
per cylinder, cfm	343	1/2 of above
Cylinder area req'd., sq.in.	129.5	Equation 31
Piston rod diameter, in.	3	Assume
Area, head end of piston, sq.in.	68.3	$A_{he}+A_{ce}-A_{rod} = A_{total}$
Area, crank end of piston, sq.in.	61.2	As above
Area, total, sq.in.	129.5	$A_{he}+A_{ce}$
Diameter of piston, inches	9.3 use 9-1/2	Calculate, round off

FRAME OR ROD LOADING		EXPLANATION
In compression (towards crank)	38319	Equation 33
In tension (away from crank)	30507	Equation 34
Frame published rating, lbs.	45000	Figure 26

Figure 2.28. Reciprocating compressor performance calculation.

does not sweep the entire volume of the cylinder, due to cylinder clearances, the actual cylinder capacity is somewhat lower than the cylinder displacement. This relation is expressed as volumetric efficiency (VE):

$$VE = \frac{\text{capacity at inlet conditions, ICFM}}{\text{cylinder displacement, CFM}} \tag{30}$$

The cylinder displacement can be calculated once the size of cylinder, speed of rotation, and piston stroke are known.

$$\text{Displacement (CFM)} = \frac{A_{he} + A_{ce}}{144} \times \frac{S}{12} \times \text{rpm} \tag{31}$$

Many formulas for volumetric efficiency have been suggested, determined from theory or from actual observed results. The following is adequate for preliminary estimates:

$$VE = 0.97 - \text{clearance}\left(\frac{r_c^{\frac{1}{k}} - 1}{Z_s/Z_d}\right) \tag{32}$$

where clearance is expressed as a decimal.

Frame Load or Rod Load

The compressor frame or piston rod has a limit to the forces that can be applied during the compression cycle. In its simplest form, the actual loadings can be computed from the known cylinder diameter and the pressures acting on the piston. This neglects the frame loading resulting from the reciprocating weights and motion of the machine.

For a double acting cylinder, when the piston is moving inward toward the crankshaft, the frame load in compression is:

$$F_c = (P_d \times A_{he}) - (P_s \times A_{ce}) \tag{33}$$

and for the piston moving outward from the crankshaft, the frame load in tension is:

$$F_t = (P_s \times A_{he}) - (P_d \times A_{ce}) \tag{34}$$

Figure 2.29 illustrates the basic relationships and terminology. Values so calculated must be checked against manufacturer's published frame limits; it is recommended

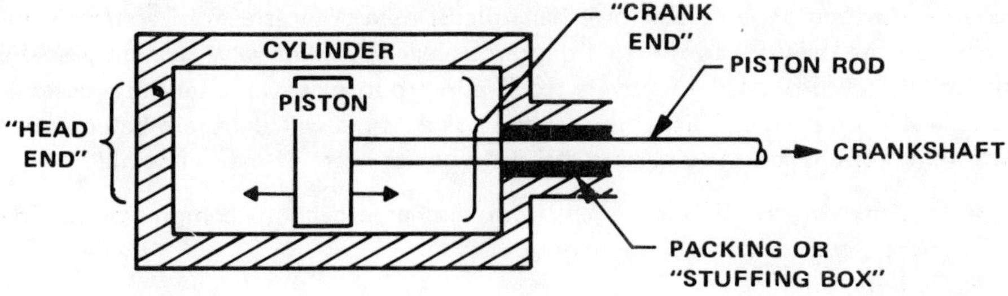

PISTON ROD MOVING TOWARDS CRANKSHAFT:

$$\text{FORCE} - \text{LB (COMPRESSION)} = \left(P_d\right)\left(A_{he}\right) - \left(P_s\right)\left(A_{ce}\right)$$

PISTON ROD MOVING AWAY FROM CRANKSHAFT:

$$\text{FORCE} - \text{LB (TENSION)} = \left(P_s\right)\left(A_{he}\right) - \left(P_d\right)\left(A_{ce}\right)$$

Figure 2.29. Diagram of cylinder.

that the actual values for a given application be not more than 75% to 85% of the published maximums. If the values calculated on a preliminary selection are too high, then the size (diameter) of the cylinder must be reduced, two cylinders used instead of one, or a larger compressor frame with greater frame rating capability must be chosen.

Compressor Speed

Limits on rotating speed and average piston speed should be specified to avoid selecting a design that operates too fast and gives excessive wear and maintenance.

$$\text{average piston speed (ft/min)} = \text{rpm} \times \frac{\text{stroke (inches)}}{12} \times 2 \qquad (35)$$

The general limit in use today for process service is 800 to 850 ft/min for lubricated pistons and 700 to 750 ft/min for non-lubricated, carbon or carbon-filled teflon ring types.

Control of Reciprocating Compressors

Because reciprocating compressors are positive displacement type, a given machine will produce an increased pressure over the rated pressure in event of a change in downstream conditions. Closing a valve, changing another process variable, shutting off a reactor or other process item downstream of a compressor can cause the pressure to rise to a dangerous level. A relief valve or high pressure shutdown may protect the machine from failure but may also upset the process or cause waste or needless plant downtime. Compressors will therefore be equipped with some form of capacity control, basically one or more of the following:

- External bypass of air or gas from discharge back to compressor suction source, with or without cooling
- Cylinder unloaders
- Cylinder clearance pockets
- Variable speed

Most reciprocating compressors are driven by electric motors so variable speed is normally ruled out. Bypass type arrangements are necessary for close regulation and minute-by-minute process variations, but are wasteful of power. More economical control methods include the use of cylinder unloaders and clearance pockets.

Cylinder unloaders are manually or automatically operated devices on one or both ends of a cylinder, designed to unload or hold open the cylinder suction valves. Thus the compressor does not work on that portion of the stroke. For example, inlet valve unloaders could be placed on the head end (outer end) of a cylinder, reducing the net output of that cylinder by approximately one-half when the devices were actuated. Unloaders are usually supplied as a means of totally unloading a compressor for startup, but prolonged operation with unloaders on one end of a cylinder may cause problems of frame load, valve life, or pulsation damping.

Clearance pockets are additional volumes of clearance built into or bolted on to a cylinder head or valve cap, either head end or crank end (or both) to increase the cylinder clearance. Reference to equation 32 will show how increased cylinder clearance results in reduced volumetric efficiency. Figure 2.30 illustrates this relationship graphically, and Figure 2.31 shows a typical cylinder with both automatically operated unloaders and a manually operated valve-cap clearance pocket.

Reflief valves must always be provided in the piping immediately downstream of any reciprocating compressor. They should be sized for the full output capacity of the compressor, set to open at 10 to 15% above the rated compressor discharge pressure.

Pulsation dampers are generally installed with reciprocating compressors on process services, to smooth out the pulsing flow generated by the reciprocating piston. These may be in the form of volume bottles or special devices with internals designed to absorb or cancel some of the pulsing flow. It is common to specify that the pulsations in the piping leading to or from a compressor shall not exceed 1% of the operating pressure in pressures up to about 400 psig; lower values are expected and required in pressures higher than this. These devices require a pressure drop as the gas passes through; careful calculations for sizing reciprocating compressors will include allowances for such pressure losses.

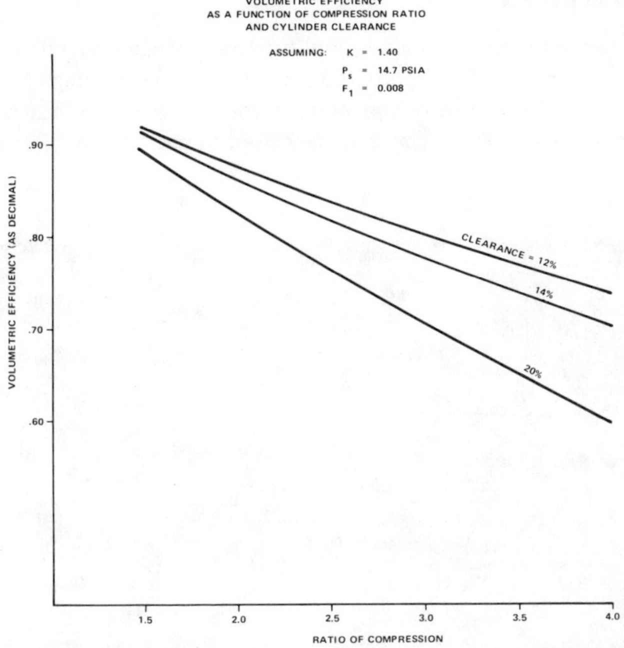

Figure 2.30. Variation in compressor capacity with increased clearance.

Figure 2.31. Compressor cylinder showing unloaders and clearance pockets.

Variable Speed Drivers

Gas engines are the most common type of variable speed driver applied to reciprocating compressors. Figure 2.32 shows a typical integral type gas engine compressor, where the engine and compressor cylinders are attached to the same frame and crankshaft. This type is widely used in gas transmission and oil and gas production.

Figure 2.32. Gas engine driven reciprocating compressor.

Steam turbines have been applied to reciprocating compressors by use of single or double-reduction gears to reduce turbine speed. Although many successful installations have been made, caution should be used when specifying this type drive arrangement as considerable additional analysis of torsional vibrations, couplings, and flywheels is required.

ROTARY COMPRESSORS

The other form of positive displacement compressor is the rotary compressor or blower, in which a rotating element within a casing displaces a fixed volume during each revolution. Many different types are in use today, which can be grouped into four basic categories:

- Lobe-type (2-lobe or 3-lobe)
- Vane-type
- Screw-type (wet-screw or dry-screw)
- Liquid-ring type

Figure 2.33. Lobe-type rotary blower.

Most widely known is the lobe-type (Figure 2.33). Two figure-8 shaped rotors (or variations from this basic concept) mesh together at relatively slow speeds, driven by timing gears attached to each shaft. This type is available in very small sizes, from approximately 2 ICFM, to largest sizes of about 20,000 ICFM. It is basically a low pressure machine, producing up to approximately 15 psig as a blower, and is widely used as also a vacuum pump. The three-lobe type will produce slightly higher pressures, to about 20 psig. Casings are normally made from cast iron, which limits this type for certain applications where steel casings may be required because of hazardous conditions. The lobe-type blower finds widest application handling air or inert gas, although in certain applications it is also widely used as a gas pump on natural gas.

The second type is the sliding vane compressor, having one offset rotor with slots in which vanes slide in and out during each revolution. Air or gas is trapped within the casing and its volume gradually reduced as the pressure rises to discharge. This type can produce up to 50 psig per stage and is also available in two-stage arrangements for pressures to about 125 psig. Volume flows range up to 2500 ICFM. A disadvantage of this type for some process applications is the fact that the sliding vanes require some lubricant to be injected into the casing and trace quantities of this oil will thus appear in the compressed air or gas.

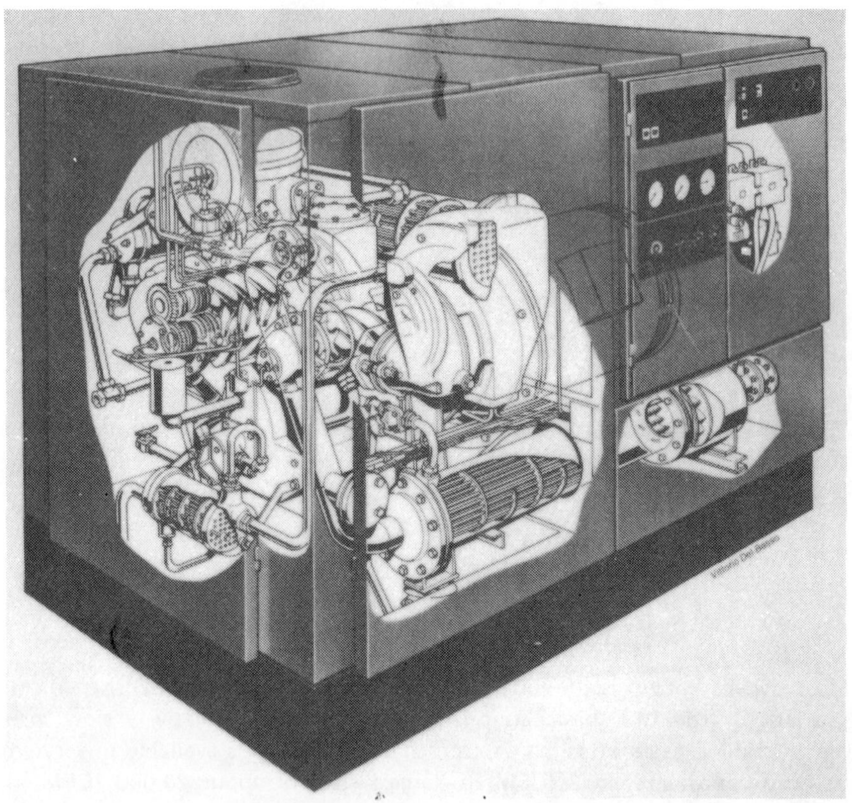

Figure 2.34. Screw-type rotary compressor.

The rotary screw type compressor has become more popular in the process industries in recent years. Although the design was introduced in Europe before World War II, U.S. manufacturers did not build this type in great quantities before the 1950s. Today the rotary screw type in the "wet-screw" or oil flooded version has taken over the portable, construction air type machine almost completely, and also occupies an important segment of the refrigeration and air conditioning compressor market. The dry-screw type has had some process applications and is gradually becoming more widespread. Both types are similar in design, utilizing twin screws which mate together and are driven thorugh timing gears. Lower speeds (to 3600 rpm) are used in the oil-flooded types, while the dry-screw types operate up to 12,000 rpm in smaller sizes. Figure 2.34 shows a sectional view of a typical screw type machine. Casing working pressures are limited to 250 psig to 400 psig, thus making this type usuable only in lower pressure applications.

The wet-screw or oil-filled machine will produce higher pressure rise per stage, with the heat of compression being absorbed by the oil which is injected into the machine for lubrication and cooling. The majority of this oil is removed after the compression cycle is over, and normally cooled and returned for re-use. The process designers should avoid this type however, if oil carryover into the process gas cannot be tolerated.

The liquid-ring pump or compressor is correctly classified as a rotary, positive displacement type. It consists of a circular or elliptical vaned type rotor turning in a circular or oval-shaped casing, in which water or other sealing liquid is also present. Centrifugal force causes the liquid to form a ring around the periphery of the casing when in operation. The air or gas travels inward towards the center of the vaned rotor, gradually decreasing in volume and increasing in pressure until it passes discharge ports and leaves the casing. Liquid still present in the air or gas is separated and either recirculated or discarded. This type is most often used as a vacuum pump, down to absolute pressures of 3 to 4 inches of mercury. It may also be used as a compressor for pressures up to about 100 psig (basis two stages in series). It has been successfully applied on certain difficult gases such as chlorine, hydrogen sulfide, acid gas and so forth. Stainless steel construction is available in most sizes.

Rotary Compressor Calculations

The rotary compressor fills only special needs such as low pressure rise, low capacity requirements. Calculation methods are not as widespread in the industry as for reciprocating and centrifugal machines. Adiabatic head, weight flow, inlet capacity calculations may be applied in a similar manner to other compressor types; however a widely recognized source of information on efficiencies for these types is not available, and most preliminary size estimates must be based on manufacturers catalog data. A sample calculation for a lobe-type gas blower, using data from a U.S. manufacturer, is shown in Figure 2.35.

CONCLUSION

A great deal of material exists in published literature today relating to compressors of all types. The reader is referred to certain references in this chapter, also to other readily available material to investigate further almost any aspect of compressor theory, design, selection or performance. References 15, 16, 17, 18, et al.

PROCESS REQUIREMENTS (GIVEN):		
Gas	Natural Gas	
Molecular weight	24	
Specific heat ratio, k	1.22	
Flow, ICFM	2000	
Suction pressure, psia	13.2	
temperature, °F/°R	100/560	
Discharge pressure, psia	21.2	
BLOWER DATA (FROM MANUFACTURER'S DATA BOOK):		
Blower size	1012J	Type RGS-J
Displacement, Cu. ft./Rev.-CFR	1.64	
Gear diameter inches	10	
Application Slip rpm	84	
Friction horsepower, maximum	18	
Temperature rise factor F_t	0.97	
Δt Maximum	205°	
Speed, maximum rpm	1800	
CALCULATION		**EXPLANATION**
Specific gravity	0.829	MW/28.96
Slip rpm	284	$\sqrt{\frac{(\Delta P)\ (14.7)\ (T_s)}{(S.G.)\ (P_s)\ (528)}}$
Speed	1504	Qs/CFR + Slip rpm
Actual gear speed, Ft/min.	3940	(0.262) (N) (Gear Dia.)
Friction HP	11	$FHP_{max}\left(\frac{\text{Act. Gear Speed}}{4720}\right)^3$
Brake horsepower	102	$\frac{(N)(FHP)}{1000} + (.00436)(CFR)(N)(\Delta p)$
Δt	86.8	$\left[\frac{(T_s)(BHP)(Ft)}{(.00436)(P_s)(Q_s)}\right]\left(\frac{k-1}{k}\right)$
Discharge temperature	186.8	$t_s + \Delta t$

Figure 2.35. Rotary lobe-type blower calculation.

ACKNOWLEDGEMENT

The writer acknowledges with thanks the information, photographs and data received from the following manufacturers of compressors and blowers: Allis-Chalmers Co., Cooper Industries, Dresser-Clark, Dresser-Roots, Elliott, Hoffman, Ingersoll-Rand, Joy Manufacturing, Nash Engineering, Transamerica DeLaval, Worthington, Atlas Copco, Sulzer Brothers Ltd.

NOMENCLATURE

A	Area
A_{tot}	Area, total
A_{he}	Area, head end

A_{ce}	Area, crank end
A_r	Area, piston rod
c	Absolute fluid velocity
CFM	Cubic feet per minute
C_p	Specific heat at constant pressure
C_v	Specific heat at constant volume
D	Diameter
F, °F	Temperature, degrees Fahrenheit
g	Gravitational constant
GHP	Gas horsepower
h	Enthalpy
H	Head
ICFM	Inlet cubic feet per minute
k	Ratio of specific heats
MC_p	Molal specific heat
MMSCFD	Million standard cubic feet per day
MW	Molecular weight
n	Polytropic exponent
N	Speed, revolutions per minute
P	Pressure
psia	Pounds per square inch absolute
psig	Pounds per square inch gauge
Q	Flow
R	Gas constant
r_c	Ratio of compression
°R	Absolute temperature, degrees Rankine
S	Stroke of reciprocating compressor
SCFM	Standard cubic feet per minute
t	Temperature, degrees F
T	Temperature, degrees R
U, u	Impeller peripheral velocity
v_s	Specific volume
v_a	Acoustic velocity of gas
V	Volume; volume flow rate
VE	Volumetric efficiency
W, w	Weight flow
Z	Compressibility factor
α, β	Impeller blade angles
Δ	Differential
μ	Head coefficient
η	Efficiency

ρ	Density
ϕ	Flow coefficient
ω	Fluid velocity relative to rotor

Subscripts

1, 2	Inlet & Discharge
s, d	Ditto
r	Reduced
c	Critical
ad, adia	Adiabatic
poly	Polytropic
p	Ditto
m	Meridional
u	Peripheral

REFERENCES

1. NGSMA Data Book, 9th Ed., 1972.
2. Compressed Air and Gas Data, C. W. Gibbs, Editor, 2nd Ed., 1971 published by Ingersoll-Rand Co.
3. "Compressibility Charts and Their Application to Problems Involving Pressure-Volume-Energy Relation for Real Gases" Research Bulletin P7637, published by Worthington Corp., 1949.
4. API 617 Centrifugal Compressors for General Refinery Services 4th Edition, 1979.
5. API 614 Lubrication, Shaft-Sealing, and Control Oil Systems for Special Purpose Applications, 1st Edition, 1973.
6. Magliozzi, T. L. "Control System Prevents Surging in Centrifugal Compressors, Chemical Engineering Magazine, May 8, 1967.
7. Staroselsky and Ladin, "Improved Surge Control for Centrifugal Compressors, Chemical Engineering Magazine, May 21, 1979.
8. API RP 541 "Recommended Practice for Form-Wound Squirrel-Cage Induction Motors 200 HP and Larger, 1972.
9. ANSI/NEMA MGI-1978 Motors and Generators
10. API 613 Special Purpose Gear Units for Refinery Services, 2nd Edition, 1977.
11. AGMA 421 "Practice for High Speed Helical and Herringbone Gear Units"
12. API 616 Combustion Gas Turbines for General Refinery Services 2nd Edition, 1980.
13. Sawyer's Gas Turbine Engineering Handbook, 3 vols., 1972 published by Gas Turbine Publications Inc.
14. API 612 Special Purpose Steam Turbines for Refinery Services 2nd Edition, 1979.
15. Scheel, L. F. "Gas and Air Compression Machinery," McGraw-Hill, 1961.
16. Scheel, L. F. "Gas Machinery" Gulf Publishing Co., 1972.
17. Stepanoff, A. J. "Turboblowers" John Wiley & Co., 1955.
18. Advanced Centrifugal Compressors, 4 papers by Turbomachinery Committee of ASME Gas Turbine Division, published by ASME, 1971.
19. Edmister, W. C. Applied Hydrocarbon Thermodynamics, 2 vols., 1961 and 1974, published by Gulf Publishing Co.

NGSMA – Natural Gas Processors Suppliers Association, Tulsa, OK 74103
API – American Petroleum Institute, 2101 L Street N.W., Washington, D.C. 20037
AGMA – American Gear Manufacturers Association, 1330 Massachusetts Ave., Washington, D.C. 20005
ASME – American Society of Mechanical Engineers, 345 East 47th Street, New York, NY 10017
NEMA – National Electrical Manufacturers Association, 2101 L Street, N.W., Washington, D.C. 20037.

CHAPTER 3

CENTRIFUGAL COMPRESSORS

M. P. BOYCE
Boyce Engineering International Inc.
Houston, Texas

INTRODUCTION

Centrifugal compressors are an integral part of the oil and chemical industry. They are used extensively because of their smooth operation, large tolerance to process fluctuations, and their higher reliability as compared to other types of compressors. Centrifugal compressors range in all sizes from pressure ratios of 1.3 per stage to pressure ratios as high as 12:1 on experimental models. We will limit ourselves to discussion of pressure ratios below 3.5:1 since these are ones used extensively in the oil and chemical industry. The proper selection of these compressors is a complex and very important decision since the successful operation of many plants depends upon the smooth and efficient functioning of these units. To ensure the best selection and consequently the proper maintenance of centrifugal compressors, the engineer must have a wide knowledge of many engineering disciplines.

SELECTION OF A COMPRESSOR

Specification Criteria

Detail compressor specifications can vary from customer to customer, some providing only basic information, such as pressure, flow rate, type of gas, driver, and site conditions, to a lengthy document detailing types of bearings, rotor response, lubrication system, acceptable tolerance on performance, etc. To do the latter, the engineer must be very conversant with comprssors and their total support systems. A starting point for the specifications are the various American Petroleum Institute specifications for turbomachinery. The following are some of the applicable publications:

API Standard 611 – General Purpose Steam Turbine for Refinery Services
API Standard 612 – Special Purpose Steam Turbine for Refinery Services
API Standard 613 – High-speed Special Purpose Gear Units for Refinery Services
API Standard 614 – Lube and Seal Oil Services
API Standard 616 – Combustion Gas Turbines for General Refinery Services
API Standard 617 – Centrifugal Compressors for General Refinery Services
API Standard 670 – Non-contacting Vibration and Axial Position Monitoring Systems

API Standard 671 – Couplings

These specifications are written by user engineers with the input of manufacturers and engineering contractors; thus, these standards represent a wealth of experience and are a very good base from which to start your turbomachinery specifications.

Many decisions, regardless of details contained in specifications, have to be made by the engineers in advance. Some of these may be his company's philosophy on various units and others could strictly be job oriented.

Layout

The general topography of the plant must be known to the engineer so that the proper site selection can be determined. Determination of whether the unit would be grade or mezzanine mounted is very important in determining the foundation characteristics. Enough space should exist for the ducting so that the inlet conditions to each stage allows the flow to enter without large distortions of velocity and pressure. Accessibility requirements should be kept in mind so that repair and maintenance work on the unit could be performed with relative ease. Location of the oil system for the unit is a very important aspect of compressor installation. It is advisable to locate the oil reservoir away from the base plate with the bottom sloped toward the low drain point, and enough space should be provided so that the return oil lines can enter the reservoir away from the oil pump suction. This would greatly reduce disturbance of the pump suction and also help in keeping the reservoir retention time to around ten minutes.

Environment

The environment in which a machine is to operate is as important a factor as any other. In many cases, this factor is often overlooked or described in a word phrase such as "extreme cold climate." The vendor needs to know much more. He must know what extreme cold means (usually below −25°F), what the transition weather is since, as a practical matter, this weather creates more problems because more equipment is exposed to it and fewer precautions are taken since operating problems are either not recognized and are glossed over. All cold weather applications need to have some icing protection, especially in the air inlet supplies. Fuel supply ventilation, pneumatic controls, actuators, all need to have some degree of de-icing. Many types of de-icing systems exist. The two most common types use exhaust gases or compressor bleed air. Tropical climate presents its own series of problems, such as excessive corrosion, high moisture content, and high ambient temperatures which increase the horsepower required and the cooling capacity of the lube oil system. Desert locations require special filterization systems so as to prevent erosion of the blades and special sealing on joints to prevent any of the small micron sand particles from entering the lube system, etc.

Machine usage is a very important environment consideration since intermittent usage can be the most severe kind of service. The aggressiveness of the gas to be handled and its temperature are very important aspects on the choice of materials to be used.

Type of Compressor

In many cases, it is not obvious what type of compressor is needed for the application. There are many types of compressors on the market as seen in Table 3.1. Some of the more significant types are the centrifugal, axial, rotary and reciprocating. The positive-displacement type compressors are used for intermittent flow in which successive volumes of fluid are confined in a closed space to increase their pressures. The other broad class of compressors are rotary type for continuous flow. In these types of compressors, rapidly rotating parts (impellers) accelerate fluid to a high speed; this velocity is then converted into additional pressure by gradual deceleration in the diffuser or volute which surrounds the impeller.

Table 3.1. Principle Types of Compressors

- COMPRESSORS
  - POSITIVE-DISPLACEMENT (Intermittent Flow)
    - RECIPROCATING
    - ROTARY
      - SLIDING VANE
      - LIQUID PISTON
      - STRAIGHT LOBE
      - HELICAL LOBE
  - CONTINUOUS FLOW
    - DYNAMIC
      - CENTRIFUGAL
      - AXIAL FLOW
      - MIXED FLOW
    - EJECTOR

The positive-displacement type of compressors can be further classified as either reciprocating or rotary types as shown in Table 3.1. The reciprocating compressor has a piston having a reciprocating motion within a cylinder. The rotary positive-displacement compressors have rotating elements whose positive action results in compression and displacement. The rotary positive-displacement can be further subdivided into sliding vane, liquid piston, straight-lobe and helical-lobe type compressors.

The continuous-flow type compressors, can be classified under dynamic or ejector type, entrain the inflowing fluid using a high velocity gas or steam jet and then convert the velocity of the mixture to pressure in a diffuser. The dynamic compressors have rotating elements which accelerate the inflowing fluid, and convert the velocity head into pressure head, partially in the rotating elements and partially in the stationary diffusers or blade. The dynamic type can be further subdivided into centrifugal, axial-flow and mixed-flow compressors. The main flow of gas in the centrifugal compressor is radial. The flow of gas in the axial compressor is axial and the mixed-flow compressor combines some characteristics of centrifugal and axial compressors.

It is not always obvious what type of compressor is needed for an application. Of the many kinds, some of the more significant are the centrifugal, axial, rotary and reciprocating. Figure 3.1 will aid in a selection of a compressor. For very high flows and low pressure ratios, an axial-flow compressor would be best. Axial-flow compressors usually have a higher efficiency but a smaller operating region than a centrifugal machine. Centrifugal compressors operate most efficiently at medium flow rates and high pressure ratios. Rotary and reciprocating compressors (positive-displacement machines) are best used for low flow rates and high pressure ratios.

This chapter deals with the centrifugal or mixed-flow type of compressors so the following discussion is concentrated on these compressors. There are many applications in a centrifugal compressor as seen in Table 3.2.

The centrifugal compressors range in size from pressure ratios of 1.3:1 per stage to as high as 12:2 on experimental models. In a typical centrifugal compressor the fluid is forced through the impeller by rapidly rotating impeller blades. The velocity of the fluid is converted into pressure, partially in the impeller and partially in the stationary diffusers. Figure 3.2 shows a section of a typical multistage centrifugal compressor used in the process industry.

A common method of classifying centrifugal compressors is based on the number of impellers and casing design. Table 3.3 shows three general types of centrifugal compressors. For each type of the compressor, approximate maximum ratings of pressure, capacity and brake horse power are also shown. Sectionalized casing types have impellers which are usually mounted on the extended motor shaft and similar sections are bolted together to obtain the desired number of stages. Casing material is either steel or cast iron. These machines require minimum supervision and maintenance and are quite economic in their operating range. The sectionalized

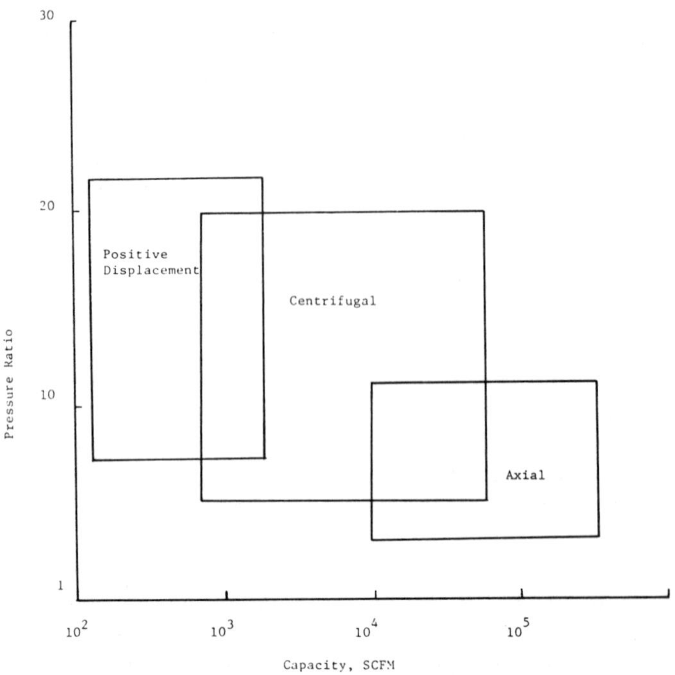

Figure 3.1. General guide to compressor selection.

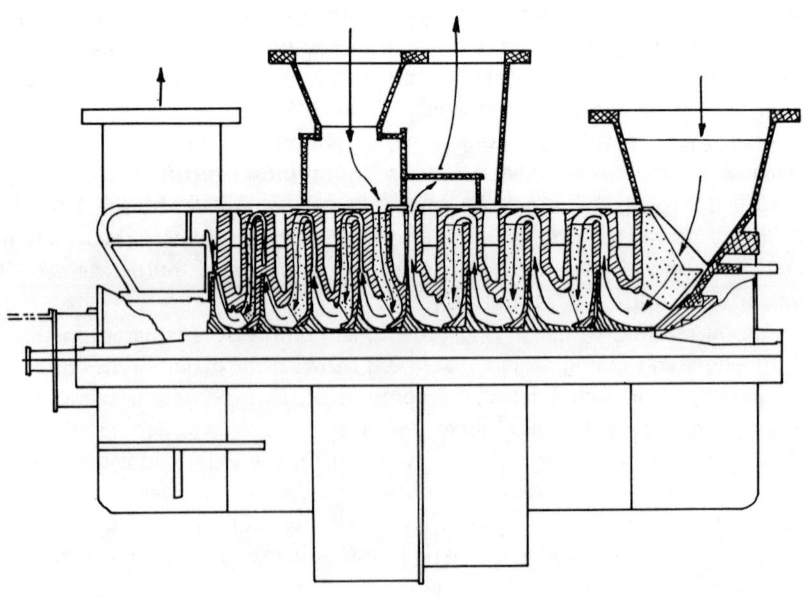

Figure 3.2. Cross section of multistage centrifugal compressor (Courtesy of Elliott)

Table 3.2. Applications of Centrifugal Compressor

Industry or Application	Service of Process	Typical Gas Handled
Iron and Steel		
Blast furnace	Combustion	Air
	Off gas	Blas furnace gas
Bessemer convertor	Oxidation	Air
Cupola	Combustion	Air
Coke oven	Compressing	Coke oven gas
	exhausting	Coke oven gas
Mining and Metallurgy		
Power	For tools and machinery	Air
Furnaces	Copper and nickel purification	Air
	Pelletizing (Iron Ore concentration)	Air
Natural Gas		
Production	Repressuring oil wells	Natural gas
Distribution	Transmission	Natural gas
Processing	Natural gasoline separation	Natural gas
	Refrigeration	Propane and Methane
Refrigeration		
Chemical	Various processes	Butane, propane, ethylene, ammonia, special refrigerants
Industrial and commercial	Air conditioning	Special refrigerants
Utilities		
Steam generators	Soot blowing	Air
	Combustion	Air
	Cyclone furnaces	Air
City gas	Manufacturing	Fuel gas
	Distribution	Fuel gas
Miscellaneous		
Sewage treatment	Agitation	Air
Industrial power	Power for tools and machines	Air
Paper making	Fourdrinier vacuum	Air and water vapor
Material handling	Conveying	
Gas engines	Supercharging	

casing design is used extensively in supplying air for combustion in ovens and furnaces.

The Horizontal Split Casting type have casings which are split horizontally at the mid-section and the top and the bottom halves are bolted and doweled together as shown in Figure 3.3. This type of design is preferred for large multistage units. The internal parts such as shaft, impellers, bearings and seals are readily accessible for inspection and repairs by removing the top half. The casing material is cast iron or cast steel. There are various types of barrel or centrifugal compressors. Low

Table 3.3. Industrial Centrifugal Compressor Classification Based on Casing Design

	Approximate Maximum Ratings		
Casing Type	Approximate Pressure PSIG	Approximate Capacity Inlet CFM	Approximate Horsepower Requirement
A. Sectionalized			
Usually Multistage	10*	20,000*	600*
B. Horizontally split			
Single stage			
(double suction)	15*	650,000*	10,000*
Multistage	1,000	200,000*	35,000
C. Vertically split			
Single stage			
(single suction)			
Overhung	30*	250,000*	10,000*
Pipeline	1,200	25,000	20,000
Multistage	over 5,500	20,000	15,000

*Based on air at atmospheric intake conditions.

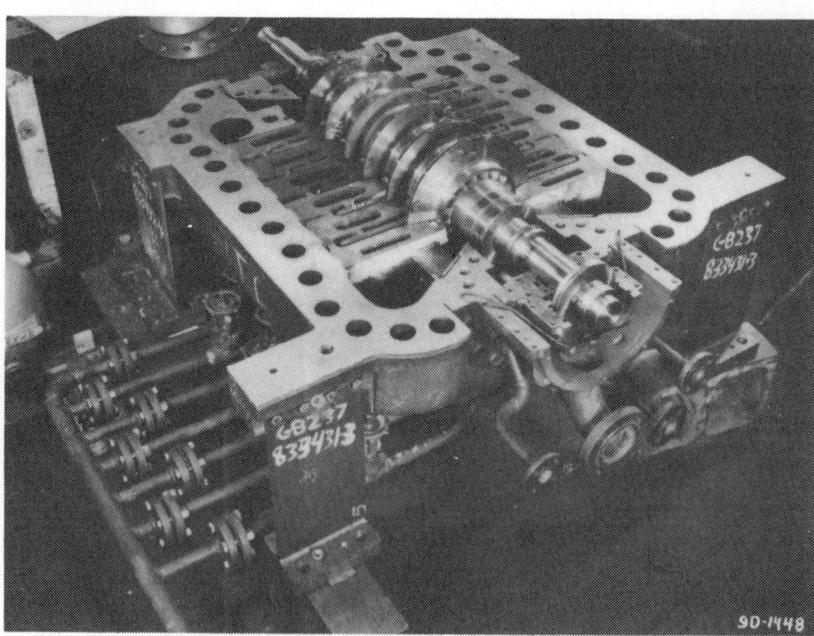

Figure 3.3. Multistage centrifugal compressor (Courtesy of Elliott)

pressure types with overhung impellers are used for combustion processes, ventilation and conveying applications. Multistage barrel casings are used for high pressures for which the horizontally split joint is inadequate. Figure 3.4 shows the barrel compressor in the background and the inner bundle from the compressor in front. Note that in most cases once the casing is removed from the barrel it is horizontally split.

Figure 3.4. Barrel type compressor showing inner casing in foreground and barrel in background (Courtesy of Elliott).

Compressor Configuration

To properly design a centrifugal compressor, the operating conditions, the type of gas, and its pressure, temperature and molecular weight must be known. The corrosive properties of the gas must be properly identified so that proper metallurgical selection can be made. Gas fluctuations due to process instabilities must be pinpointed so that the compressor can operate without surging.

Centrifugal compressors for industrial applications have relatively low presure ratios per stage. This is necessary so that the compressors can have a wide operating range, and stress levels can be kept at a minimum. Due to the low pressure ratios for each stage, a single machine may have a number of stages in one "barrel" to achieve the desired overall pressure ratio. Figure 3.5 shows some of the many config-

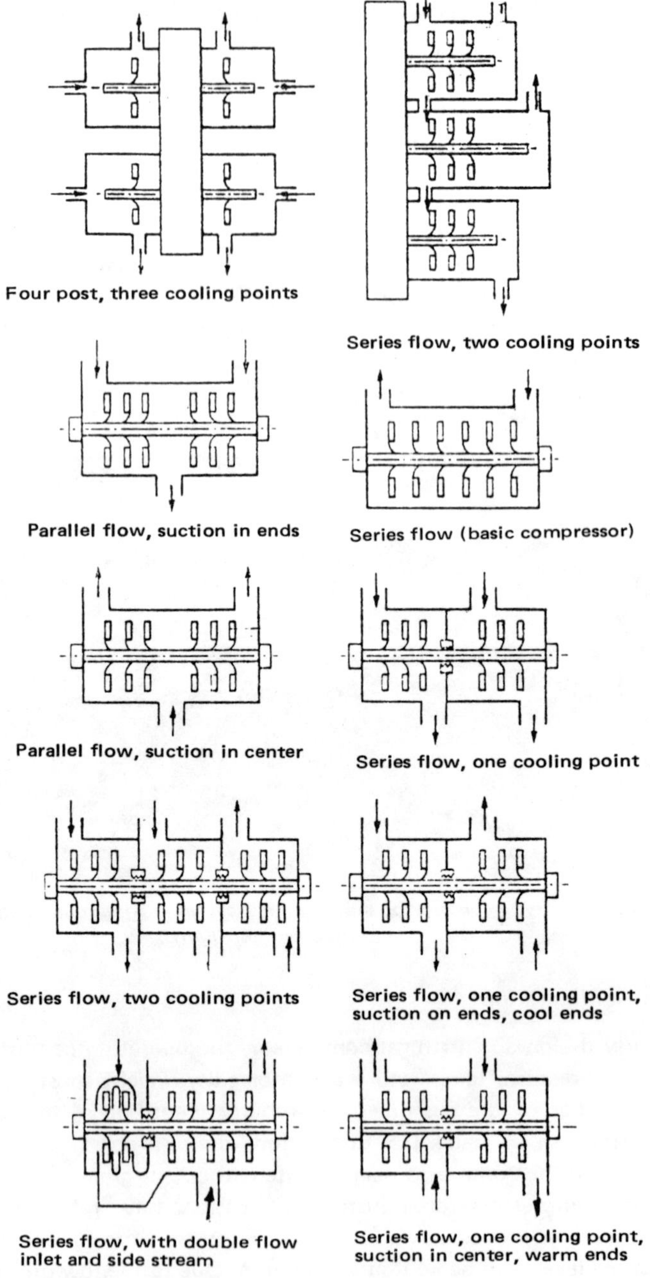

Figure 3.5. Various configurations of centrifugal compressors to meet needs of the process plants.

urations. Some of the factors to be considered when selecting a configuration to meet plant needs are:

1. Intercooling between stages can considerably reduce the power consumed.
2. Back-to-back impellers allow for a balanced rotor thrust, and minimize overloading the thrust bearings.
3. Cold inlet or hot discharge at the middle of the case reduces oil-seal and lubrication problems.
4. Single inlet or single discharge reduces external piping problems.
5. Balance planes that are easily accessible in the field can appreciably reduce field-balancing time.
6. Balance piston with no external leakage will greatly reduce wear on the thrust bearings.
7. Hot and cold sections of the case that are adjacent to each other will reduce thermal gradients, and thus reduce case distortion.
8. Horizontally split casings are easier to open for inspection than vertically split ones, reducing maintenance time.
9. Overhung rotors present an easier alignment problem because shaft-end alignment is necessary only at the coupling between the compressor and driver.
10. Smaller, high-pressure compressors that do the same job will reduce foundation problems but will have greatly reduced operational range.

Arrangements

The general configuration of the compressors and their drive trains must be evaluated to fit the location, environment, and type of compressor selected. A decision must be made as to whether the units are to be in series or parallel. This decision requires a knowledge of the flow and discharge pressure required. In most cases, a number of casings are connected together to form a "compressor train." This is nothing else but the connection of various compressors in series. The limit to the number of connected casings is due to the rotor dynamics of the coupled train. In the arrangement, one must also decide what type of mounting is desirable. Most turbomachinery is mounted on structural steel platforms which are usually referred to as base plates or skids. These platforms are then mounted on a mass of concrete at the job site by mounting them on sole plates or through direct grouting. Platforms should be considered part of the foundations and great care should be placed in their design. Insufficient rigidity of these platforms can allow the rotating machinery to excite them.

Driver Selection

The three main types of drives on compressors are: 1) steam turbines, 2) gas turbines, and 3) electric motors. The decision of which drive is the best suited is not always an easy decision. The selection of the drive depends on many factors such as location, process, and unit size. In remote locations, gas turbines are mostly used due to their low maintenance and the ability to prepackage the units. Their light weight makes them also a must for offshore platforms. In petrochemical plants, due to the process steam created, steam turbines are widely used. In this manner, plants can utilize their energy — thus operating at an overall high plant efficiency. Electric motors drive most of the smaller flow compressors, usually through a speed increasing gear system. Figure 3.6 shows the typical ranges for the various drives. Also from this figure, one can note that the higher the flow, the lower the speed. This is due to the fact that at high flows the compressor diameter must be large; therefore, the speed must be reduced to maintain the same stress levels.

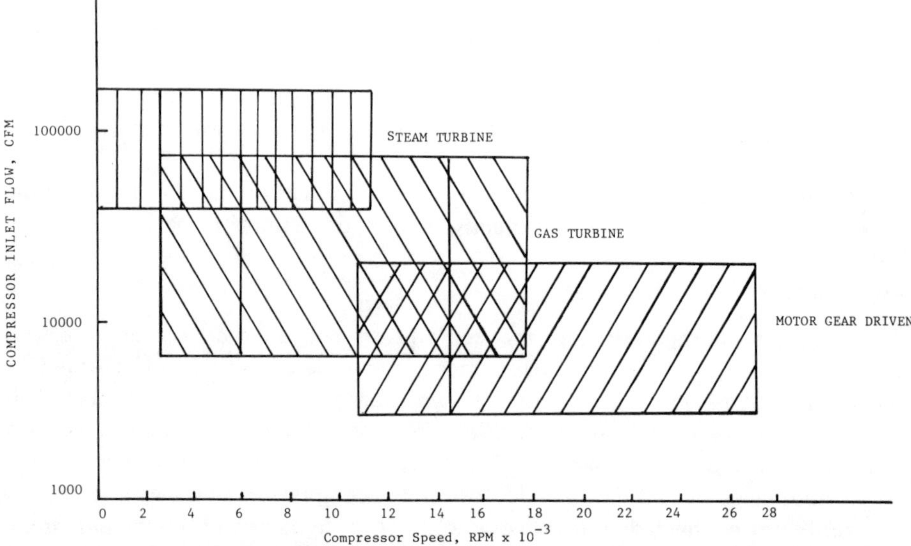

Figure 3.6. Application chart for centrifugal compressors.

AERO-THERMODYNAMICS OF CENTRIFUGAL COMPRESSORS

Thermodynamics

The energy transfer to the rotor can be obtained from the first law of thermodynamics. The first law applied between two stations can be written as follows:

$$\dot{q} + h_1 + \frac{V_1^2}{2g_c J} = h_2 + \frac{V_2^2}{2g_c J} + \dot{W} \tag{1}$$

where:

- $\dot{q}$ is the rate of heat transfer per unit mass flow
- $\dot{W}$ is the rate of shaft work per unit mass flow
- h is the enthalpy of the fluid
- V is the velocity of the fluid

If the energy transfer to the fluid is considered as being reversable and adiadabatic (i.e. losses such as due to friction in the flow passage being absent, along with no heat transfer) the work energy utilized to raise the enthalpy of the fluid from its initial to final state can be written as:

$$\dot{W} = h_1 + \frac{V_1^{\,2}}{2gJ} - h_2 + \frac{V_2^{\,2}}{2gJ} = h_{1t} - h_{2t} \qquad (2)$$

The above relationships will give a negative value in the case of a compressor and a positive value in the case of a turbine thus indicating that input or output shaft work respectively is equal to the increase in the total enthalpy rise across the turbomachine.

The mechanical energy available to the rotor of the centrifugal compressor from its prime mover is given by the Euler Turbine equation and can be written as:

$$\dot{W} = U_1 V_{\theta 1} - U_2 V_{\theta 2}$$

where U_1 and U_2 are the peripheral velocities at the inlet (mean) and exit diameters of the rotor, and $V_{\theta 1}$ and $V_{\theta 2}$ are the tangential components of the absolute velocities at the inlet and exit respectively.

Efficiencies

1. Adiabatic Efficiency — The work in a centrifugal compressor under ideal conditions occurs at constant entropy as shown in Figure 3.7. The actual work done is indicated by the dotted line. Therefore the isentropic efficiency of the compressor can be written in terms of the total changes in enthalpy:

$$\eta_{ad} = \frac{\text{Isentropic Work}}{\text{Actual Work}} = \left(\frac{h_{2t} - h_{1t}}{h_2{}'_t - h_{1t}}\right)\begin{matrix}\text{id}\\ \\ \text{act}\end{matrix} \qquad (3)$$

This equation can be rewritten for a thermally and calorically perfect gas in terms of total pressure and temperature as follows:

$$\eta_{ad} = \frac{\left\{\left(\frac{P_{2t}}{P_{1t}}\right)^{\frac{\gamma-1}{\gamma}} - 1\right\}}{\left\{\frac{T_2{}'_t}{T_{1t}} - 1\right\}} \tag{4}$$

The process between 1 & 2′ can be defined by the following equation of state:

$$P/\rho^n = \text{Const.}$$

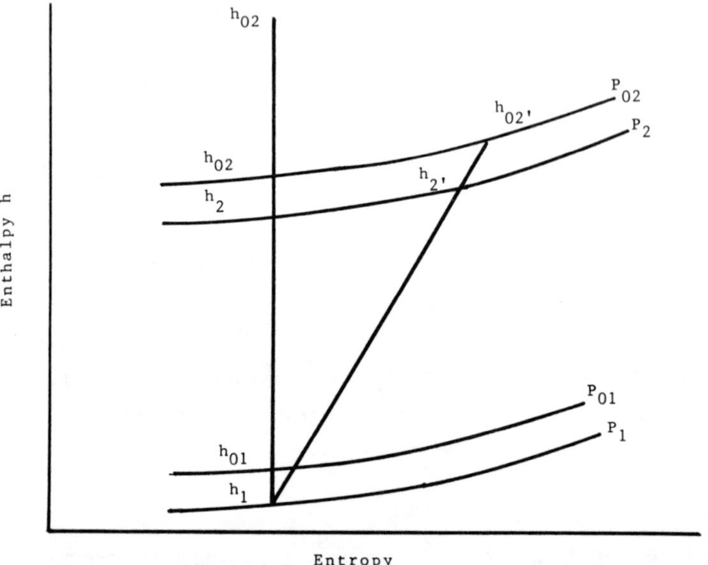

Figure 3.7. Entropy – enthalpy diagram for a centrifugal compressor.

Where n is some polytropic process. The adiabatic efficiency can then be represented by:

$$\eta_{ad} = \frac{\left\{\left(\frac{P_{2t}}{P_{1t}}\right)^{\frac{\gamma-1}{\gamma}} - 1\right\}}{\left\{\left(\frac{P_{2t}}{P_{1t}}\right)^{\frac{n-1}{1}} - 1\right\}} \tag{5}$$

2. Polytropic Efficiency – Polytropic efficiency is another concept of efficiency. It is often referred to as the small stage frequency i.e. when there is a very small pressure rise across the compressor:

$$\eta_p = \frac{\left[1 + \frac{dP_{2t}}{P_1}\right]^{\frac{\gamma - 1}{\gamma}} - 1}{\left[1 + \frac{dP_{2t}}{P_1}\right]^{\frac{n - 1}{n}} - 1} \tag{6}$$

Which can be expanded assuming that $dp_{2t}/P_1 \ll 1$ and neglecting second order terms as:

$$\eta_p = \frac{\frac{\gamma - 1}{\gamma}}{\frac{n - 1}{n}} \tag{7}$$

From this it is obvious that polytropic efficiency is the limiting value of the isentropic efficiency as the pressure increase approaches zero, and that the value of the polytropic efficiency is higher than the corresponding adiabatic efficiency. Figure 3.8 shows the relationship between the adiabatic and polytropic efficiency as the pressure ratio across the compressor increases.

Another characteristic of the polytropic efficiency is that the polytropic efficiency of a multistage compressor is equal to the stage efficiency if each stage has the same efficiency.

Dimensional Analysis of a Centrifugal Compressor

Turbomachines can be compared with each other by the use of dimensional analysis. This type of analysis produce various types of geometric similar parameters. Dimensional Analysis is a procedure where variables representing a physical situation are reduced into groups which are dimensionless. These dimensionless groups can then be used to compare performance of various types of machines with each other. Dimensional Analysis as used in turbomachines can be used to a) compare data from various types of machines, and is a very useful technique in development of blade passages, blade profiles, etc. b) for selection of various types of units based on maximum efficiency and pressure head required. c) prediction of a prototype's performance from test conducted on a smaller scale model or at lower speeds etc.

Dimensional Analysis as used in turbomachines is a very useful tool and leads to

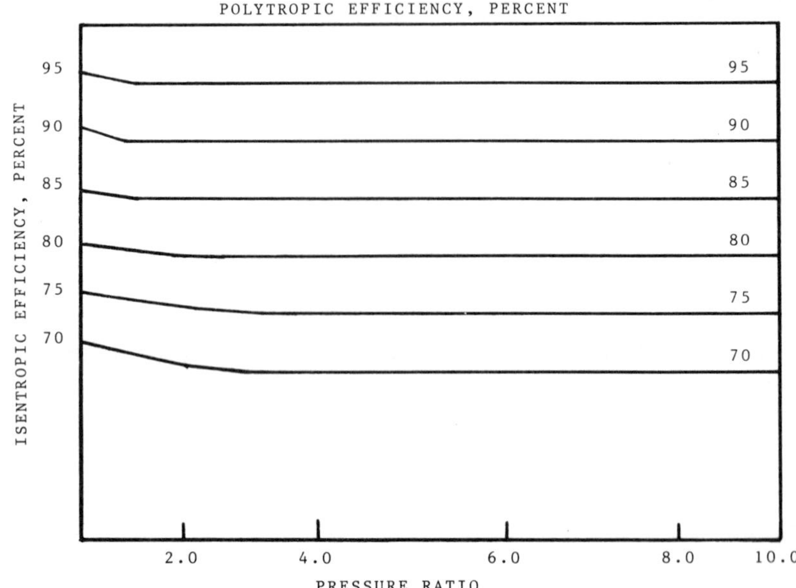

Figure 3.8. Relationship between adiabatic and polytropic efficiency.

various dimensionless parameters which are based on the dimensions, such as Mass (M), Length (L) and Time (T). Based on these we can obtain various independent parameters such as density (ρ), viscosity (μ), speed (N), diameter (D) and velocity (V). This leads to forming various dimensionless groups which are used in fluid mechanics of turbomachines.

$$Re = \frac{\rho V D}{\mu} \quad \text{(Reynolds Number)} \tag{8}$$

$$N_s = \frac{N\sqrt{Q}}{H^{3/4}} \quad \text{(Specific Speed)} \tag{9}$$

Where H is the adiabatic head, Q is the volume rate, and N the speed

$$D_s = \frac{D\,H^{1/4}}{Q} \quad \text{(Specific Diameter)} \tag{10}$$

Where D is the diameter of the impeller.

$$\phi = \frac{Q}{ND^3} \quad \text{(Flow Coefficient)} \tag{11}$$

$$\psi = \frac{H}{N^2\,D^2} \quad \text{(Head Coefficient)} \tag{12}$$

The above are some of the major dimensionless parameters. In many cases for the flow to remain dynamically similar all the above parameters must remain constant, however, this is not possible in a practical sense so one must make choices.

In seclecting turbomachines the choice of specific speed and specific diameter determines the type of compressor which is most suitable as seen in Figure 3.9. It is obvious from this figure that high head and low flow require a positive displacement type unit, while a medium head and medium flow require a centrifugal type unit, and high flow and low head an axial flow type unit. Figure 3.10 shows a blow-up of the centrifugal compressor section and can be used as a reference for selection of centrifugal compressor units.

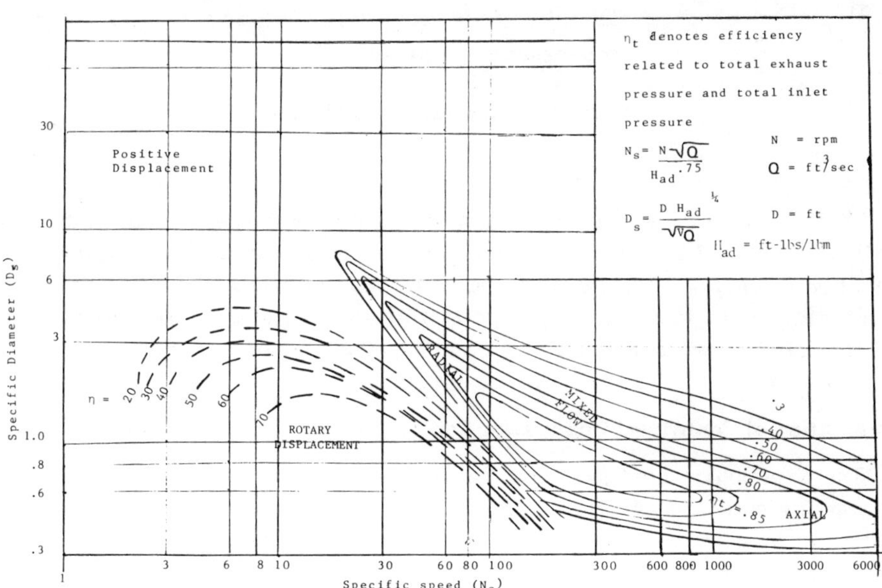

Figure 3.9. Compressor map.

Flow coefficients, pressure coefficients can be used to determine various off design characteristics. Reynolds number effects the flow calculations as far as skin friction and velocity distribution are concerned.

It must be remembered when using dimensional analysis in computing performance

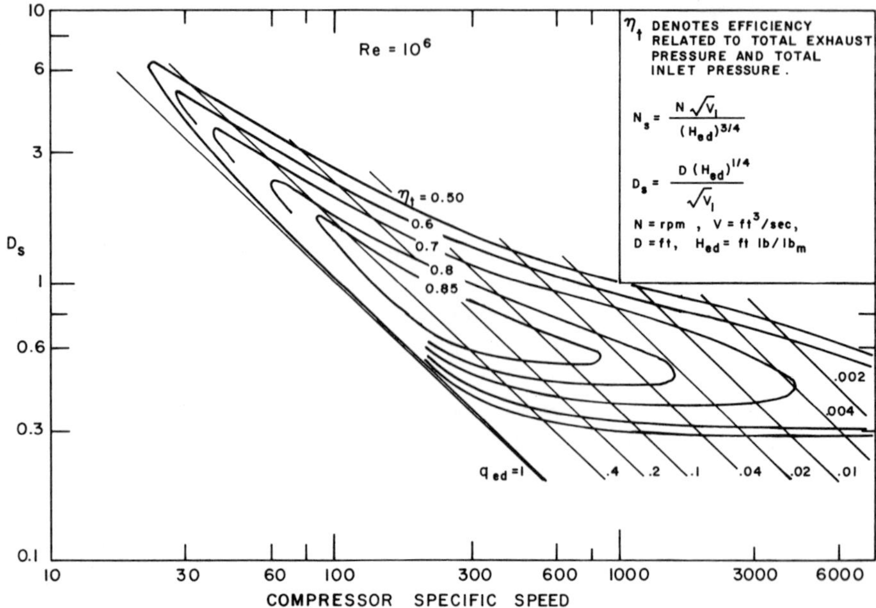

Figure 3.10. Compressor specific speed.

or predicting performance based on tests performed on smaller scale units that due to the fact that it is not physically possible to keep all parameters constant, the variation of the final results will depend on the scale up factor, the difference in the fluid medium etc. It is thus very important in any type of dimensionless study to understand the limit of the parameters and thus the geometrical scale up.

Description of Centrifugal Compressor Components

The terminology used to define the components of a centrifugal compressor is shown in Figure 3.11. A centrifugal compressor is usually composed of inlet guide vanes, an inducer, an impeller, a diffuser and a scroll. Generally the inlet guide vanes (I.G.V.) are used only for a high pressure-ratio transonic compressor.

The fluid comes into the compressor through an intake duct, and is given prewhirl by an I.G.V. It then flows into an inducer almost without any incidence angle from a design standpoint, where its flow direction is changed from the axial to the radial direction. At this stage, the fluid is given energy by the rotor as it goes through the impeller while compressing. It is then discharged into a diffuser where the kinetic energy is converted into static pressure. The flow then enters the scroll, from which the compressor discharge is taken. Figure 3.12 shows the variations in pressure and velocity through a compressor.

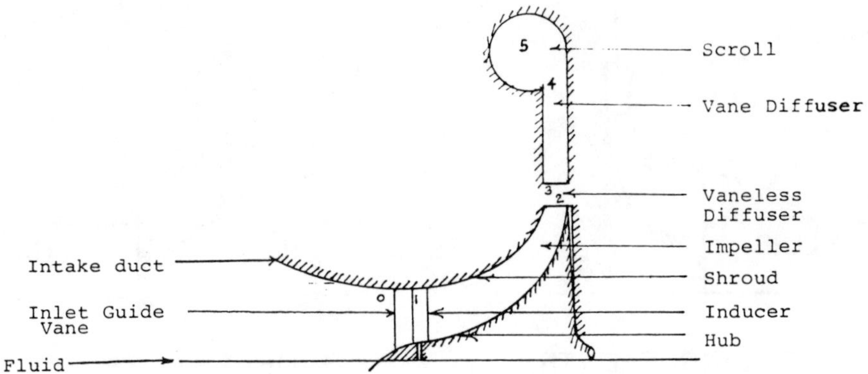

Figure 3.11. Schematic of a centrifugal compressor.

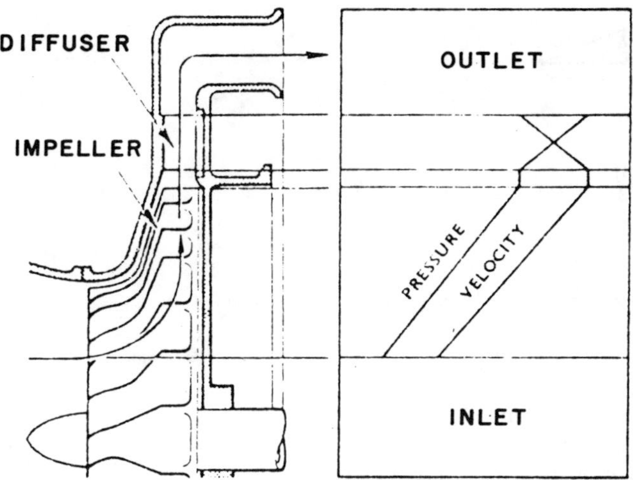

Figure 3.12. Pressure and velocity through a centrifugal compressor.

There are two kinds of entry inducer systems, namely, a single-entry inducer and a double-entry inducer as shown in Figure 3.13.

A double-entry inducer system essentially halves the inlet flow so that a smaller inducer tip diameter can be used reducing the inducer tip Mach number; however, the configuration is difficult to integrate into many configurations.

There are three types of impeller vanes as shown in Figure 3.14. These are defined with respect to the exit blade angles. Impellers with exit blade angle $\beta_2 = 90^\circ$ are said to have radial vanes. Impellers with $\beta_2 < 90$ are said to have backward-curved or backward-swept vanes and for $\beta_2 > 90$, the vanes are said to be forward-

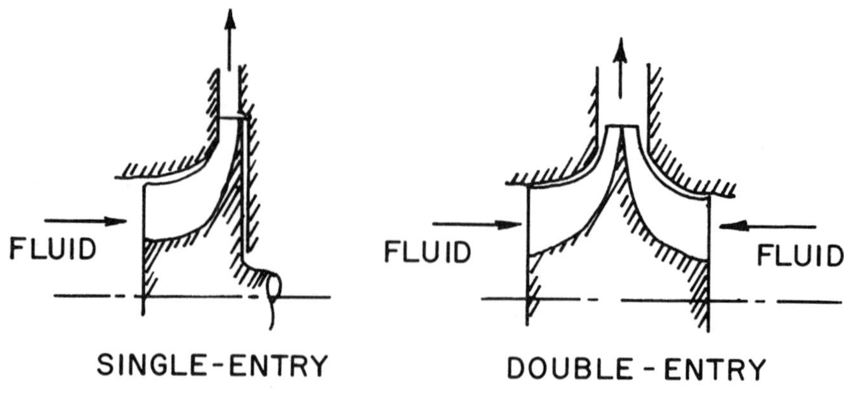

Figure 3.13. Types of entry inducer system.

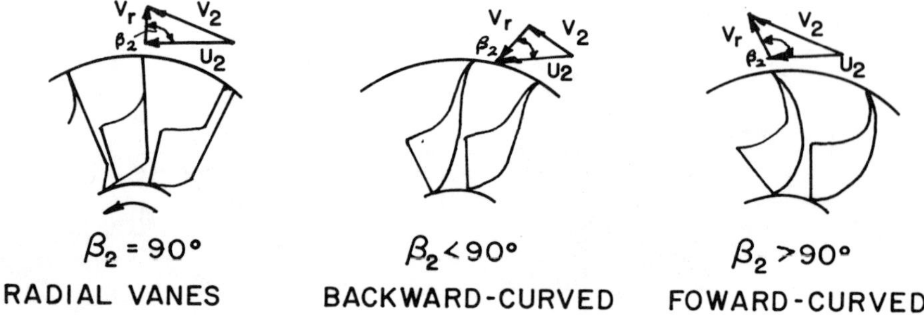

Figure 3.14. Various types of impeller blading.

curved or forward-swept. They have different characteristics of a theoretical head-flow relationship to each other as shown in Figure 3.15. It should be pointed out that even though in Figure 3.15 the forward curve head is the largest, in actual practice the head characteristics of all the impellers is similar to the backward curve impeller. Table 3.4 shows the merits and demerits of various impellers.

The Euler equation, assuming simple one-dimensional flow theory, is the theoretical amount of work imparted to each pound of fluid as it passes through the impeller and is given by:

$$E = U_2 V_{\theta 2} - U_1 V_{\theta 1} \tag{13}$$

where:

E — Work per lb of fluid
U_2 — Impeller peripheral velocity

U_1 — Inducer velocity at the mean radial station
$V_{\theta 2}$ — Absolute tangential fluid velocity at impeller exit
$V_{\theta 1}$ — Absolute tangential air velocity at inducer inlet

For the axial inlet, $V_{\theta 1} = 0$
Then:

$$E = U_2 V_{\theta 1} = 0 \qquad (14)$$

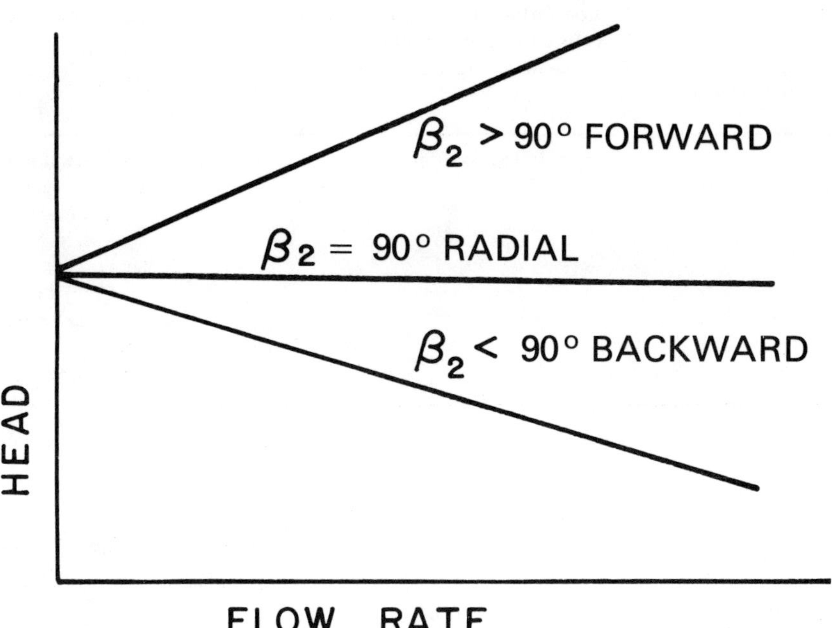

Figure 3.15. Head-flow rate characteristics for various blade angles.

Supposing constant rotational speeds, no slip and axial inlet, the velocity triangle becomes one as shown in Figure 3.16. For the radial vane, the absolute tangential fluid velocity at the impeller exit is constant even if the flow rate is decreased. Therefore:

$$E = U_2 V_{\theta 2} = U_2 V'_{\theta 2} = E' \qquad (15)$$

For the backward-curved vanes, the absolute tangential fluid velocity at impeller exit increases according to the reduction of flow rates as shown in the following equation:

$$E = U_2 V_{\theta 2} = U_2 V'_{\theta 2} = E \qquad (16)$$

Table 3.4. The Merits and Demerits for Various Impellers

Types of Impellers	Merits	Demerits
Radial Vanes	1. Reasonable Compromise between low energy transfer and high absolute outlet velocity 2. No complex bending stress 3. Easy Manufacturing	1. Surge Margin is relatively narrow
Backward Curved Vanes	1. Low outlet kinetic energy = low diffuser inlet mach number 2. Surge margin is wide	1. Low energy transfer 2. Complex bending stress 3. Hard manufacturing
Forward	1. High energy transfer	1. High outlet kinetic energy = High diffuser inlet mach number 2. Surge margin is less than radial vanes 3. Complex bending stress 4. Hard manufacturing

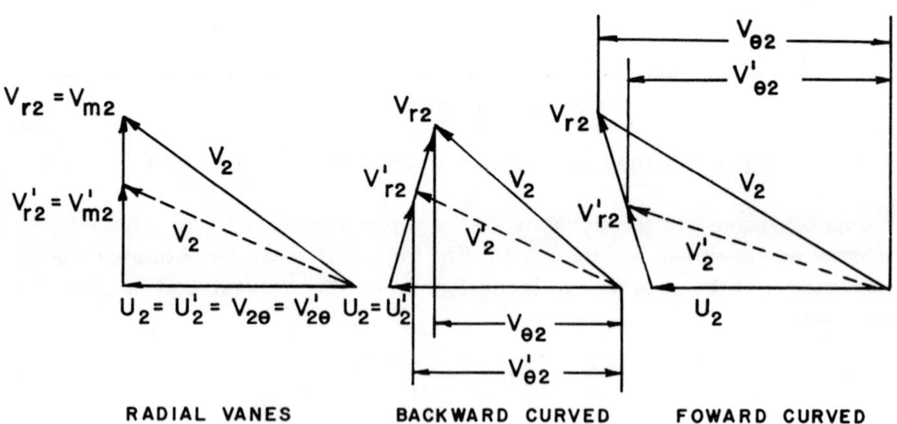

Figure 3.16. Velocity triangle.

For the Forward-Curved vanes, the absolute tangential fluid velocity at impeller exit decreases according to the reduction of flow rates as shown in the following equation:

$$E = U_2 \, V_{\theta 2} > U_2 V'_{\theta 2} = E' \tag{17}$$

Inlet Guide Vanes

The inlet guide vanes give a circumferential velocity component to the fluid at the inducer inlet. This is called pre-whirl. Figure 3.17 shows the inducer inlet velocity diagrams with and without I.G.V.

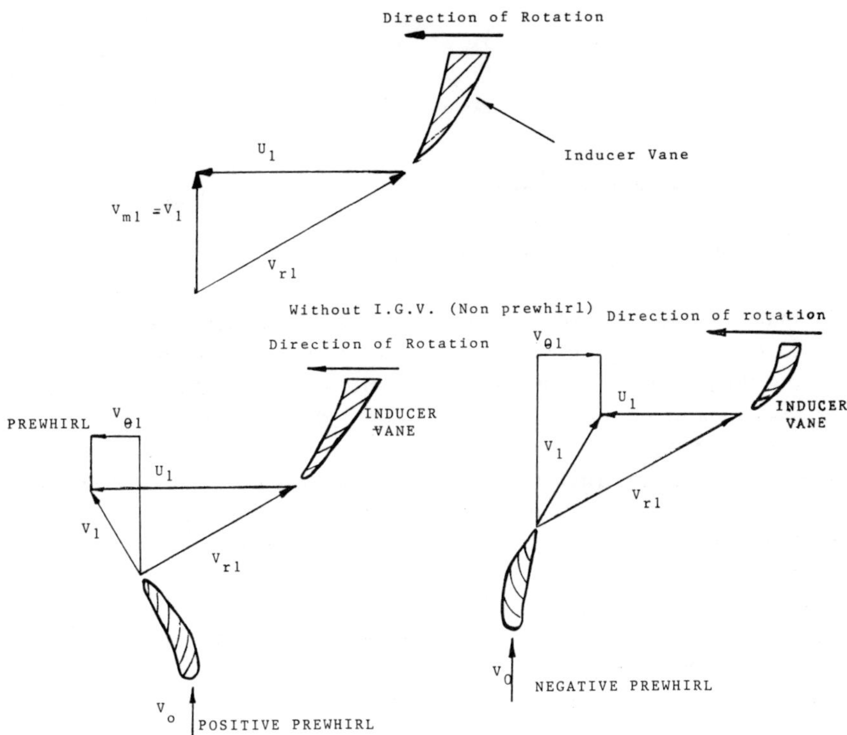

Figure 3.17. Inducer inlet velocity diagrams.

I.G.V. may be installed directly in front of the inducer or where an axial entry is not possible, located radially in an intake duct.

A positive vane angle produces prewhirl in the direction of the impeller rotation and a negative vane angle produces prewhirl in the opposite direction. The disadvantage of positive prewhirl is that a positive inlet whirl velcoity reduces the energy transfer. The Euler equation is defined by:

$$E = U_2 V_{\theta 2} - U_1 V_{\theta 1} \tag{18}$$

For non-prewhirl (without I.G.V.) $V_{\theta 1}$ is equal to zero. Then the Euler work is $E = U_2 V_{\theta 2}$.

In the case of positive prewhirl, the second term of the Euler equation remains, that is $E = U_2 V_{\theta 2} - U_1 V_{\theta 1}$. Therefore, Euler work is reduced by the use of positive prewhirl. On the other hand, negative prewhirl increases the energy transfer by the amount $U_1 V_{\theta 1}$.

The positive prewhirl decreases the relative Mach number at the inducer inlet. However, negative prewhirl increases it. A relative Mach number is defined by:

$$M_{rel} = \frac{V_{r1}}{a_1} \tag{19}$$

where:

M_{rel} = Relative Mach number
V_{r1} = Relative velocity at an inducer inlet
a_1 = Sonic velocity at an inducer inlet condition

The main purpose of installing the I.G.V. is to decrease the relative Mach number at the inducer tip inlet because the highest relative velocity at the inducer inlet is the tip section. When the relative velocity is close to the sonic velocity or greater than it, a shock wave takes place at the inducer section. This gives us not only the shock loss but also chokes the inducer.

The main purpose to install the I.G.V. is to decrease this relative Mach number at an inducer inlet, so prewhirl will be mentioned mainly with respect to the positive case. Figure 3.18 shows the effect of inlet prewhirl with respect to a compressor efficiency.

There are three kinds of prewhirl. These are as follows:

1. Free Vortex Type – This type is represented by $r_1 V_{\theta 1}$ = constant with respect to the inducer inlet radius. This prewhirl distribution is shown in Figure 3.19. In this type $V_{\theta 1}$ is a minimum at the inducer inlet shroud radius. Therefore, it is not good to decrease the relative Mach number in this manner.

2. Forced Vortex Type – This type is shown as $V_{\theta 1}/r_1$ = constant. This prewhirl distribution is also shown in Figure 3.19. $V_{\theta 1}$ is a maximum at the inducer inlet shroud radius, so this contributes to decreasing the inlet relative Mach number.

3. Control Vortex Type – This type is represented by $V_{\theta 1} = AR_1 + B/r_1$ where A and B are constants. This equation shows the first type with $A = 0$, $B \neq 0$ and the second type with $B = 0$, $A \neq 0$.

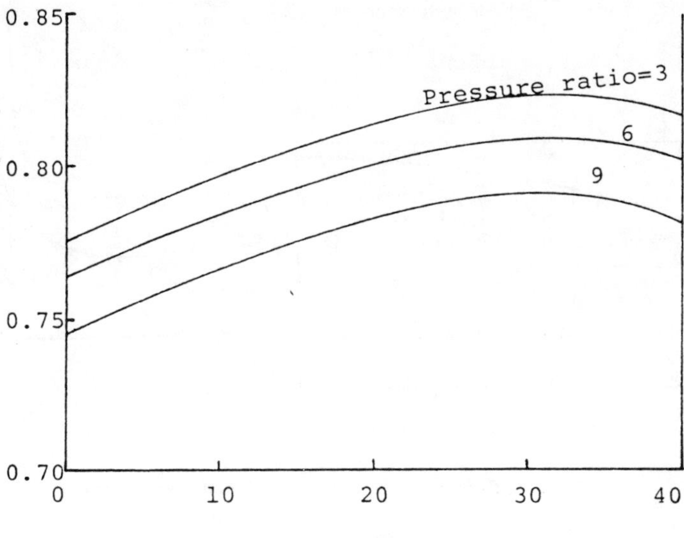

Figure 3.18. Estimated effect of inlet prewhirl by C. Rogers & L. Sapiro.

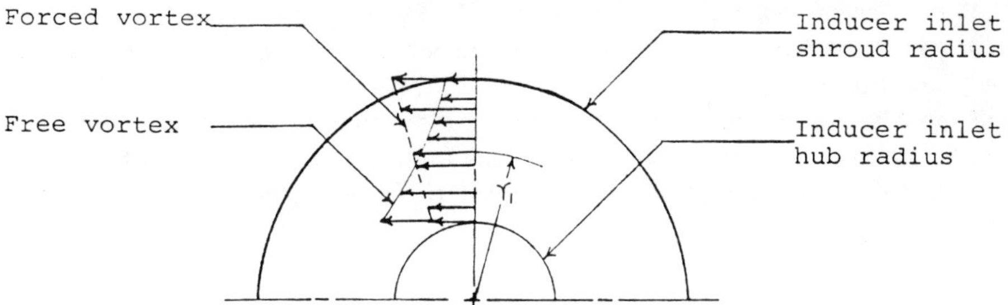

Figure 3.19. Prewhirl distribution patterns.

Euler work distributions at an impeller exit with respect to the impeller width are shown in Figure 3.20. From Figure 3.20 it is noted that the prewhirl distribution should be made not only from the relative Mach number at the inducer inlet shroud radius, but also from Euler work distribution at the impeller exit. This is because uniform impeller exit flow conditions, considering the impeller losses, are important factors to obtain good compressor performance.

Impeller

An impeller in a centrifugal compressor imparts energy to the fluid. The impeller

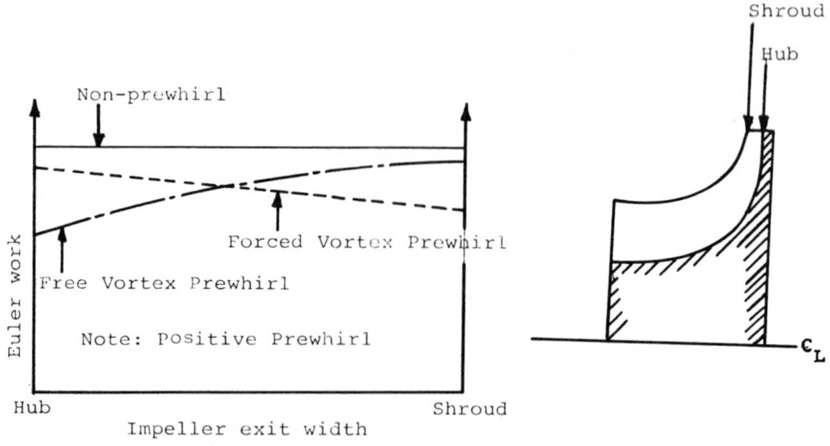

Figure 3.20. Euler work distribution at an impeller exit.

consists of two basic solutions 1) the inducer which is very much like an axial flow rotor, and 2) the radial blades where energy is imparted by centrifugal force. The flow enters the impeller usually in the axial direction and leaves in the radial direction. The velocity variations from hub to shroud resulting from these changes in flow directions complicate the design procedure for centrifugal compressors. C. H. Wu has presented the three dimensional theory in an impeller, but it is difficult to solve for the flow in an impeller using the above theory without certain simplified conditions. J. D. Stanitz, T. Katsanis, F. Dallenback, M. P. Boyce, Y. Senoo and so on have dealt with it as a quasi-three dimensional solution. It is composed of two solutions, one in the meridional surface (hub-to-shroud) and the other in the stream surface of revolution (blade-to-blade). These surfaces are illustrated in Figure 3.21.

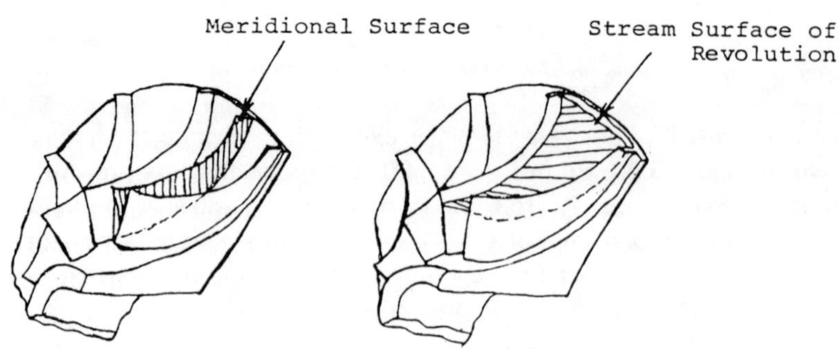

Figure 3.21. Two dimensional surface for flow analysis.

By the application of the previous mentioned method using a computer program, it is possible to achieve impeller efficiencies of more than 90%. The actual flow phenomena in an impeller is much more complicated than the one calculated. One example of this complicated flow is shown in Figure 3.22. The stream lines observed in Figure 3.22 do not cross, but are actually in different planes. Figure 3.23 shows the flow in the meridional plane, note separation regions at the inducer section and at the exit.

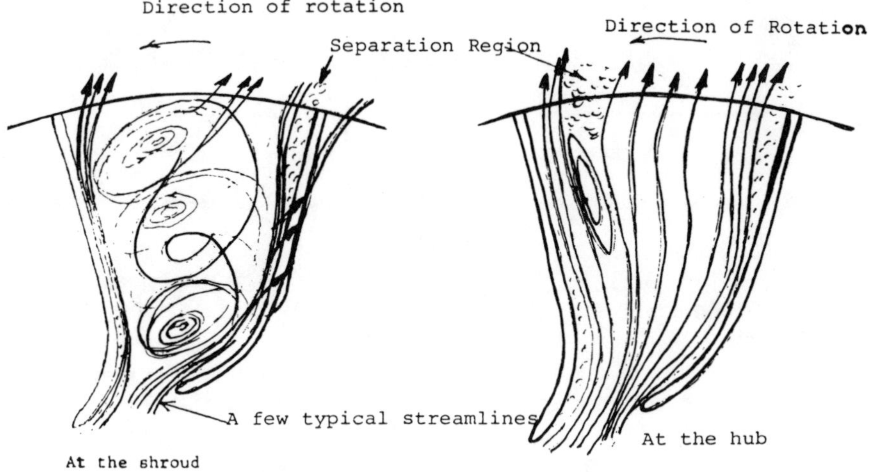

Figure 3.22. Flow map of impeller plane.

As mentioned previously, experimental studies of the flow within impeller passages have shown that the distribution of velocities on the blade surfaces is rather different from the distributions predicted theoretically. It is likely that the discrepancies between theoretical and experimental results are due to secondary flows from the pressure losses and boundary layer separation in the blade passages. High performance impellers should be designed, when possible, with the aid of theoretical methods for determining the velocity distributions on the blade surfaces.

Examples of the theoretical velocity distributions in the impeller blades of a centrifugal compressor are shown in Figure 3.24. The blades should be so designed that there are no large decelerations or accelerations of flow in the impeller since this would lead to high losses and separation of the flow. Potential flow solutions predict the flow well in regions away from the blades where boundary layer effects are negligible. In a centrifugal impeller, the viscous shearing forces create a boundary layer with reduced kinetic energy. If the kinetic energy is reduced below a certain limit, the flow in this layer becomes stagnant then reverses.

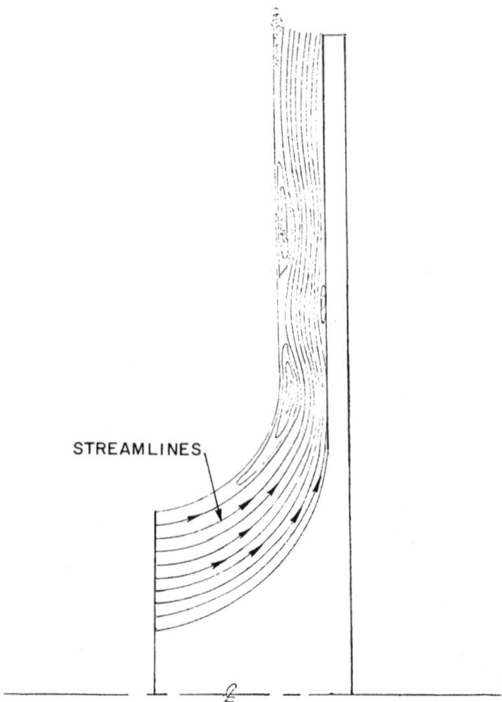

Figure 3.23. Flow map as seen in meridional plane.

Inducer

The function of the inducer is to increase the fluid's angular momentum without increasing its radius of rotation. In the inducer section the blades bend toward the direction of rotation as shown in Figure 3.25. The inducer is very much an axial rotor and changes the flow direction from the inlet flow angle to the axial direction. It has the largest relative velocity in the impeller and if not properly designed can lead to choking conditions at its throat as shown in Figure 3.25.

There are usually three forms of inducer camber lines in the axial direction. These are circular arc, parabolic and elliptical. Circular arc camber lines are used in compressors with low pressure ratios while the elliptical produces good performance at high pressure ratios where the flow has tronsonic mach numbers.

Due to choking conditions in the inducer many designs incorporate a splitter blade design. The flow pattern in such an inducer section is shown in Figure 3.26a. This flow pattern indicates a separation on the suction side of the splitter blade. Other designs include tandem inducers. This consists of the inducer section being slightly rotated as shown in Figure 3.26b. This modification by Boyce and Nishida gives additional kinetic energy to the boundary which is otherwise likely to separate.

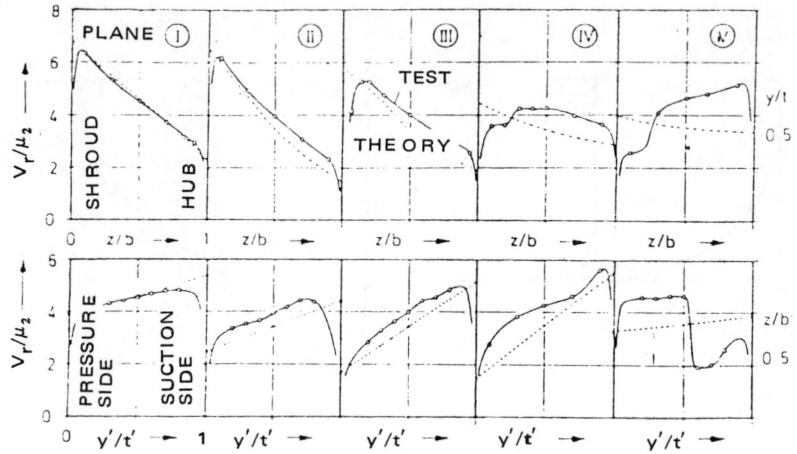

z/b = relative meridional channel width

y'/t' = relative blade spacing

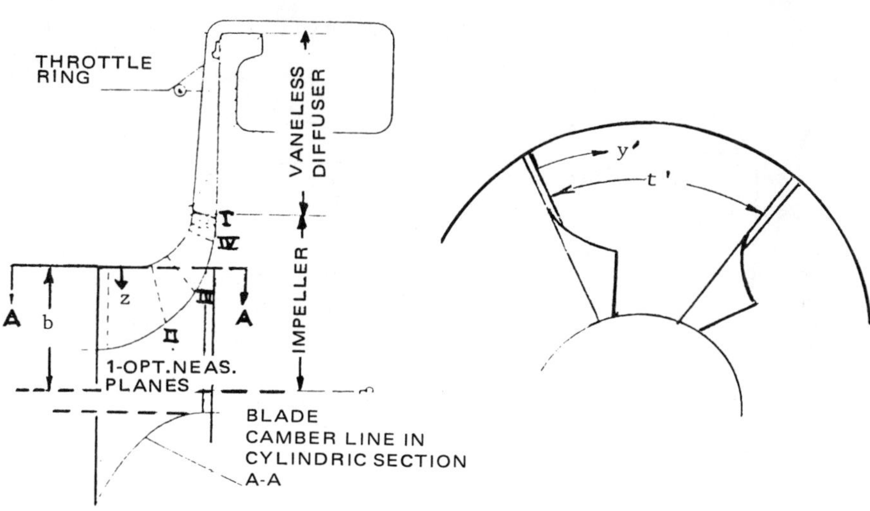

Figure 3.24. Velocity profiles through a centrifugal compressor by Senoo.

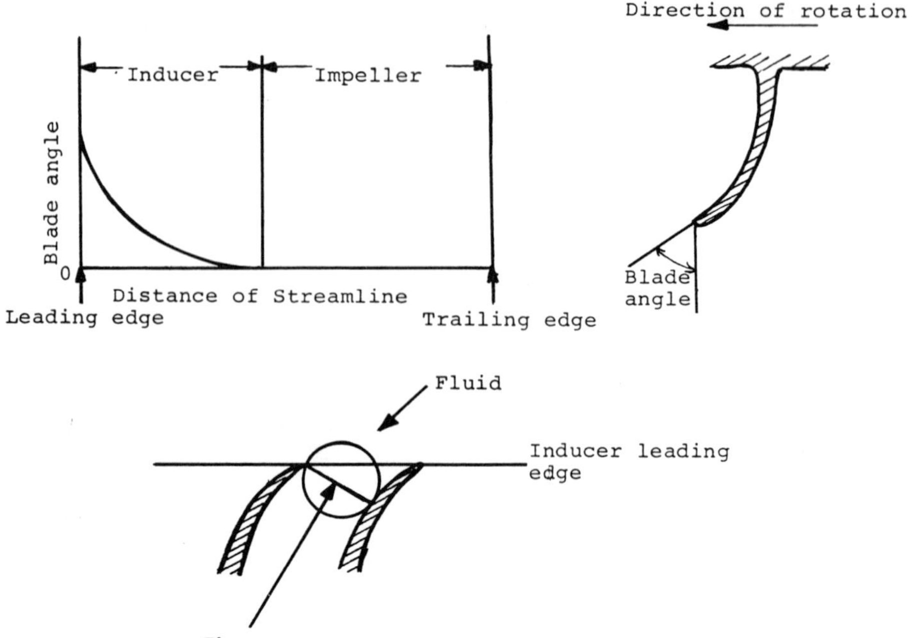

Figure 3.25. Inducer centrifugal compressor.

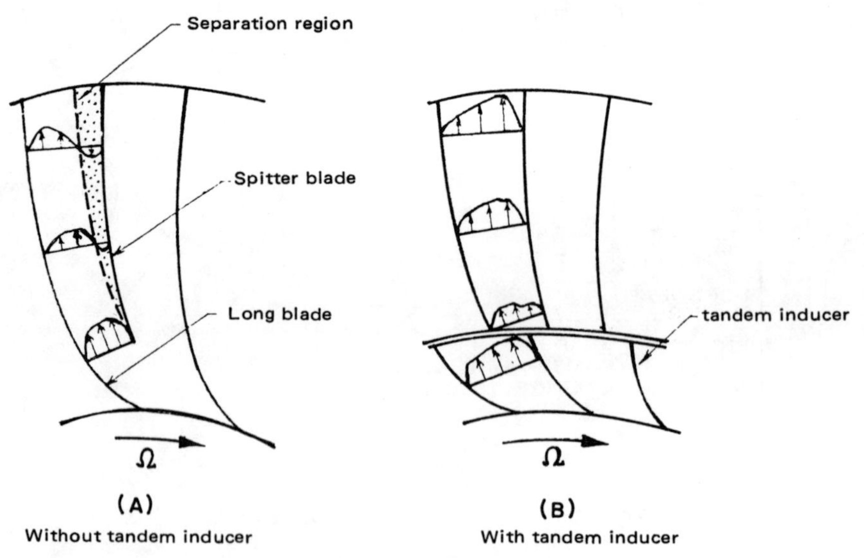

Figure 3.26. Impeller channel flow.

Centrifugal Section of an Impeller

The flow in this section of the impeller enters from the inducer section and leaves the impeller in the radial direction. The flow in this section is not completely guided by the blades and hence the effective fluid outlet angle does not equal the blade outlet angle.

To account for the flow deviation (which is similar to the effect accounted for by the deviation angle in axial flow machines), a factor known as the slip factor is used, defined as:

$$\mu = \frac{V_{\theta 2}}{V_{\theta 2 \infty}} \tag{20}$$

Where $V_{\theta 2}$ is the tangential component of the absolute exit velocity with a finite number of blades and $V_{\theta 2 \infty}$ is the tangential component of the absolute exit velocity, if the impeller were to have an infinite number of blades (i.e., no slipping back of the relative velocity at outlet). For the case of radial blades at the exit, we have:

$$\mu = \frac{V_{\theta 2}}{U_2} \tag{21}$$

Flow in a rotating impeller channel (i.e., blade passage) would be a vector sum of that with the impeller stationary and the flow due to rotation of the impeller as seen in Figure 3.27.

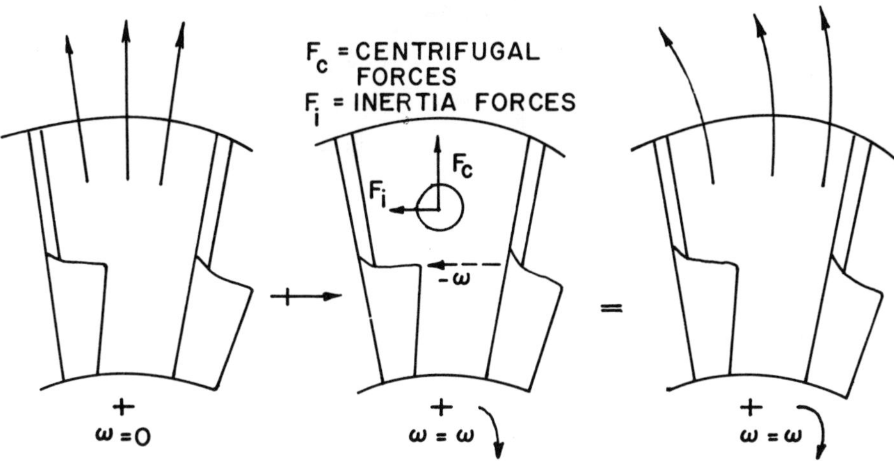

Figure 3.27. Forces and flow characteristics in a centrifugal compressor.

In a stationary impeller the flow could well be expected to follow the blade shape and exit tangentially to it. Here, of course, a very high adverse pressure gradient along the blade passage and subsequent flow separation are not considered to be general possibilities.

Inertia and centrifugal forces would cause the fluid elements to move closer to and along the leading surface of the blade, towards the exit, and once out of the blade passage where there is no positive impelling action present, these fluid elements tend to slow down.

Causes of Slip

The cause of the slip factor phenomenon that occurs within an impeller is not known exactly. However, some general reasons can be used to explain why the flow is changed. These are:

1. Coriolis Circulation – due to the pressure gradient between the walls of two adjacent blades, Coriolis forces, centrifugal forces, and follows the Helmholtz vorticity law. The combined gradient that results causes a fluid movement from one wall to the other and vice versa. This sets up a circulation within the passage as seen in Figure 3.28. Because of this circulation, a velocity gradient results at the impeller exit with a net change in the exit angle.

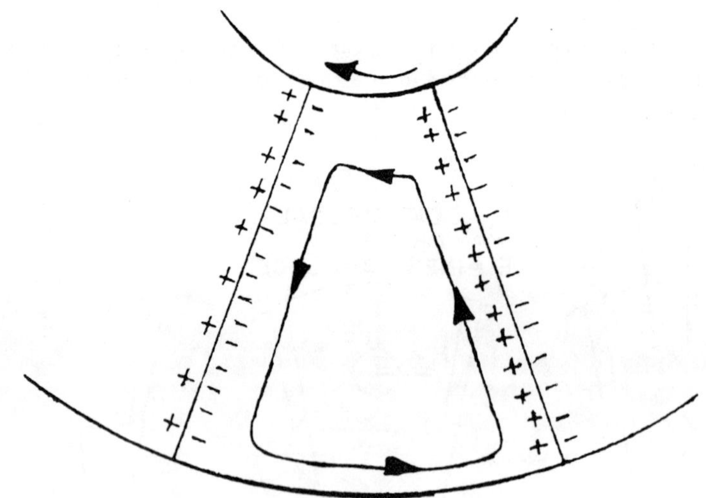

Figure 3.28. Coriolis circulation.

2. Boundary Layer Development – The boundary layer that develops within an impeller passage causes the flowing fluid to experience a smaller exit area as shown in Figure 3.29. This is due to small, if any, flow within the boundary-layer. For the

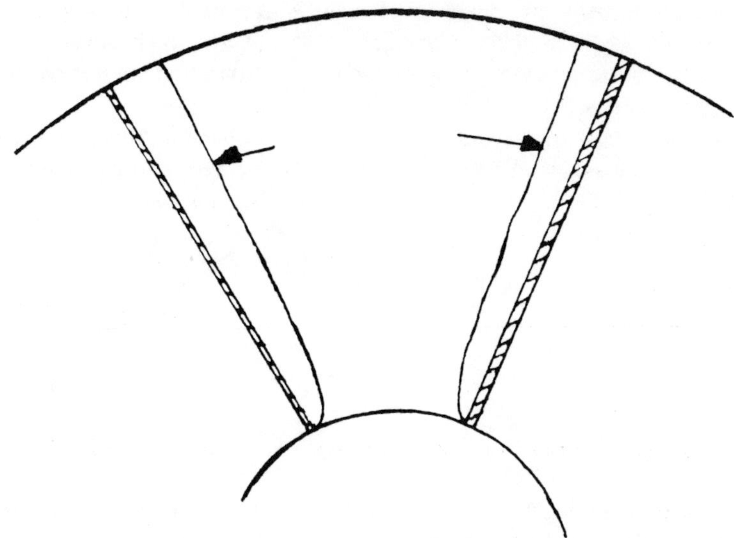

Figure 3.29. Boundary-layer development.

fluid to exit this smaller area, its velocity must increase. This gives a higher relative exit velocity. Since the meridional velocity remains constant, the increase in relative velocity must be accompanied with a decrease in absolute velocity.

3. Leakage – Fluid flow from one side of a blade to the other side is referred to as leakage. Leakage reduces the energy transfer from impeller to fluid and decreases the exit velocity angle.

4. Number of Vanes – The higher the number of vanes, the lower the vane loading and the closer the fluid follows the vanes. With higher vane loadings, the flow tends to group upon the pressure surfaces and introduces a velocity gradient at the exit.

5. Vane Thickness – Because of manufacturing problems and physical necessity, impeller vanes have some thickness. When the fluid exits the impeller, the vanes no longer contain the flow and the velocity is immediately slowed. Because it is the meridional velocity that decreases, both the relative and absolute velocities decrease, thus changing the exit angle of the fluid.

To combine all these effects, consider a backward curved blade impeller. The exit velocity triangle for this impeller, with the different slip phenomenon changes is shown in Figure 3.30. As this shows the actual conditions when the compressor is running may be far removed from the design condition.

Several theoretical and empirical equations have been derived for the slip factor. Some of the widely used of these are the theoretically based equation derived by Stodola and Stanitz, all of which assume the inviscid fluid-flow through the impeller.

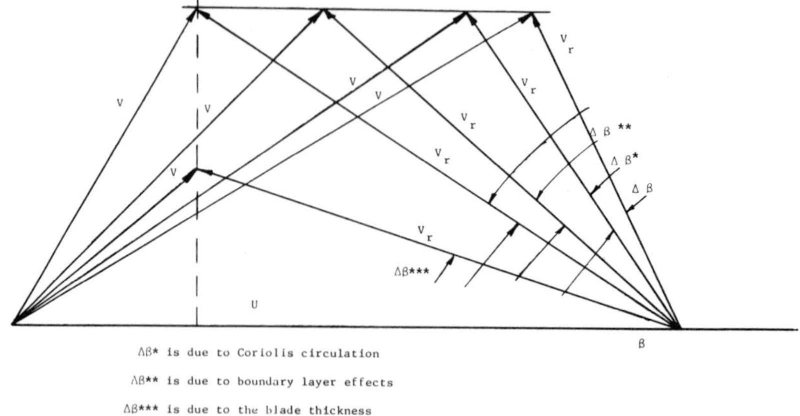

Figure 3.30. Effect on exit velocity triangles by various parameters.

6. Slip Factor Due to Stodola – The second Helmholtz law states that the vorticity of a frictionless fluid does not change with time. Hence, if the flow at the inlet to an impeller is irrotational, the absolute flow must remain irrotational throughout the impeller. As the impeller has an angular velocity ω, the fluid must have an angular velocity $-\omega$ relative to the impeller. This fluid motion is called the relative eddy. Thus if there were no flow through the impeller, the fluid in the impeller channels would rotate with an angular velocity equal and opposite to the impeller's angular velocity.

The approximate the flow Stodola's theory assumes that the slip is due to the relative eddy.

The relative eddy is considered as a rotation of a cylinder of fluid at the end of the blade passage (shown as a shaded circle) at an angular velocity of $-\omega$ about its own axis. The Stodola slip factor is given by:

$$\mu = 1 - \frac{\pi}{Z}\left[1 - \frac{\sin\beta_2}{\frac{W_{m2}\,\mathrm{Cot}\beta_2}{U_2}}\right] \tag{22}$$

Where β_2 is the blade angle, Z the number of blades, Wm_2 is the meridional velocity and U_2 the blade tip speed. Calculations using this equation have been found to be generally lower than experimental values.

7. Stantiz Slip Factor – Stanitz calculated blade-to-blade solutions for eight impellers and concluded that, for the range of conditions covered by the solutions, U is a function of the number of blades (Z) and the blade exit angle (β_2) is approximately the same whether the flow is compressible or incompressible.

$$\mu = 1 - \frac{0.63\pi}{Z}\left[\frac{1}{1 - \dfrac{W_{m2}}{U_2}\cot\beta_2}\right] \tag{23}$$

Stanitz's solutions were for $\pi/4 < \beta_2 < \pi/2$. This equation compares well with experimental results for radial or near radial blades.

Diffusers

Diffusing passages have always played a vital role in obtaining good performance from turbomachines. Their role is to recover the maximum possible kinetic energy leaving the impeller at a minimum expense of loss in total pressure. The efficiency of centrifugal compressor components has been steadily improved by advancing their performance. Significant further improvement in efficiency, however, will only be gained by improving the pressure recovery characteristics of the diffusing elements of these machines, since these elements have the lowest efficiency.

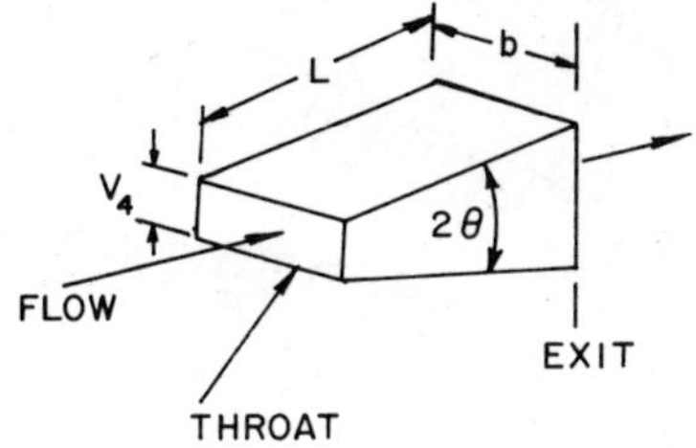

(A) STRAIGHT-WALL, RECTANGULAR DIFFUSER

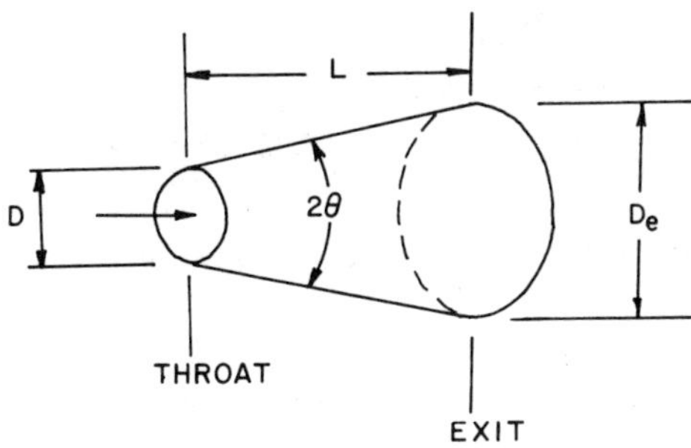

(B) STRAIGHT-WALL, CONICAL DIFFUSER

Figure 3.31. Geometric classification of diffusers.

The performance characteristics of a diffuser are complicated functions of the diffuser geometry, inlet flow conditions and in some cases exit flow conditions. Figure 3.31 shows typical diffusers classified based on their geometry. The selection of an optimum channel diffuser for a particular task is difficult since it must be chosen from an almost infinite number of cross sectional shapes and wall configurations. In radial and mixed-flow compressors the requirement of high performance and compactness leads to the use of vaned diffusers as shown in Figure 3.32. Figure 3.32 also shows the flow regime of a vane-island diffuser.

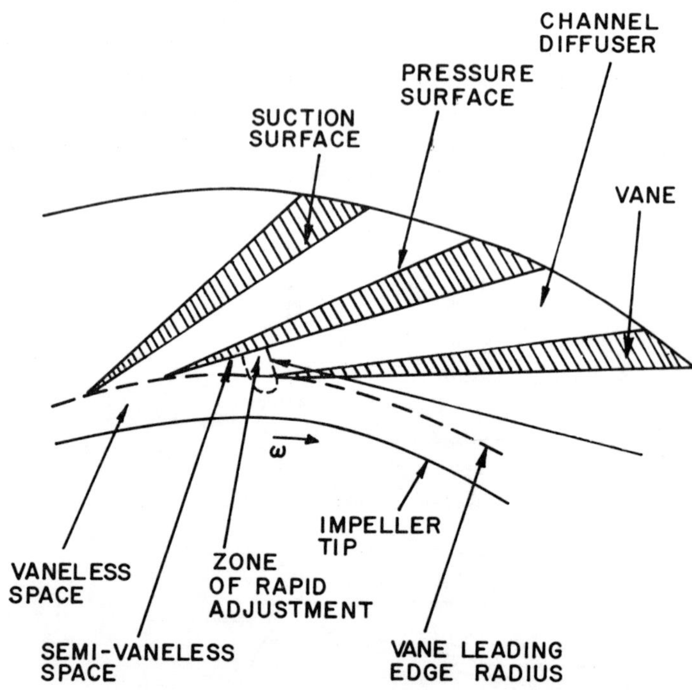

Figure 3.32. Flow regions of the vaned diffuser.

Matching the flow between the impeller and the diffuser is very complex because the flow path changes from a rotating system into a stationary system. This complex unsteady flow is strongly affected by the jet-wake of the flow leaving the impeller as seen in Figure 3.33. The three-dimensional boundary layers and secondary flows in the vaneless region, and flow separation at the blades also effect the overall flow in the diffuser.

The flow in the diffuser is in many cases assumed to be of a steady nature, in order to obtain the overall geometric configuration of the diffuser. In a

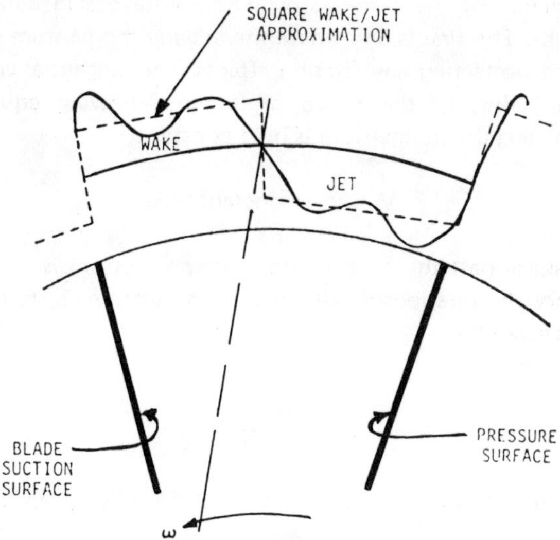

Figure 3.33. Jet/wake flow distribution from impeller.

channel-type diffuser the viscous shearing forces create a boundary layer with reduced kinetic energy. If the kinetic energy is reduced below a certain limit, the flow in this layer becomes stagnant and then reverses. This flow reversal causes separation in a diffuser passage, as mentioned above, which results in eddy losses, mixing losses and changed flow angles. The separation should be avoided or delayed to improve compressor performance.

The high-pressure ratio centrifugal compressor has a narrow stable operating range. This operating range is due to the close proximity of the surge and choke flow limits. The word, surge, is widely used to express unstable operation of a compressor in general. "Surge" will be defined as the flow breakdown period during unstable operation. The unsteady flow phenomena during the onset of surge in a high-pressure ratio centrifugal compressor caused the mass flow throughout the compressor to oscillate during supposedly "stable" operation. The throat pressure, in the diffuser increases during the precursor period almost up to collector pressure, P_{col}, at the beginning of surge. All pressure traces, except plenum pressure, suddenly drop at the surge point. The sudden change of pressure can be explained by the measured occurrence of back flow from the collector through the impeller during the period between the two sudden changes.

Scroll or Volute

The purpose of the volute is to collect the fluid leaving the impeller or diffuser and deliver it to the compressor outlet pipe. The volute has an important effect on the overall efficiency of the compressor. The volute design considers two major schools of thought. The first is one where the angular momentum of the flow in the volute is constant neglecting any friction effects. The tangential velocity $V_{5\theta}$ is the velocity at any radius in the volute thus the following equation shows the relationship if the angular momentum is held constant:

$$V_{5\theta}\ r = \text{constant} = K \tag{24}$$

Assuming no leakage past the tongue and that the pressure is constant around the impeller periphery the relationship of flow at any section Q_θ to the overall flow in the impeller Q is given by:

$$Q_\theta = \frac{\theta}{2\pi}\ Q \tag{25}$$

Thus the area distribution at any section θ can be given by the following relationship:

$$A_\theta = Qr\ X\ \frac{\theta}{2\pi}\ X\ \frac{L}{K} \tag{26}$$

where:

r = radius to the center of gravity
L = Volume width

The second technique is to design the volute by assuming that the pressure and velocity are independent of θ. Then the area distribution in the volute is given by:

$$A_\theta = K\ \frac{Q}{V_{5\theta}}\ \frac{\theta}{2\pi} \tag{27}$$

To define the volute section at a given θ the shape of the section must be decided as well as the area. Flow patterns in various types of volute are shown in Figure 3.34, note that the flow in the assymetrical volute has a single vortex instead of the double vortex in the symmetrical volute. In the case where the impeller is discharging directly into the volute it is better to have the volute width larger than the impeller width this is due to the flow from the impeller being bounded by the vortex generated from the gap between the impeller and the casing.

At different flows from design conditions there exists a circumferential pressure gradient, at the impeller tip and in the volute at a given radius. At low flows the pressure rises with the peripheral distance from the volute tongue and at high flow

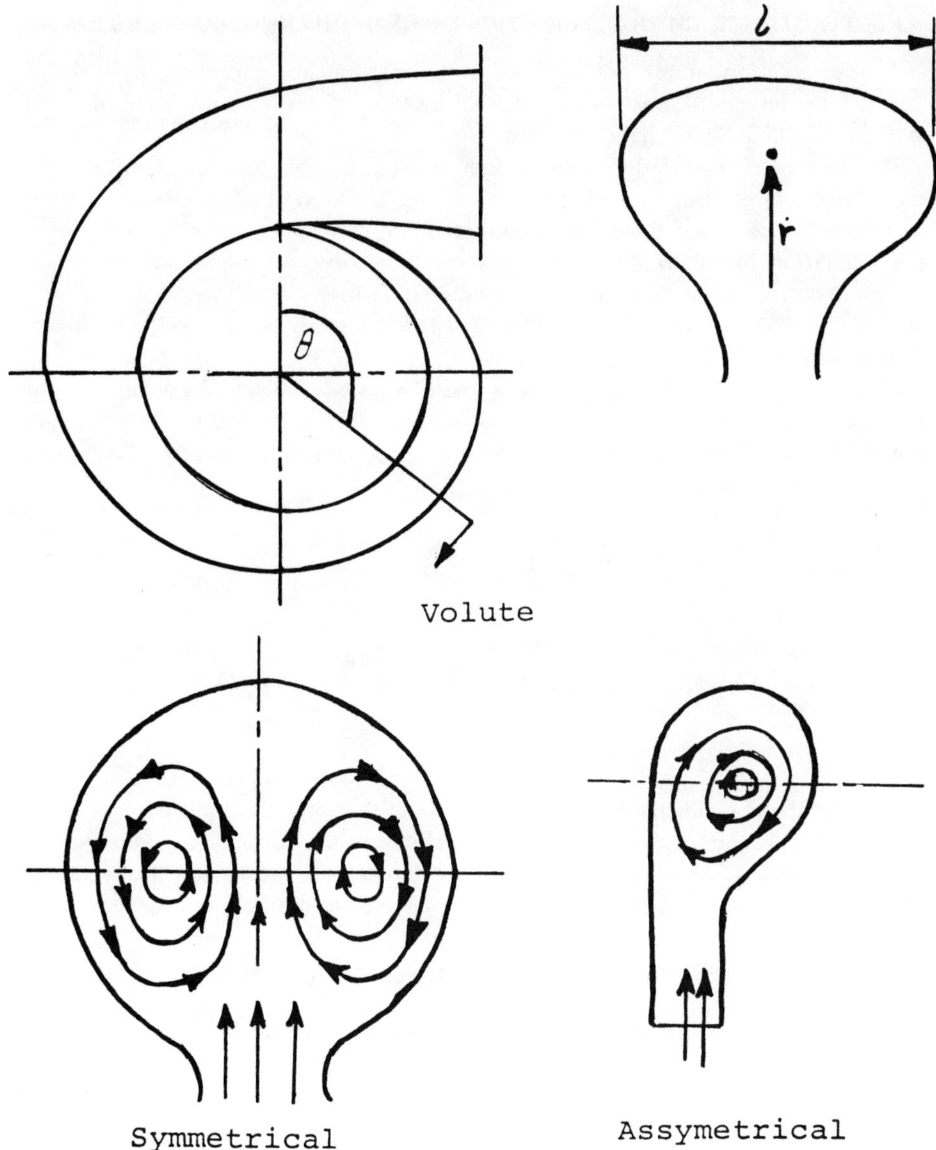

Figure 3.34. Flow patterns involute.

the pressure falls with distance from the tongue. This is due to the fact that near the tongue the flow is guided by the outer wall of the passage. The circumferential pressure gradients reduce the efficiency away from the design point, as non uniform pressure at the impeller discharge results in unsteady flows in the impeller passage and causes flow reversal and separation in the impeller.

PERFORMANCE CHARACTERISTICS OF CENTRIFUGAL COMPRESSORS

The calculation of the performance of a centrifugal compressor at both design and off design conditions require the knowledge of the various types of losses encountered in a centrifugal compressor.

The accurate calculation and proper evaluation of the losses within the centrifugal compressor is as important as the calculation of the blade loading parameter, since unless the proper parameters are controlled, the efficiency drops, the evaluation of the various losses is a combination of experimental results and theory. The losses are divided into two groups: 1) losses encountered in the rotor and, 2) losses encountered in the stator. The losses are usually expressed as a loss of heat or enthalpy.

The losses are usually expressed as a loss of heat or enthalpy. A convenient way to express them is in a non-dimensional manner with reference to the exit blade speed. The theoretical total head available (q_{tot}) is equal to the head available from the energy equation:

$$q\ th = \frac{1}{U_2^2}\ (U_2 V_{\theta 2} - U_1 V_{\theta 1}) \tag{28}$$

plus the head which is lost due to disk friction (Δq_{DF}) and due to any recirculation (Δq_{rc}) of the air back into the rotor from the diffuser.

$$q_{tot} = q_{th} + \Delta q_{DF} + \Delta q_{rc} \tag{29}$$

The adiabatic head that is actually available at the rotor discharge is equal to the theoretical head minus the heat due to the shock in the rotor (Δq_{sh}), the inducer loss (Δq_{in}), the blade loadings, (Δq_{b1}) the clearance (Δq_c) between the rotor and the shroud, and the viscous losses (Δq_{sf}) encountered in the flow passage.

$$q_{ia} = q_{th} - \Delta q_{in} - \Delta q_{sh} - \Delta_{b1} - \Delta q_c - \Delta q_{sf} \tag{30}$$

Therefore, the adiabatic efficiency in the impeller is:

$$\eta_{imp} = \frac{q_{ia}}{q_{tot}} \tag{31}$$

The calculation of the over-all stage efficiency must also include the losses encountered in the diffuser. Thus, the overall actual adiabatic head attained would be the actual adiabatic head of the impeller minus the head losses encountered in the diffuser due to wake caused by the impeller blade (Δq_w), the loss of part of the kinetic head at the exit of the dffuser (Δq_{ed}), and the loss of head due to the frictional forces (Δq_{osf}) encountered in the vaned or vaneless diffuser space.

$$q_{oa} = q_{ia} - \Delta q_w - \Delta q_{ed} - \Delta q_{osf} \tag{32}$$

Thus the overall adiabatic efficiency in impeller is given by the following relationship:

$$\eta_{ov} = \frac{q_{oa}}{q_{tot}} \tag{33}$$

The individual losses can now be computed. These losses are broken up into two major categories: 1) losses in the rotor, and 2) losses in the diffuser.

Rotor Losses

The rotor losses as mentioned previously are divided further into various categories. The following is the analysis of each of these losses.

1. Shock in Rotor Losses — This loss is due to the shock occurring at the rotor inlet. The inlet of the rotor blades should be wedge-like so as to obtain a weak oblique shock and then should gradually be expanded to the blade thickness so as to avoid another shock. If the blades were blunt, this would lead to a blow shock which would cause the flow to detach from the blade wall and the loss to be much higher.

2. Incidence Loss — At off design conditions, flow enters the inducer at an incidence angle that is either positive or negative, as shown in Figure 3.35. A positive incidence is that which causes a reduction in flow. Fluid approaching a blade with incidence suffers an instantaneous change of velocity at the blade inlet to comply with the blade inlet angle. Separation of the blade also creates a loss associated with this phenomenon.

3. Disk Friction Loss — This is the loss due to the frictional torque on the back surface of the rotor as seen in Figure 3.36. This loss is the same for a given size disk whether it is used for a radial inflow compressor, or a radial inflow turbine. In many cases, the losses in the seals, bearings, and gear box are also lumped in with this loss, and the entire loss can be called an external loss. In this loss unless the gap is of the order of magnitude of the boundary layer, the effect of the gap size is negligible. A point of interest that should be indicated here is that the disk friction in a housing is less than that on a free disk. This is due to the existence of a "Core" which rotates at half the angular velocity.

4. Diffusion Blading Loss — This loss arises because of negative velocity gradients in the boundary layer. This deceleration of the flow increases the boundary layer and gives rise to separation of the flow. The adverse pressure gradient, which a compressor normally works against increases the chances of separation and gives rise to a rather significant loss.

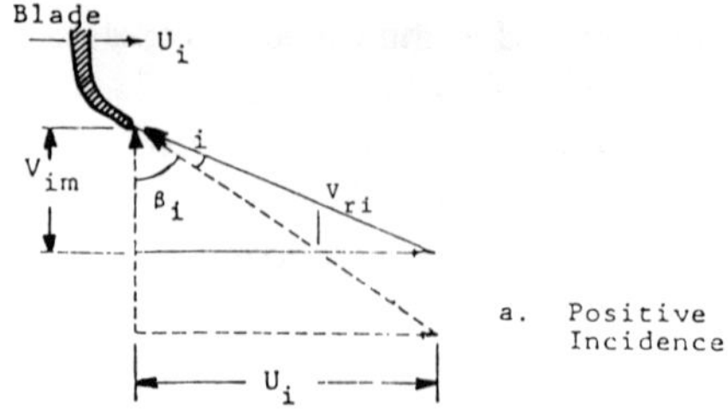

Design Triangles Are Shown Dotted

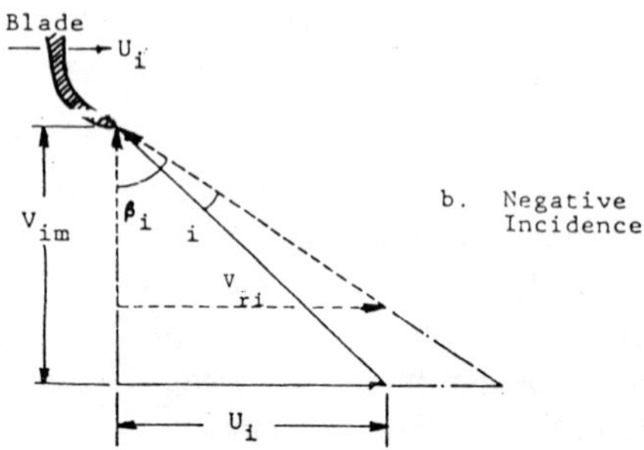

Figure 3.35. Inlet velocity triangles at non-zero incidence.

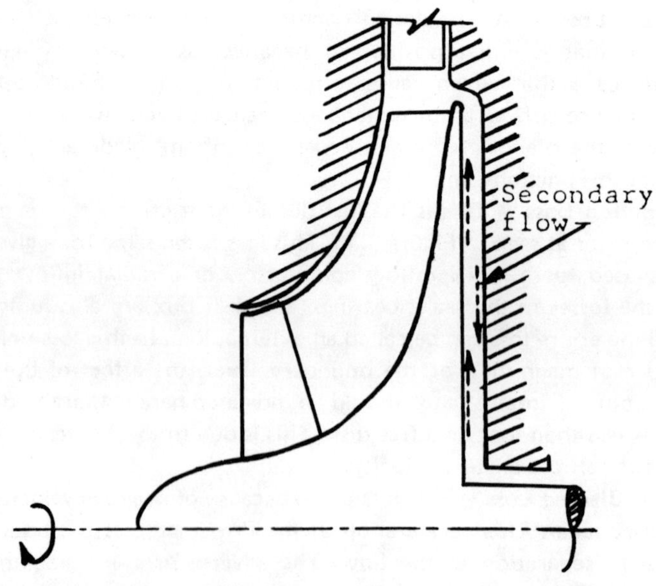

Figure 3.36. Secondary flow at the back of impeller.

5. Clearance Loss – When a fluid particle has a translatory motion relative to a non-inertial rotating coordinate system, it experiences, besides other forces, a force known as the Coriolis force. A pressure difference exists between the driving and trailing faces of an impeller blade, due mainly to Coriolis acceleration. The shortest path, and generally that of least resistance, for the fluid to flow and neutralize this pressure differential, is provided by the clearance between the rotating impeller and the stationary casing. In the case of the shrouded impellers, such a leakage from the pressure side to the suction side of an impeller blade is not possible. Instead, the existence of a pressure gradient in the clearance between the casing and the impeller shrouds, predominant along the direction shown in Figure 3.37, accounts for the clearance loss.

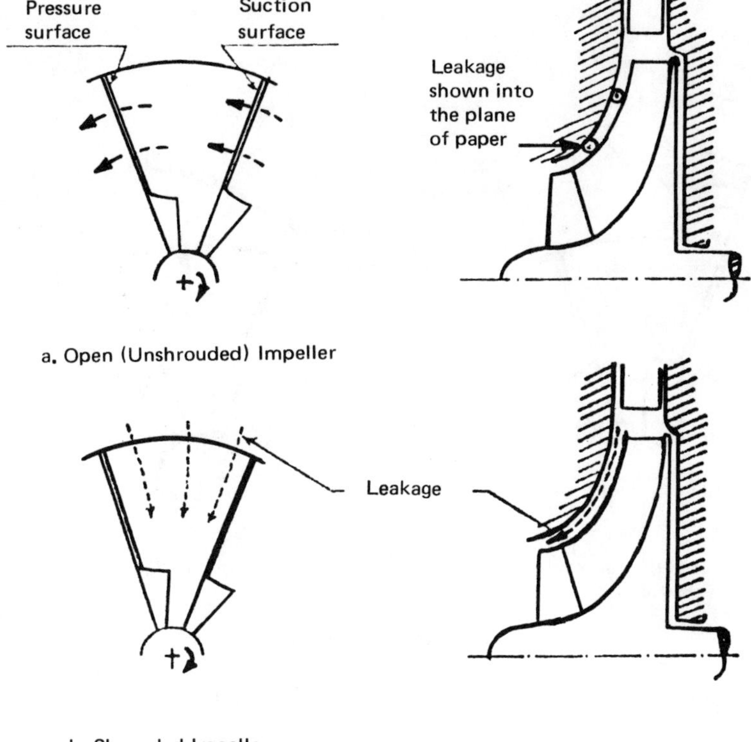

Figure 3.37. Leakage affecting clearance loss.

This loss may be quite substantial. The leaking flow undergoes a large expansion and contraction due to temperature variation across the clearance gap affecting both the leaking flow and the stream into which it discharges.

6. Skin Friction Loss – Skin friction loss is defined as the loss due to the shear forces on the impeller wall which are mostly due to turbulent friction. This type of loss is usually determined by considering the flow as an equivalent circular cross section with a hydraulic diameter. The loss is then computed based on the well known pipe flow pressure loss equations.

Stator Losses

1. Recirculating Loss – This loss occurs due to the back flow into the impeller exit of a compressor and is a direct function of the air exit angle. As the flow through the compressor reduces, there is an increase in the absolute flow angle at the exit of the impeller as seen in Figure 3.38. Part of the fluid is recirculated from the diffuser to the impeller and its energy is returned to the impeller.

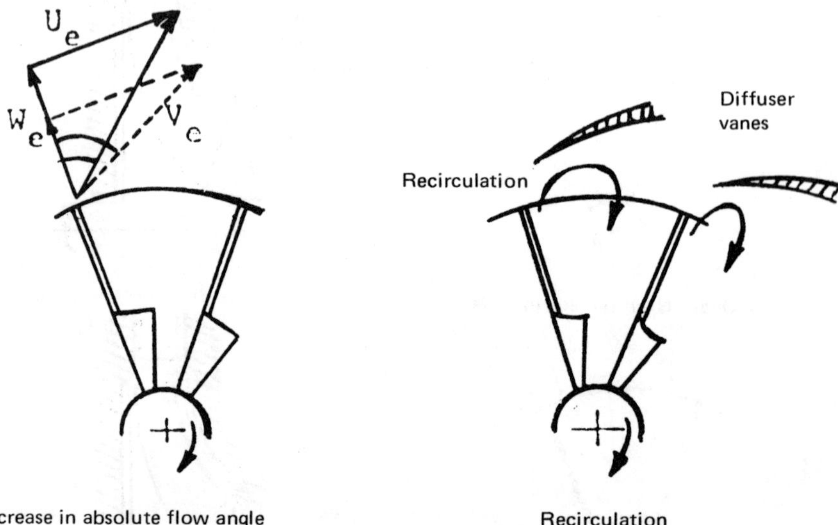

Figure 3.38. Recirculating loss.

2. Wake Mixing Loss – This loss is due to the impeller blades, causing a wake in the vaneless space behind the rotor. This loss is minimized in a diffuser which is symmetric around the axis of rotation.

3. Vaneless Diffuser Loss – This loss is experienced in the vaneless diffuser due to the friction and the absolute flow angle.

4. Vaned Diffuser Loss – Vaned diffuser losses are based on the conical diffuser test results. They are a function of the impeller blade loading and the vanlesss space radius ratio. They also take into account the blade incidence angle and the skin friction due to the vanes.

5. Exit Loss – The exit loss assumed that one half of the kinetic energy leaving the vaned diffuser is lost.

Losses are a complex phenomena and as discussed are a function of many parameters such as inlet conditions, pressure ratios, blade angles, flow etc. Figure 3.39 shows the loss distributed in a typical and centrifugal stage of pressure ratio below 2:1 with backward curved blades. This figure is just a guide line and should be used as such.

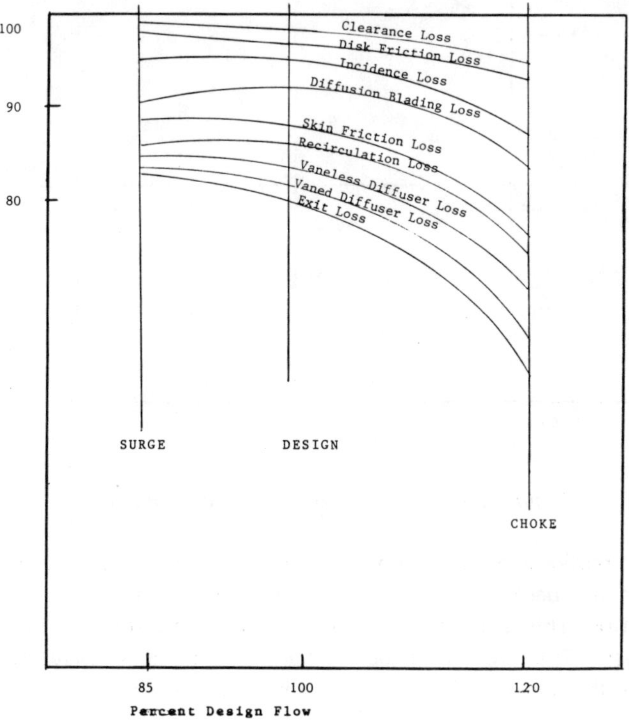

Figure 3.39. Losses in a centrifugal compressor.

Performance Characteristics

A plot showing the variation of total pressure ratio across a compressor as a function of the mass flow rate through it at various speeds is known as the performance characteristics of that compressor. Figure 3.40 shows such a plot.

The actual mass flow rates and speeds are corrected by factor $(\sqrt{\theta}/\delta)$ and $(1/\sqrt{\theta})$ respectively, in order to take into account the variation in the inlet conditions of temperature and pressure. The surge line is the line which joins the points on different speed lines where the compressor's operation begins to be unstable. A compressor is said to be in surge when the main flow through the compressor

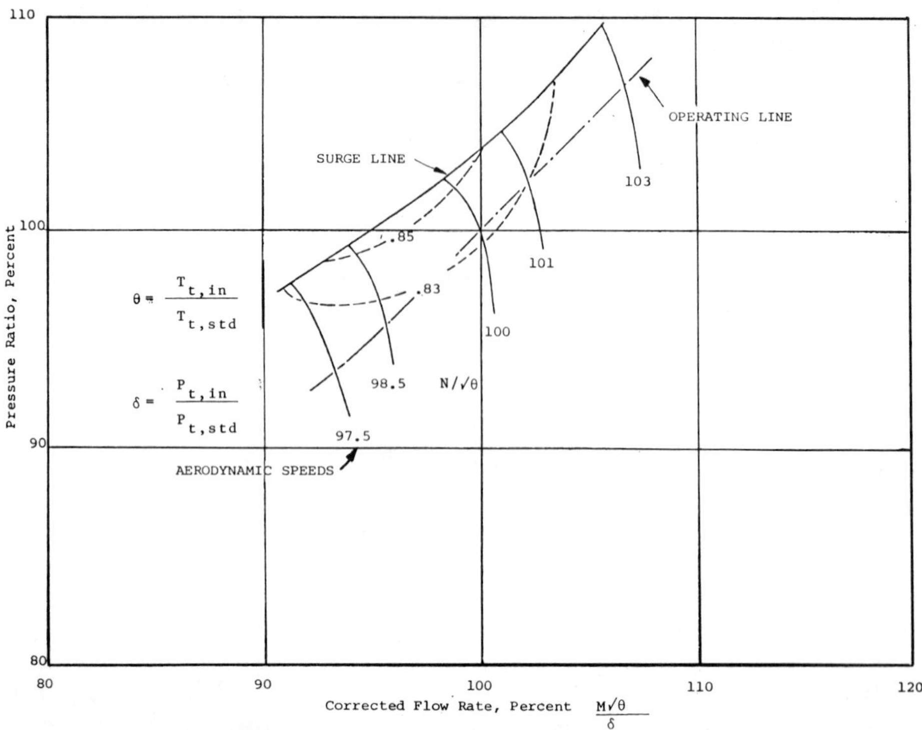

Figure 3.40. Typical compressor performance map.

reverses its direction and flows from the exit to the inlet, for a short time interval during which the back (exit) pressure drops and then the main flow assumes its proper direction. This is followed by the rise in back pressure causing the main flow to reverse again. This unsteady process, if allowed to persist, may result in irreparable damage to the machine. Lines of constant adiabatic efficiency (sometimes called the efficiency islands) are also plotted on the compressor map. A condition known as choke or commonly known as "Stonewall" is indicated on the map which shows the maximum mass flow rate possible through the compresor at that operating speed.

Surge

Compressor surge is a phenomenon of considerable interest and is not yet fully understood. Essentially, it is a situation of unstable operation and should therefore be avoided in both design and operation. Surge has been traditionally defined as the lower limit of stable operation of a compressor and involves the reversal of flow. This reversal of flow occurs because of some kind of aerodynamic instability within the system. Usually it is part of the compressor that is the cause of the aerodynamic instability though it is possible that the system arrangement could be capable of magnifying this instability. Figure 3.40 shows a typical performance map for a centrifugal compressor showing efficiency islands and constant aerodynamic speed lines. The total pressure ratio can be seen to change with flow and speed. Usually compressors are operated at a working line separated by some safety margin from the surge line.

Usually, surge is linked with excessive vibration and an audible sound; yet, there have been cases in which surge problems which are not audible have caused failures. Extensive investigations have been conducted on surge. Poor quantitative universality of aerodynamic loading capacities of different diffusers and impellers and an inexact knowledge of boundary layer behavior makes the exact prediction of flow in turbomachines at the design stage difficult. It is, however, quite evident that the underlying cause of surge is aerodynamic stall. The stall may occur in either the impeller or the diffuser.

When the impeller is the cause of surge, the inducer is what actually causes the stall. Either a decrease in the mass flow rate or an increase in the rotational speed of the impeller or both can cause stall if the compressor is operating at the surge line.

Stalling the diffuser occurs in basically the same way as in the inducer. A diffuser usually consists of a vaneless diffuser, a pre-diffuser section before the throat containing the initial portion of the vanes, the throat and a diffusion passage.

The pre-diffuser accepts the velocity generated by the centrifugal impeller and turns the flow from the vaneless space to the restriction of the diffuser passage at the throat and beyond. When the pre-diffuser stalls, the flow will not enter the throat. The discharge volume senses this drop in flow pressure and discharges through the diffuser, thus causing flow reversal and surge. As before, stalling of the pre-diffuser can be accomplished in two ways — by increasing impeller speed or decreasing the flow rate.

Whether surge is caused by a decrease in flow velocity or an increase in rotational speeds, either the inducer or pre-diffuser stalls. Which stalls first is difficult to determine, but findings have shown that for low pressure ratio compressor the surge initiates in the diffuser section while for units with single stage pressure ratios above 3:1 the indication is that surge is initiated in the inducer.

Surge Detection and Control

Surge detection devices may be broken into two groups: 1) static surge detection devices, and 2) dynamic surge detection devices. To this date, static surge detection devices have been widely used and more research work is still to be done before a dynamic detection device can be used. It is probably the dynamic surge detection device that will meet the requirements and hopes of many engineers for a control device that could anticipate stall and surge and hence prevent its occurrence. Obviously, detection devices must be linked with a control device which would prevent an unstable operation of a compressor.

Static surge detection devices are those which attempt to avoid stall and surge by the measurement of some compressor parameters and ensure that a pre-decided value is not exceeded. When the parameters meets or exceeds the limit, some control action is taken. A typical pressure oriented anti-surge control system is shown in Figure 3.41. The pressure transmitter monitors the pressure and controls a device which might open a blow-off valve. A temperature sensing device corrects the readings for the effect of flow and speed for the effect of temperature. A typical flow oriented device is also shown in Figure 3.41.

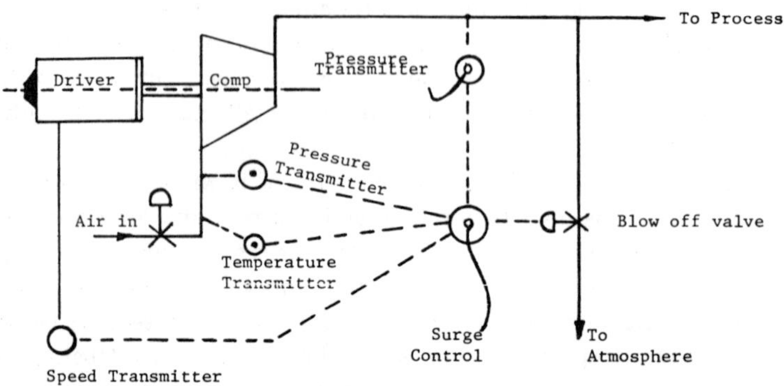

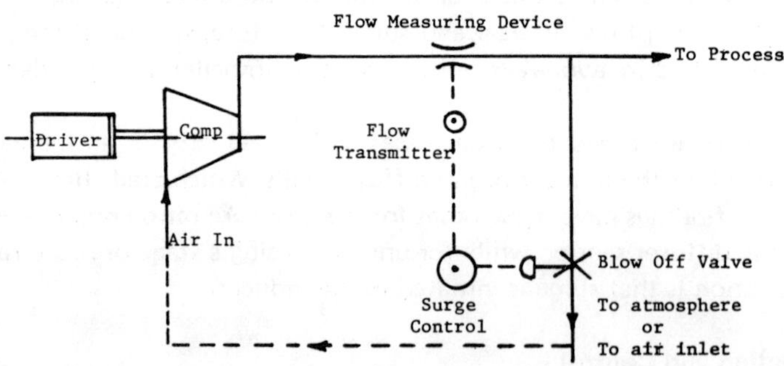

Figure 3.41. Pressure and flow – oriented anti-surge control system.

The important point for all static surge detection devices is that the actual phenomena of flow reversal (surge) is not being directly monitored. What is being monitored are other parameters that are related to surge and the control limits are set from past experience and a study of the compressor characteristics.

Dynamic surge detection and control methods are currently being researched today. Here, an attempt is made to detect the *start* of a reversal of flow before it reaches the critical situation of surge. This is done using a boundary layer probe.

Boyce has obtained a patent for a dynamic surge detection using a boundary layer probe which is presently undergoing some actual field tests. This system consists of specially mounted probes in the compressor to detect boundary layer flow reversal, as shown in Figure 3.42. The concept being that the boundary layer would reverse initially before the entire unit would be in surge, and since it is measuring an actual onset by monitoring the flow reversal it is not dependent on the molecular weight of the gas and is not effected by the movement of the surge line.

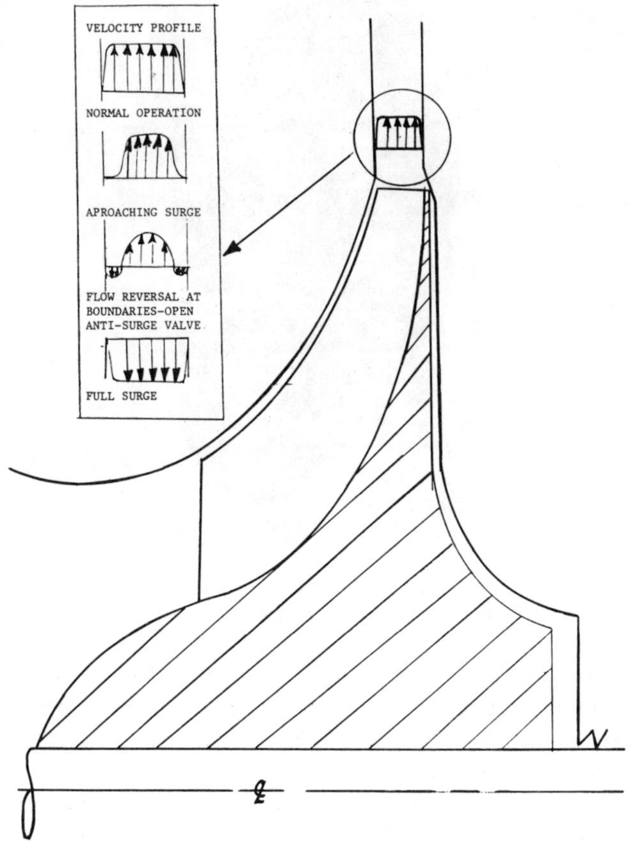

Figure 3.42. Boundary layer surge prediction technique.

IMPELLERS AND THEIR FABRICATION

Centrifugal compressor impellers are either shrouded or unshrouded as seen in Figure 3.43 and 3.44. The blading for the impellers can have one of three configurations most commonly radial, followed by backward-curved and, in some rare cases, forward curved. Backward curved blades have the largest operating range and are usually higher in efficiency. The radial bladed impeller produces a higher head but has a smaller operating range and a slightly lower efficiency than one with backward curved blades.

Figure 3.43. Closed impeller. (Courtesy of Elliott)

Open, shrouded impellers that are mainly used in single stage applications are made by investment casting techniques or by three dimensional milling. Such impellers are used, in most cases, for the high pressure ratio stages. In the CPI, the shrouded impeller is the most common.

Figure 3.45 shows several fabrication techniques. The most common type of construction is seen in Figure 3.45a and where the blades are fillet-welded to the hub and shroud. In Figure 3.45b the welds are full penetration. The disadvantage of this type of construction is the obstruction to the aerodynamic passage. In Figure 3.45c the blades are partially machines with the covers, and then butt-welded down the middle. For backward-lean-angle blades, this technique has not been very suc-

Figure 3.44. Open faced impeller.

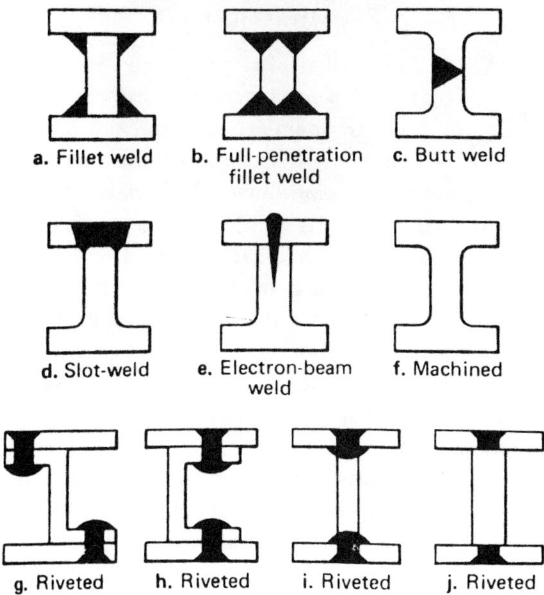

Figure 3.45. Techniques of blade attachment.

cessful, and there has been difficulty in achieving a smooth contour around the leading edge.

Figure 3.45d illustrates a slot-welding technique and is used where blade-passage height is too small, or the backward-lean angle too high, to permit conventional fillet welding. In Figure 3.45e, an electron-beam technique is used to weld on the shroud or the hub. This technique is still in its infancy and work needs to be done to perfect it. Its major disadvantage is that electron-beam welds should preferably be stressed in tension, but for the configuration of Figure 3.45e, they are in shear. The configurations of Figures 3.45g through 3.45j use rivets. Where the rivet heads protrude into passage, aerodynamic performance is reduced.

Materials for fabricating these impellers are usually low-alloy steels, such as AISI 4140 or AISI 4340. AISI 4140 is satisfactory for most applications; AISI 4340 is used for large impellers requiring higher strengths. For corrosive gases, AISI 410 stainless steel (about 12% chromium) is used. Monel K-500 is employed in halogen gas atmospheres and in oxygen compressors because of its resistance to sparking. Titanium impellers have been applied to chlorine service. Aluminum-alloy impellers have been used in great numbers especially at lower temperatures (below 300°F). With new developments in aluminum alloys, this rage is increasing. Aluminum and titanium are sometimes selected because of their low density. This can cause a shift in the critical speed of the rotor, which may be advantageous.

ROTOR DYNAMICS

The movement of the rotor and its effect on the entire performance of the unit is the most important aspect of centrifugal compressor design. Most compressors today, which are used in the petrochemical industry, are built in accordance with API 617 specifications. The natural frequency of the rotor cannot occur in the variable speed range of the compressor. Many of the newer high speed compressors, operate above their first critical. Shafts which operate above their critical are said to be "flexible shafts." API specifications call for the first critical to be at least 15 percent below any operating speed and the second critical to be at least 20 percent over the maximum continuous speed. It is desirable that the first critical not be around half the design speed; otherwise, a problem known as "oil whirl" may be induced. Oil whirl is a major cause of instability in turbomachines. It may occur in the journal bearings or in the seals in which the shaft and the stationary seal are separated by a film of fluid.

In the case of the newer flexible rotors which operate in many cases above the first critical and in many cases above the second or third criticals, balancing is a major problem. Figure 3.46 shows the various modes that the rotor shaft undergoes as it passes through these criticals, note that the mode shapes are also effected by the bearing stiffness. High-speed balancing of these rotors is sometimes a must for smooth operation. This inevitably means field balancing since there are only a few rigs which can balance these rotors at design speed, and these also operate in a vacuum chamber. Figure 3.47 is a typical rotor response curve for a four stage rotor. Here the rotor is operating above the first critical, but the steepness of the curve near the design point is a cause for concern. Modification of the rotor and change in bearing stiffness moved the slope from design point.

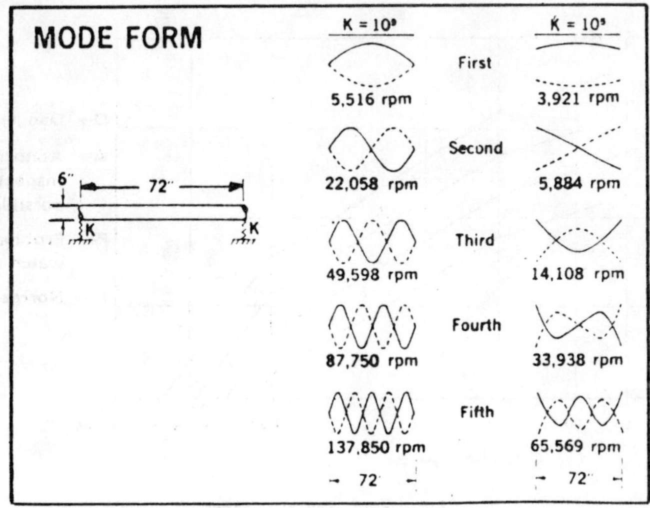

Figure 3.46. Mode forms for various bearing stiffness.

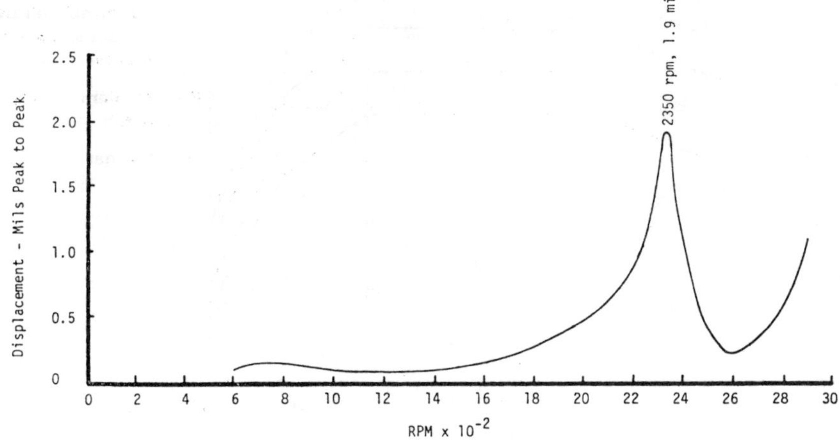

Figure 3.47. Rotor response curve.

All rotating machines vibrate when operating, but the failure of the bearings is due mainly to their inability to resist cyclic stresses. The level of vibration a unit can tolerate is shown in the severity charts as shown in Figures 3.48, 3.49, and 3.50. These charts are modified by many users to reflect their critical machines in which they would like to maintain much lower levels. These charts should only be used as guidelines, note that the absolute levels are effected by the speed of the

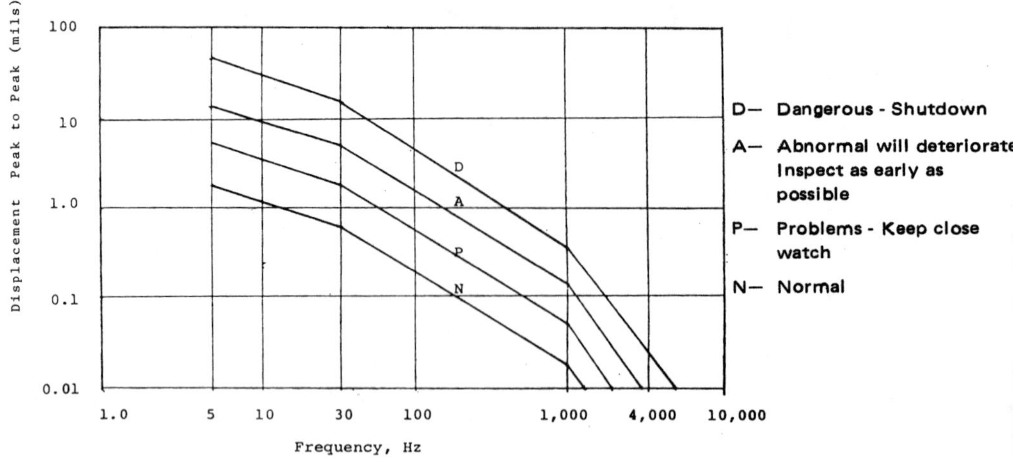

Figure 3.48. Severity chart.

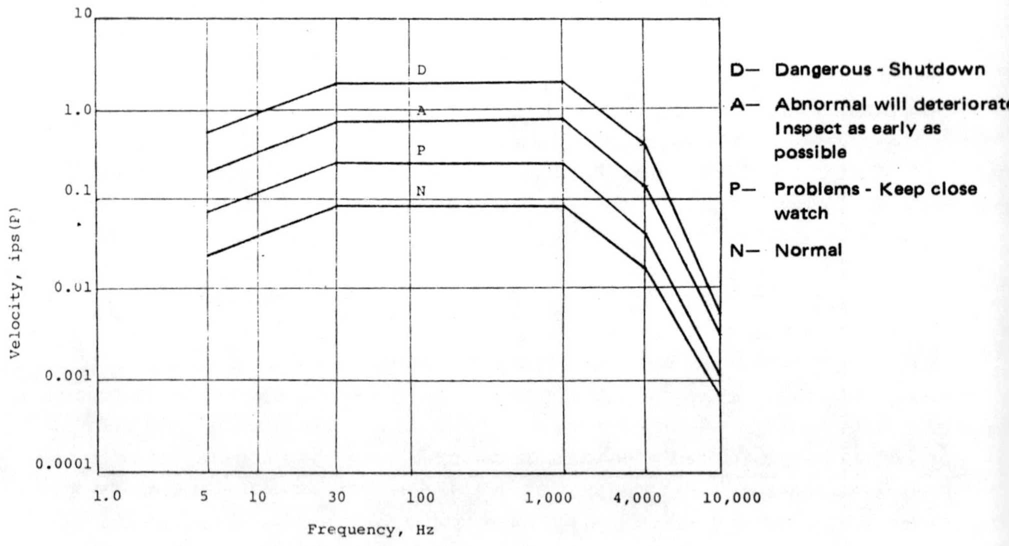

Figure 3.49. Severity chart.

machine. The velocity chart figure is the only one which is least effected by this problem.

Forces Acting On A Rotor Bearing System

There are many types of forces which act on a rotor bearing system. The forces

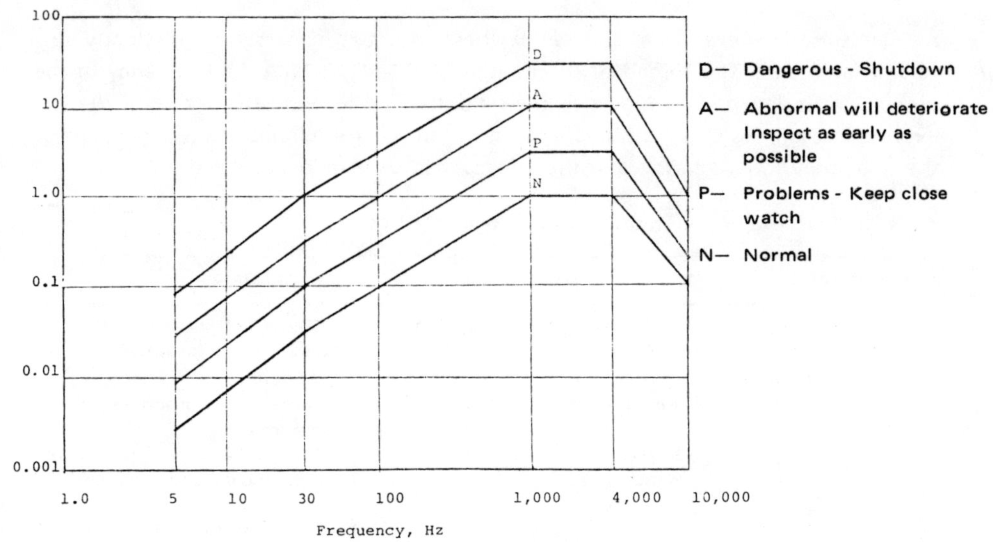

Figure 3.50. Severity chart.

can be classified into three major categories: 1) casing and foundation, 2) forces generated by rotor motion, 3) forces applied to a rotor. Table 5 by Reiger (Ref. 2) is an excellent compilation of these forces.

1. Casing and Foundation Forces — These forces can be due to foundation instability, other nearby unbalanced machinery, piping strains, rotation in gravitational or magnetic fields, and excitation of casing or foundation natural frequencies. These forces can be constant or variable with impulse loadings. The effect of these forces on the rotor bearing system can be large for example piping strains can cause major misalignment problems and unwanted forces on the bearings.

2. Forces Generated by Rotor Motion — These forces can be classified into major categories: 1) forces due to mechanical and material properties, 2) forces due to various loadings of the system. The forces due to mechanical and material properties are unbalanced which can be caused by lack of homogenity of materials rotor bow and elastic hysterisis of the rotor. The forces due to loadings of the system are viscous and hydrodynamic forces in the rotor bearing system and various blade loading forces which vary in the operational range of the unit.

3. Forces Applied to a Rotor — These forces can be due to drive torques, couplings, gears misalignment and axial forces such as due to balance piston and thrust unbalance. These forces can be very destructive and in many cases result in total destruction of a machine.

Rotor Bearing System Instabilities

Instabilities in rotor bearing systems may be the results of different forcing mechanisms. However, one can divide these into two general and distinctly different categories. The forced or resonant type is one in which the frequency of the oscillations is dependent on outside mechanisms. The second category being the self excited instabilities these are independent of outside stimuli and are independent of the frequency. Characteristics of these forces can be seen in Table 3.6.

Table 3.5. Forces Acting on Rotor-Bearing Systems (Ref. 115).

Source of Force	Description	Application
1. Forces transmitted to foundations, casing, or bearing pedestals	Constant, unidirectional force	Constant linear acceleration
	Constant force, rotational	Rotation in gravitational or magnetic field
	Variable, unidirectional	Impressed cyclic ground—or foundation-motion
	Impulsive forces	Air blast, explosion or earthquake.
	Random forces	Nearly unbalanced machinery. Blows, impact
2. Forces generated by	Rotating unbalance: Residual, or bent shaft.	Present in all rotating machinery.
	Coriolis forces	Motion around curve of varying radius. Space applications. Rotary-coordinated analyses.
	Elastic hysteresis of rotor	Property of rotor material which appears when rotor is cyclically deformed in bending, torsionally or axially.
	Coulomb friction	Construction damping arising from relative motion between shrunk fitted assemblies. Dry-friction bearing whirl.
	Fluid friction	Viscous shear of bearings. Fluid entrainment in turbomachinery. Windage.
	Hydrodynamic forces, static	Bearing load capacity
	Hydrodynamic forces, dynamic.	Bearing stiffness and damping properties.
	Dissimilar elastic beam	Rotors with differing rotor lateral stiffnesses
	Stiffness reaction forces	Slotted rotors, electrical machinery, Keyway Abrupt speed change conditions
	Gyroscopic moments	Significant in high-speed flexible rotors with disks.
3. Applied to rotor	Drive torque	Accelerating or constant-speed operation
	Cyclic forces	Internal combustion engine torque and force components.
	Oscillating torques	Misaligned couplings. Propellers. Fans. Internal combustion engine drive.
	Transient torques	Gears with indexing or positioning errors
	Heavy applied rotor force	Drive gear forces

(Continued)

Table 3.5. Forces Acting on Rotor-Bearing Systems (Ref. 115) (Continued)

Source of Force	Description	Application
		Misaligned 3-or-more rotor-bearing assembly.
	Gravity	Non-vertical machines. Non-spatial applications.
	Magnetic field, stationary or rotating	Rotating electrical machinery
	Axial forces	Turbomachine balance piston. Cyclic forces from propeller, or fan. Self-excited bearing forces. Pneumatic hammer.

(Concluded)

Table 3.6. Characteristics of Forced and Self Excited Vibration

	Forced or Resonant Vibration	Self Excited or Instability Vibration
Frequency/RPM Relationship	$N_F = N_{RPM}$ or N or rational fraction	Constant and relatively independent of rotating speed
Amplitude/RPM Relationship	Peak in narrow bands of RPM	Blossoming at onset and continue to increase with increasing RPM
Inf. of Damping	Add. damping Reduce amplitude No change in RPM at which it occurs	Add. dampoing may defer to a higher RPM. Will not materially affect amplitude
System Geometry	Lack of axial sym. external forces	Independently of symmetry small deflectin to an axisymetric system. Amplitude will self propagate.
Vibration Frequency	At or near shaft Critical or Natural frequency	same
Avoidance	1. Critical Freq. above running speed. 2. Axisymetric 3. Damping	1. Operating RPM below onset. 2. Eliminates instability Introduce damping

1. Forced or Resonant Vibration — In forced vibration the most usual drying frequency in rotating machinery is the shaft speed or multiples of this speed. This becomes critical when the frequency of excitation is equal to one of the natural frequencies of the system. In forced vibration the system is function of the frequencies that can also be multiples of rotor speed which can be exicted by frequencies other than the frequency of the speed such as blade passing frequencies, gear mesh frequencies, and other component frequencies. Figure 3.51 shows that for forced vibration the critical frequency remains constant at any shaft speed. The

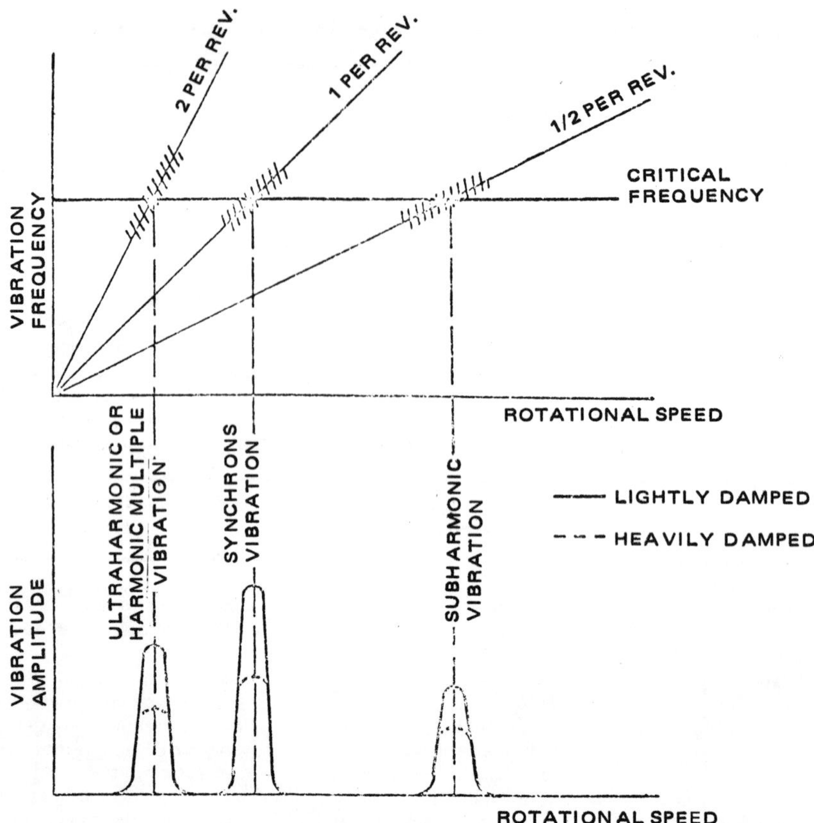

Figure 3.51. Characteristics of forced vibration or resonance in rotating machinery. (Ref. 118)

critical speeds occur at one-half, one and two times the rotor speed. The effect of damping in forced vibration is to reduce the amplitude but does not effect the frequency at which this phenomenon occurs.

Typical Forced Vibration Stimuli are as follows:

1. Unbalance – in this case due to material imperfections, tolerances, etc. the mass center of gravity is different than the geometric case, thus leading to a centrifugal force acting on the system.

2. Asymetric flexibility – the sag in a rotor shaft will cause a periodic excitation force twice every resolution.

3. Shaft misalignment – when the rotor center line and the bearing support line are not true. Misalignment may also be caused by an external piece such as the driver to a centrifugal compressor. In these cases, flexible couplings can be used and better alignment techniques to reduce the large reaction forces.

2. Self Excited Instabilities – The self excited instabilities are characterized by mechanisms which will whirl at its own critical frequency essentially independent of external stimuli. These types of self excited vibrations can be very destructive since they induce alternating stress which can lead to fatigue failures of rotating equipment. The whirling type of motion which characterize this type of instability generates a tangential force normal to the radial deflection of the shaft and whose magnitude is proportional to that deflection. The type of instabilities which fall under this category are usually called whirling or whipping. At some rotational speed where such a force is started it will overcome the external stabilizing damping force and will induce a whirling motion of every increasing amplitude. Figure 3.52 shows the onset speed note that damping in this case results mainly in a shift of this frequency not in the lowering of the amplitude as is the case in forced vibration. Important examples of such instabilities include hysteretic whirl, dry friction whip, oil whip, aerodynamic whirl and whirl due to fluid trapped in rotor. In these self excited systems friction or fluid energy dissipations generate the distabilizing force.

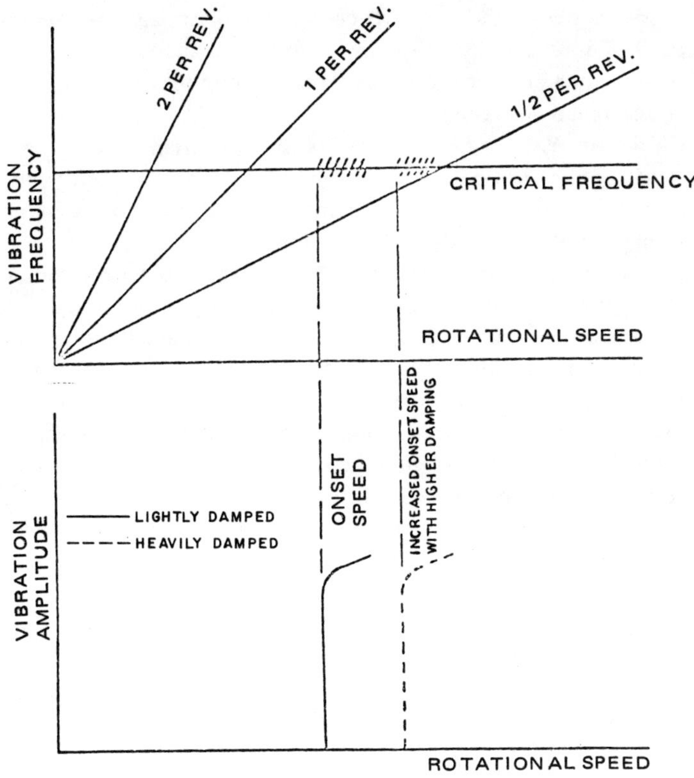

Figure 3.52. Characteristics of instabilities or self excited vibration in rotating machiner. (Ref. 118).

a. Hysteretic Whirl — This type of whirl has been diagnosed as occurring in flexible rotors due primarily to shrink fits.

When a radial deflection is imposed on a shaft, a neutral strain axis is induced normal to the direction of flexure. From first order considerations, the neutral axis of stress is coincident with the neutral axis of strain, and a restoring force is developed perpendicular to the neutral stress axis. The restoring force is then parallel and opposing the induced force. In actuality, internal friction exists in the shaft which causes a phase shift in the stress. The result is that the neutral strain and neutral stress axis are displaced so that the resultant force is not parallel to the deflection. The tangential component which is normal to the deflection results in a whirl instability. As whirl begins, the centrifugal force component increases thus causing larger deflections which result in larger stresses and still larger whirl forces. This type of increasing whirl motion may eventually be destructive.

It often requires some unbalance initial impulse to start the whirl motion. The essential effect is caused by interfaces of joints in a rotor (shrink fits) rather than defects in the material of the rotor. This type of whirl phenomenon occurs only at rotational speeds above the first critical, it may disappear and then reappear at a higher speed. To reduce this type of whirl some success has been achieved by reducing the number of separate parts, restricting the shrink fits, providing some lock-up of assembled elements.

b. Dry Friction Whirl — Dry friction whip is experienced when the surface of a rotating shaft comes into contact with an unlubricated stationary guide. This can take place in an unlubricated journal, contact in radial clearance of labyrinth seals, and loss of clearance in hydrodynamic bearings.

This phenomenon occurs when the contact is made between the surface and the rotating shaft, the coluomb friction will induce a tangential force on the rotor. This friction force is aprpoximately proportional to the radial component of the contact force thus creating a condition for instability. It should be noted that the whirl detection is counter to the shaft direction.

c. Oil Whip — This instability begins when fluid entrained in the space between the shaft and bearing surfaces begins to circulate with an average velocity of one-half of the shaft surface speed. The pressures developed in the oil are not symmetric about the rotor. Due to viscous losses of the fluid circulating through the small clearance, higher pressure exists on the upstream side of the flow than on the downstream side. Again a tangential force results. A whirl motion exists when the tangential force exceeds any inherent damping. It has been shown that the shafting must rotate at approximately twice the critical speed for this to occur. Thus the ratio of frequency to RPM is close to 0.5 for oil whirl. It should be pointed out here that this phenomenon is not restricted to the bearing but also can occur in seals.

The most obvious way to prevent oil whirl is to restrict the maximum rotor speed to less than twice its critical. Bearing designs incorporating grooves or tilting pads have been found effective in inhibiting oil whirl instability. Changing the oil temperature can also sometimes get the machine out of an oil whirl condition.

d. Aerodynamic Whirl – Although the mechanism is not clearly understood, it has been shown that aerodynamic components such as compressor wheels and turbine wheels can create cross-coupled forces due to the motion of the wheel.

The acceleration or deceleration of the process fluid imparts a net tangential force on the blading. If the clearance between the wheel and housing varies circumferentially, a variation of the tangential forces on the blading may also be expected, resulting in a net destabilizing force. The resultant force from the cross-coupling of angular motion and radial forces may destabilize the rotor causing a whirl motion.

e. Whirl Due to Fluid Trapped in a Rotor – This type of whirl occurs when liquids are entrapped inadvertently in internal cavities of rotors. The fluid does not remain in a radial direction but has a component in the tangential direction. The onset of this type of instability occurs just above the first critical and below twice the critical speed.

BEARINGS FOR HIGH SPEED MACHINERY

Journal and thrust bearings are among the most important components to assure maintenance free running of high speed turbomachines. Bearings in these machines range from simple journal bearings and flat thrust bearing to multiwedge designs for both thrust and journal bearings. Some of the many factors that enter into the selection of such bearings are:

- Speed range of shaft
- Maximum misalignment that can be tolerated by the shaft
- Loading of the compressor inlets
- Oil temperature and viscosity
- Foundation stiffness
- Axial movement that can be tolerated
- Type of lubrication system and its contamination
- Maximum vibration levels that can be tolerated

All rotating machines vibrate when operating, but failure of the bearings is mainly due to their inability to resist cyclic stresses. The level of vibration that a unit can tolerate is shown in the severity charts as shown in Figures 3.48, 3.49, and 3.50. These charts are modified by many users to reflect the critical values for their machines.

Journal Bearings

The journal bearings for turbomachinery has a fluid film that carries the load. Film thickness in most applications range from 0.0003 in. for gases to 0.008 in. for hydrostatic oil lubricated bearings.

Figure 3.53 shows a number of journal bearings in which a positive supply of lubricant is fed to the bearing at all times. The circumferential grooved bearing normally has the oil groove at half the bearing length. This provides better cooling, but reduces load capacity by dividing the bearing into two parts. The cylindrical bearing, used in turbines, has a split construction with two axial oil feed grooves at the split. The pressure dam bearing is used where bearing stability is required.

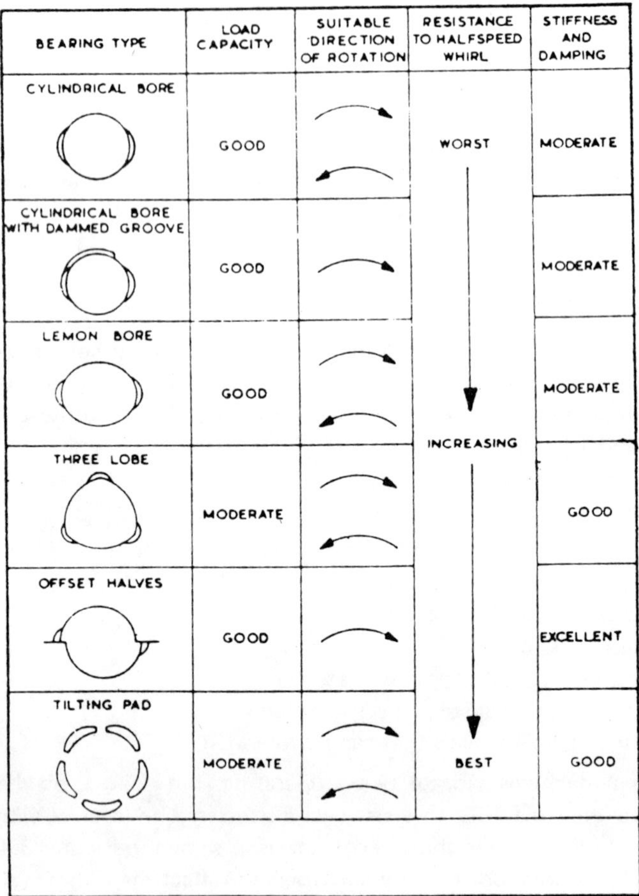

BEARING TYPE	LOAD CAPACITY	SUITABLE DIRECTION OF ROTATION	RESISTANCE TO HALFSPEED WHIRL	STIFFNESS AND DAMPING
CYLINDRICAL BORE	GOOD		WORST	MODERATE
CYLINDRICAL BORE WITH DAMMED GROOVE	GOOD			MODERATE
LEMON BORE	GOOD			MODERATE
THREE LOBE	MODERATE		INCREASING	GOOD
OFFSET HALVES	GOOD			EXCELLENT
TILTING PAD	MODERATE		BEST	GOOD

Figure 3.53. Comparison of journal type bearings.

The most common bearing is the tilting pad type, whose most important feature is self alignment when the bearing is used with spherical pivots. This bearing offers the greatest increase in fatigue life because of these advantages:

1. Self aligning for optimum alignment and minimum limit.

2. Thermal conductivity backing material to dissipate heat developed in the oil film.
3. Thin babbitt layer, centrifugally cast to a uniform thickness of about 0.005 in. Thick babbitts greatly reduce bearing life. Babbitt thickness of about 0.01 in. reduces bearing life by more than half.
4. Oil film thickness can be varied by changing the number of pads, directing the load onto or in between the pads, or changing the axial length of the pad. Oil film thickness is critical when making bearing stiffness calculations.

Thurst Bearings

The most important function of a thrust bearing is to resist the unbalanced force developed in the working fluid of the machine, and to maintain the rotor in its position within the prescribed limits. A complete analysis of the thrust load must be conducted. Compressors with back to back rotor greatly reduce the load on thrust bearings.

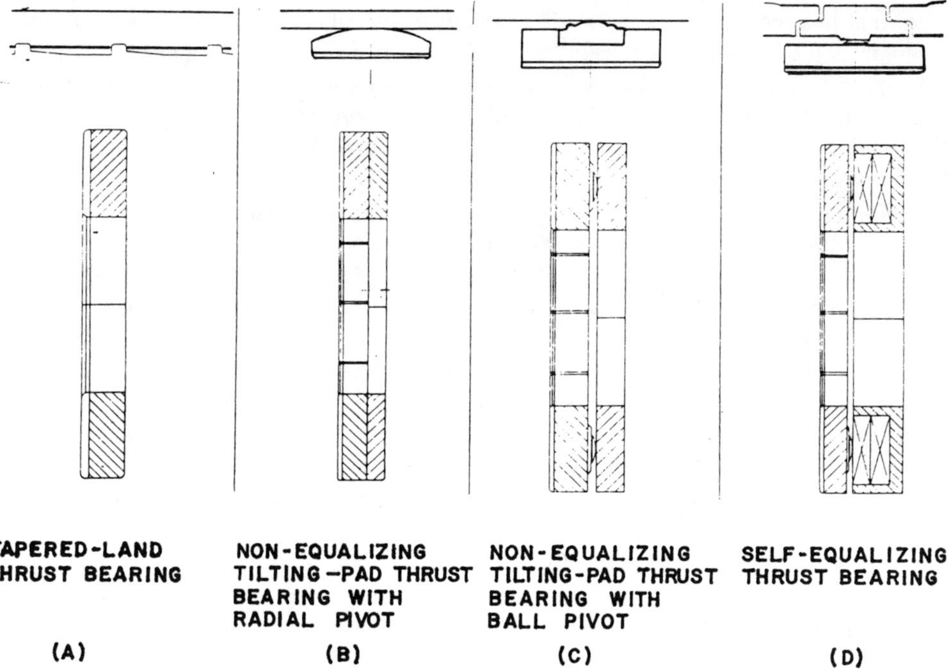

TAPERED-LAND THRUST BEARING (A)

NON-EQUALIZING TILTING-PAD THRUST BEARING WITH RADIAL PIVOT (B)

NON-EQUALIZING TILTING-PAD THRUST BEARING WITH BALL PIVOT (C)

SELF-EQUALIZING THRUST BEARING (D)

Figure 3.54. Types of thrust bearings.

Figure 3.54 shows a number of different types of such bearings. When properly designed, the tapered-land thrust bearing as seen in Figure 3.54a can take and support a load equal to that of a tilting pad thrust bearing. With perfect alignment, it can match the load of even a self equalizing tilting pad thrust bearing. Figure 3.54b is a non equalizing tilting pad thrust bearing that pivots on the back of the pad along a radial line. Figure 3.54c is a non equalizing tilting pad bearing whose pads are supported on spherical pivot points. Since this allows the pads to pivot in any direction, alignment is not as serious a problem as in the other two types. Figure 3.54d is the Kingsbury type self equalizing thrust bearing. This bearing virtually eliminates the problem of misalignment. The major drawback is that standard designs require more axial space than do a non equalizing type.

MISALIGNMENT

The amount of misalignment which can be tolerated depends on the types of journal and thrust bearings used. As previously mentioned, the tilting pad type bearings greatly reduce the problem of misalignment. Figure 3.55 shows misalignment in both the journal and thrust bearing. The effect of misalignment on a journal bearing is that of the shaft contacting the end of the bearing. Thus, journal length is a criteria in the amount of misalignment a bearing can tolerate; a shorter length bearing obviously can tolerate more misalignment. The effect on the thrust bearing is to load up one segment of the thrust bearing arc and unload the opposite segment. This effect is more pronounced at the higher loads and less flexible bearings.

To adjust for misalignment, beyond correcting by using tilting pad bearings, various misalignment techniques have been used. Initially, the simplest and most common type of alignment technique is used, which is the so called "cold alignment" method. This is also usually referred to as a base alignment. Once this is accomplished hot alignment checks are required; thus, hot alignment is carried out. Hot alignment techniques measure the changes when the unit is operational and temperature growth stabilized so that accurate alignment data is done. The most recommended technique is to do the cold alignment, using the "reverse indicator graphical plotting" and the hot alignment by the use of mechanical technqiues such as "Acculign" or "DOD Bars" or other hot alignment techniques such as optical or laser techniques.

The reverse indicator graphical plotting technique is normally done when the unit is cold. This is done by first laying out the desired hot operating line on a graph paper. This is the line which shows the final desired operating equilibrium conditions. Then the desired cold position of the shaft is plotted. Figure 3.56 shows such a chart. The actual positions of the shaft in the field are then taken. This information is plotted and the difference computed and the shims added to the supports. This procedure is then repeated after hot checking the alignment. To make the hot alignment checks, a mechanical alignment procedure is recommended.

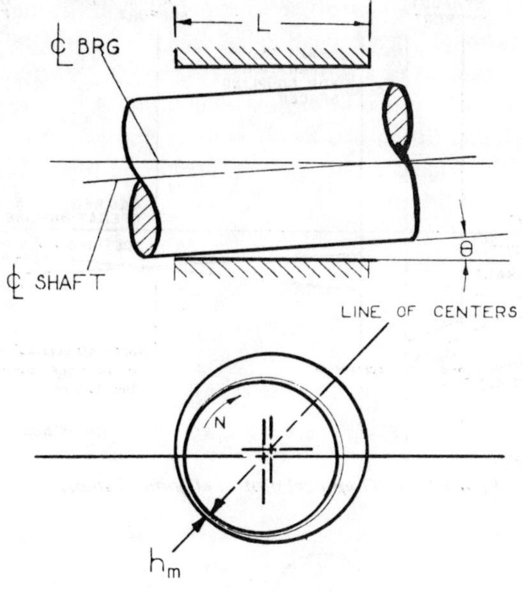

Journal Bearing

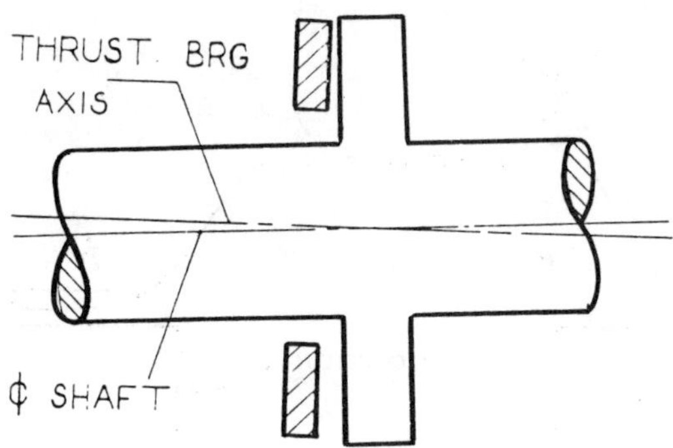

Thrust Bearing

Figure 3.55. Journal misalignment in thrust bearings.

The heart of this technique is a mechanical instrument with built in dial indicator to measure displacement from precise and reference marks. Thus, after the initial cold alignment, the train is ready for start up. Bench marks are established at each end and each side of each unit of the train as close to the couplings as possible, on most units, this is the bearing housing as seen in Figure 3.57. The train is then

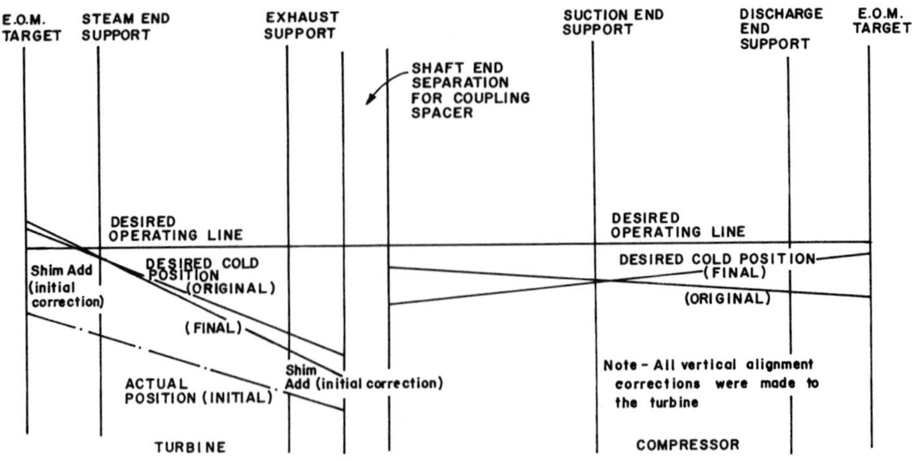

Figure 3.56. Graphical plot of alignment steps.

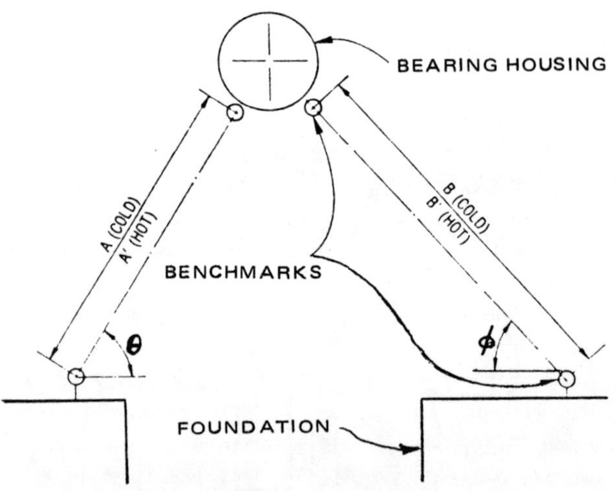

Figure 3.57. Typical placement of benchmarks on foundation and bearing housing.

started up and operated at design or near design conditions as much as possible and temperatures are allowed to stabilize. Another set of readings are taken as shown in Figure 3.58. Thus, the actual thermal growth can then be plotted on the graph and new corrections can be computed. The technique outlined above is used on new machines, and on old machines, a reverse technique can be used. First, a hot alignment check is made, then a cold check, followed by the mechanical reverse indicator readings. This information is plotted and realignment measurements taken.

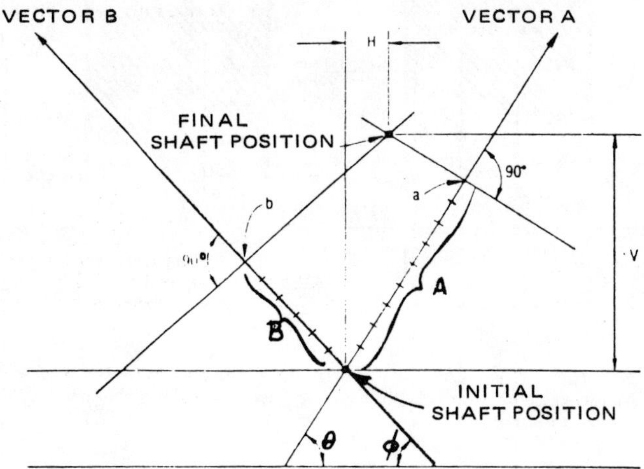

Figure 3.58. Graphical determination of shaft in hot position relative to cold position.

The above outline is simple, but in actual practice, one must develop this skill. Some of the major problems encountered in alignment are caused by pipe strain. This is caused by the piping being off from the intake or exhaust from a few thousandths to several inches. Many engineers take the attitude that they cannot understand how a small pipe like this is able to move a large piece of machinery. The results are very surprising. Tension on pipe hangers can change the vibration level considerably. Another contributor to the alignment problem is the gear casing. Thermal growth in the new fabricated cases has been unpredictable in many applications. In some cases, gear cases rise with a twist. This creates another problem which is very hard to correct. Short couplings also present a problem and magnify misalignment problems. Some users are now specifying that the coupling spacer will be at least 18 inches long. The above has been a general cursory outlook on alignment techniques. Alignment of high speed machinery must be accurate; otherwise, major problems will arise. Sometimes, to prevent major shutdowns for correction, heaters are added to one or the other legs of the unit to align them while running. This technique is not advised as a cure, but as a temporary relief when shutdown is possible.

COMPRESSOR SEALS

The internal seals that prevent leakage around the impellers are usually labyrinth type, as shown in Figure 3.59. They have a series of circumferential knife edges that are positioned closely to the rotating impeller. If damaged by rubbing erosion or corrosion, these knife edges will lose their effectiveness. In some

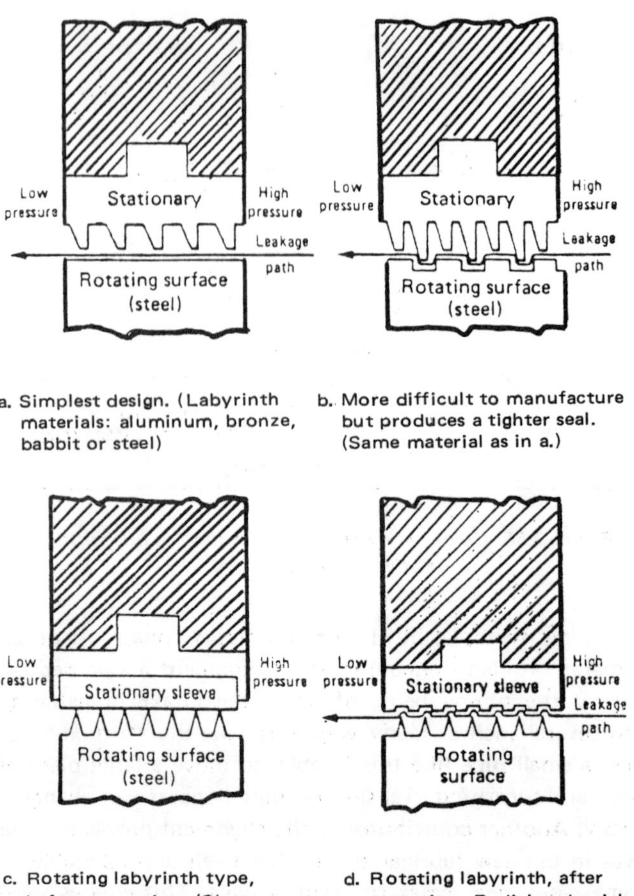

Figure 3.59. Labyrinth seals.

cases, the knife edges are machined onto the rotating part, while a sleeve of soft material is positioned on the stationary part. The rotating part then cuts a groove into the stationary sleeve, reducing leakage considerably and thus increasing the compressor efficiency. The stationary sleeve can be manufactured in babbitt lined steel, ceramic, compressed steel fiber material, etc., while the knives are made of a high quality steel.

Shaft seals are usually mechanical contact or liquid film types as seen in Figure 3.60. In some cases where a small amount of leakage can be tolerated, labyrinth

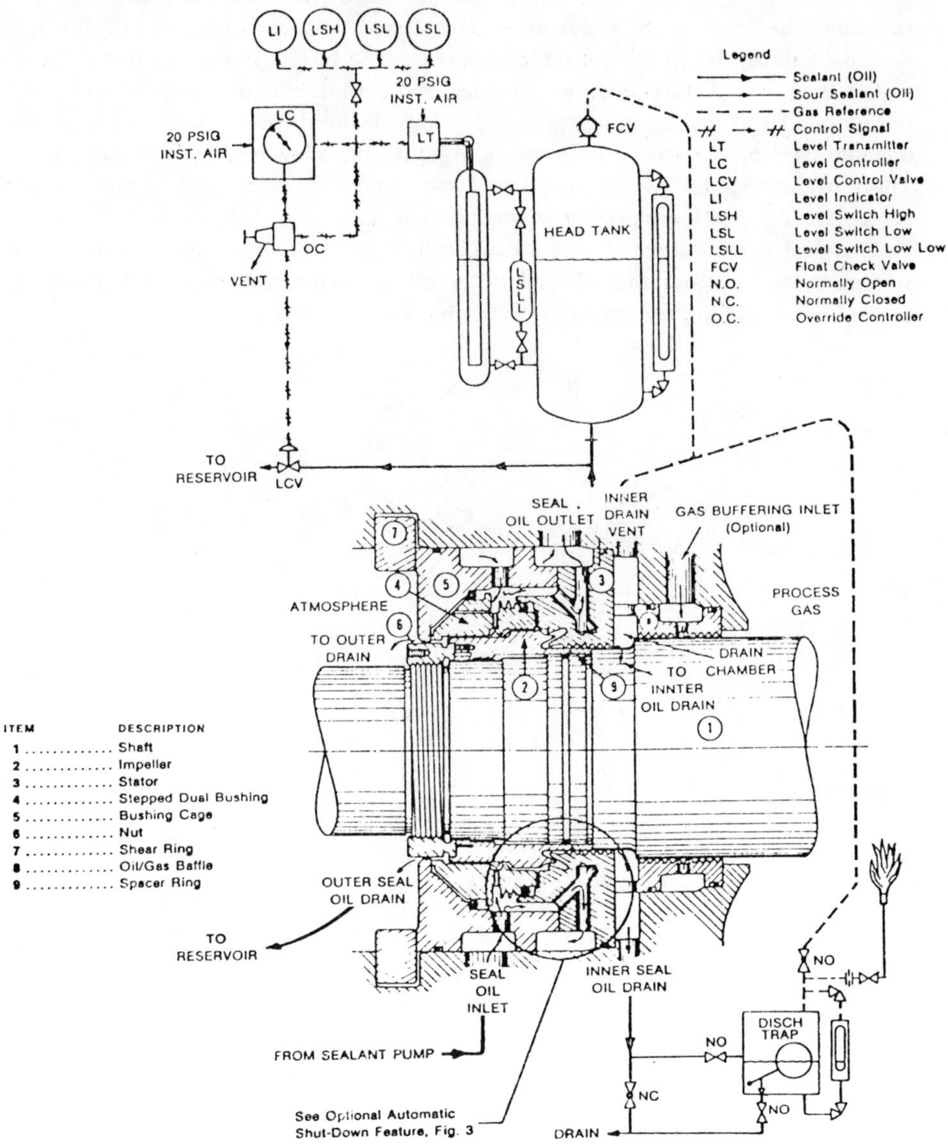

Figure 3.60. Liquid film seal for compressor shaft provides continuous flow of seal oil.

seals are used. The oil or liquid film seal consists of two stationary bushings that surround the shaft with a clearance of a few thousandths of an inch. Oil at a nominal rate of 10 gpm is introduced between the bushings at a positive pressure higher than that of the process gas, and leaks in both directions along the shaft. The oil is retained in the seal housing by "O" rings. To limit the inward oil leakage, the differential pressure across the inner bushing is only a few pounds per square inch. The inner leakage rate varies from 1 to 4 gph, depending on the size of the seal, but is independent of the gas pressure being contained.

This leakage is collected in a chamber that is usually separated from the gas stream by a labyrinth seal. Overflow of oil from the leakage chamber and its subsequent entering the compressor is the biggest problem.

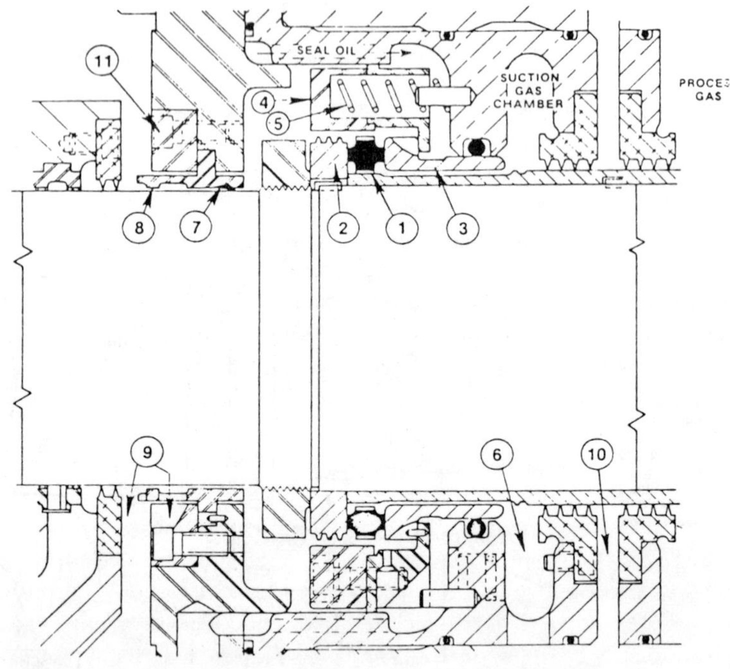

1. ROTATING CARBON RING
2. ROTATING SEAL RING
3. STATIONARY SLEEVE
4. SPRING RETAINER
5. SPRING
6. GAS AND CONTAMINATED OIL DRAIN
7. FLOATING BABBITT-FACED STEEL RING
8. SEAL WIPER RING
9. SEAL OIL DRAIN LINE
10. BUFFER GAS INJECTION PORT
11. BYPASS ORIFICE

Figure 3.61. Mechanical contact shaft seal.

Mechanical contact seals have two major elements as shown in Figure 3.61. These are the oil to process gas seal, or carbon ring, and the oil to uncontaminated seal oil drain seal, or breakdown bushing. This seal can maintain a lower inner leakage ratio with higher oil to gas differential pressure.

In operation, the seal oil pressure is maintained at about 25 to 50 psia over the process gas pressure against which the seal is sealing. High pressure oil enters the seal cavity, completely filling it. Some of the oil (ranging from 2 to 8 gph) is forced across the carbon seal face, and flows out the contaminated oil drain. The mechanical seal's great advantage over the oil film seal is that it has a minimum effect on rotor dynamics. On the other hand, when the oil film bushings lose their free floating feature, they can upset the stability of the rotor when operating at high speeds.

BALANCING

In large rotors operating at super critical speeds, balancing of the rotor becomes very important. This is particularly true since stresses caused by unbalanced rotors are proportional to sequence of the frequency of rotation. In perfectly balanced rotors, the vertical axis of the rotor should be located on the axis of rotation. In reality no rotor is perfectly balanced and the unbalance is caused by various factors such as non-homogeneous material, misalignment of bearings and couplings, thermal gradients, hydraulic and aerodynamic faces, etc.

In the rotor, the exact location of the unbalance cannot always be found. The only way to locate any unbalance is to study the vibration of the rotor. Balancing of the rotor is done in two stages: 1) the static balance and, 2) the dynamic balance.

The purpose of static balance is to make the center of gravity of the rotor approach the center line. Static balance is usually performed by placing the rotor on a set of frictionless supports; the heavy point usually has a tendency of rolling down. Noting the location of this point helps determine the amount of unbalance.

Dynamic balancing is more complex. The rotor is placed on its supports and by noting the vibration pattern, the location and amount of unbalance can be determined. Balancing is then done by placing correction weights at the appropriate locations in plates perpendicular to the rotor. A better balance is obtained by placing weights in as many planes as possible.

It is common practice to balance individual components such as impeller wheels, couplings, etc. and to then mount them on the shaft and the final balancing performed, if then an unbalance exists, correction is then usually made to the last impeller installed. The best technique for high speed rotors is to balance them — not in low speed machines, but at their rated speed. This is not always possible in the shop; therefore, it is often done in the field. New facilities are being built which can run a rotor in an evacuated chamber at running speeds in a shop. Figure 3.62 shows the evacuation chamber and Figure 3.63 the control room.

Figure 3.62. Evacuation chamber for high speed testing (Courtesy of Transamerica DeLaval)

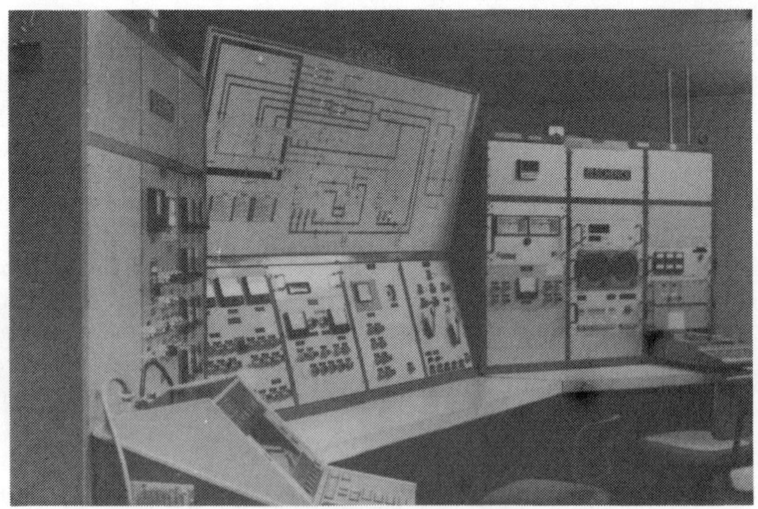

Figure 3.63. Control room for high speed balancing test (Courtesy of Transamerica DeLaval)

High speed balancing should be considered for one or more of the reasons listed below:

1. The actual field rotor operates with characteristic mode shapes significantly

different than those which occur during a standard production balance.

This is a primary consideration since flexible rotor balancing must be performed with the rotor whirl configuration approximating the mode in question.

2. The oeprating speed(s) is in the vicinity of a major flexible mode resonance (damped critical speed).

As these two speeds approach one another, a tighter balance tolerance will be required. Of special concern are those designs which have a lower rotor bearing stiffness ratio or bearings in the vicinity of mode nodal points.

3. The predicted rotor response of an anticipated unbalance distribution is significant.

This type of analysis may indicate a sensitive rotor which should be balanced at rated speed. It will also indicate which components need to be carefully balanced prior to assembly.

4. The available balance planes are far removed from locations of expected unbalance and thus relatively ineffective at the operating speed.

The rule of balancing is to compensate in the planes of unbalance when possible. A low speed balance utilizing inappropriate planes can have an adverse effect on the high speed operation of the rotor.

In many cases, implementation of an incremental low speed balance as the rotor is assembled will provide an adequate balance, since compensations are being made in the planes of unbalance. This is particularly effective with designs incorporating solid rotor construction.

5. A very low production balance tolerance is needed in order to meet rigorous vibration specifications.

Vibration levels below those associated with a standard production balanced rotor are often best obtained with a multiple plane balance at the operating speed(s).

6. The rotor on other similar designs have experienced field vibration problems.

Even a well designed and constructed rotor may experience excessive vibrations due to improper or ineffective balancing. This situation can often occur when the rotor has had multiple rebalances over a long service period and thus contains unknown balance distributions. A rotor originally balanced at high speed should not be rebalanced at low speed.

Influence coefficient methods present by far the best method for balancing in the field or even shop balancing. An additional advantage of this method is the fact that rotor can be balanced at many speeds and planes in one attempt.

The method of influence coefficients is based on the principle that the deflection, Z_i, at plane i is a function of the unbalanced forces or:

$$Z_i = e_{ij} P_j \tag{34}$$

The matrix of e_{ij}, is called the influence coefficient matrix or compliance matrix. Once the compliance matrix is obtained then unbalanced forces can be determined and corrections made.

In this type of balancing, the initial unbalanced amplitudes and phases are recorded. Next trial weights are placed in the required planes and the unbalanced amplitudes and phases are recorded. The influence coefficients are then calculated. From these the correction weights are determined. Usually the balancing is done over a number of speeds so that the system is balanced over the whole operating range rather than at the running speed only. Figure 3.64 shows the rotor response curves before and after balancing using influence coefficient techniques.

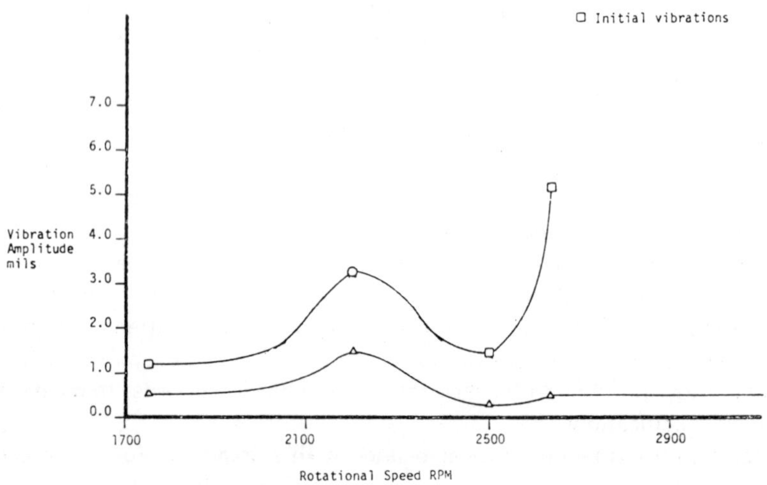

Figure 3.64. Rotor response curves before and after balancing.

Though the manufacturers may have balanced the rotor, it is often the responsibility of the end user who has to balance the machines finally in the field. Everytime the setup is changed, the system has to be balanced, since any addition or removal of parts, even though these may be balanced individually, will alter the unbalance of the system as a whole.

COUPLINGS

Flexible gear couplings have been the most widely used. In many new machines, they are being replaced by disk-type couplings. Gear-type couplings consist of a hub gear and a sleeve gear. In most cases, male teeth are integral with the hub, but some couplings have male teeth integral with the sleeve. Most of these couplings are mounted on a taper shaft end. In many instances, a key and keyway are provided to transmit torque; in others, the coupling is shrink fitted on the tapered shaft.

Lubrication of the gear coupling is accomplished by a continuous or batch packed technique. With continuous lubrication, nozzles spray oil into the coupling's teeth. The oil is recirculated through the machine's lubricant system, which dissipates the heat generated in the coupling, and thus maintains a relatively constant temperature. With batch-packed or seal lubrication, a recommended grease or oil is sealed in the coupling, which is changed as necessary or during scheduled shutdown. The major advantages of the batch packed technique are that the lubricant is the best available for the application and that the lubricant does not get contaminated. For high temperature applications, continuous lubrication is recommended. Such lubrication requires a good filtration system; otherwise, a centrifugal of contaminants in the gear teeth may occur.

The disc type or metal flexible coupling as seen in Figure 3.65 consists of basically two hubs rigidly mounted by interference fit of flange bolting to the driven and driving shaft. Two flexible elements, one attached at each hub (these compensate for misalignment) are then connected by a spacer, which is usually tubular, and spans the gap between shafts. Its major problem is its potential to vibrate when excited at its resonsnace frequency. Its major advantage is that it needs no lubrication system and can tolerate a much higher degree of misalignment. The old gear type couplings are rapidly being replaced.

When selecting a coupling there are some very important areas of evaluation. One is sliding friction coefficient. This produces a resistance to the axial movement necessary as rotors heat and expand. For the same catalog size rating, different coupling types can have about the same pitch line diameter, so the unit load on teeth would be the same. This load and a selected friction coefficient result in a force which must be handled by the thrust bearings and/or, one helix of a double helix gear before the teeth will slide and the normal operating position is achieved. A reaction will occur during every thermal change the system goes through.

This is a subject of more interest for the user than it may be for the designer, since the user must contend with tooth surface finishes other than new or as manufactured. This is the reason the words "a selected friction coefficient" were used.

This characteristic of the gear coupling therefore requires a serious study of the capabilities of the driven shaft thrust bearing and the loading service factors used in the gear design "in order to suit the coupling."

Intensive design efforts and complete design changes have been made in recent years to gear couplings to diminish the sliding friction coefficient, with some success, but again the burden is on the user to understand and maintain ideal conditions to insure continuous production.

Closely associated with this latter problem is the matter of lubrication required by gear couplings. A failure in this area results in the infinite sliding friction factor — the locked coupling.

Another area the user must be aware of is the tolerance permitted for axial growth or axial growth transients. This can be a more serious problem with disc

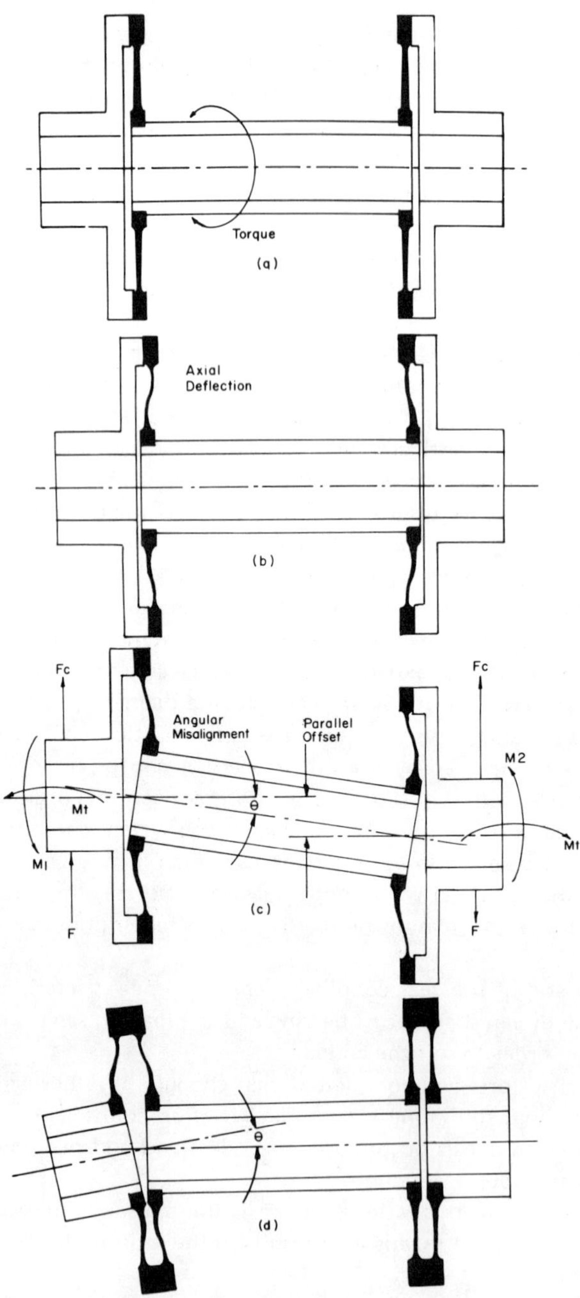

Figure 3.65. Flexible coupling (Courtesy of Bendix).

couplings than with gear couplings on certain types of units. A rather specific control is placed on the disc deflection range, and the equipment has to be adjusted axially to suit with more accuracy than with gear couplings. Some units are very difficult to move once they have been set. A gas expander, for instance. Where this is a problem, however, arrangements can be made to permit repositioning the coupling by remachining the components for one time per coupling fitup, or by adding a spacer plate.

The application of a disc coupling therefore almost completely eliminates the very generation reeducation of coupling users in the intricacies of the design itself – once the coupling is designed, because in order to design it, the coupling characteristics must be determined and adjusted to suit the connecting rotors. Furthermore, there are no changes or wear taking place in the coupling characteristics throughout the life of the unit.

One of the main features of the disc coupling is its known axial deflecting load value. When compared to the varying unknown sliding friction factor in the gear coupling, this feature eliminates the greatest concern of the user and to an equal extent the designers of the rotating equipment.

In most cases, both the disc, and the gear couplings can be applied. The difference between them can be noticed, but even though one has advantages over the other, both will succeed as couplings. The one overriding difference to users is the obvious promise that the disc coupling will be less susceptible to usage than the gear coupling, and that it will require less attention.

LUBRICATION SYSTEMS

API Standard 614 covers in detail the minimum requirements for lubrication sytems, oil type shaft sealing systems, and control oil supply systems for special purpose applications.

The major components of a typical oil system are reservoir, oil pumps, oil coolers, oil filters, alarms and shutdowns, and thermometers.

The reservoir should be separate from the equipment base plate. It should be sealed against the entrance of dirt and water. The bottom should be sloped to the low drain point and the return oil lines should enter the reservoir away from the oil pump suction to avoid disturbances of the pump suction. The working capacity should be at least five minutes based on normal flow. Reservoir retention time should be 10 minutes, based on normal flow and total volume below minimum operating level. Facilities for heating the oil should also be provided. If thermostatically controlled electrical emersion heating is provided, the maximum watt density should be 15 watts per square inch. When steam heating is used, the heating element should be external to the reservoir.

The oil system should be equipped with a main oil pump, a standby and for critical machines, an emergency pump. Each pump must have its own driver and check valves must be installed on each pump discharge to prevent reverse flow through idle pumps. The pump capacity should be 10–15% greater than maximum system usage. The standby and emergency pumps shall be equipped with automatic controls to start the pumps on failure of the main oil pumps. This control system should have manual reset. If steam turbine and electric motors are to be used for the main and standby oil pumps, the steam turbine should drive the standby pump. The emergency oil pump can be driven with an AC motor but from a power source that is different to the standby pump. DC electric motors can also be used when DC power is available. Process gas or air driven turbines and quick-start steam turbines are often used to drive the emergency pumps.

Pressure gauges should be provided at the discharge of the pumps, the bearing header, the control oil line and the seal oil system.

Each atmospheric oil drain line should be equipped with steel nonrestrictive bull's-eye type flow indicators positioned for viewing through the side.

Lubricant Selection

The characteristics of turbomachinery oil should be correct viscosity, rust inhibitor, oxidation inhibitor, non-sludging, form resistant and good demulsibility. Besides lubrication, oil has to cool bearing and gears, prevent excessive metal to metal contact during starts, transmit pressure in control systems, and carry away foreign materials, reduce corrosion, and resist degradation.

As a general rule most turbomachines are lubricated with premium quality turbine grade oil. However, under certain environmental conditions, it may be advantageous to consider another type oil. For example, if a machine is subject to the exposure of low concentration of chlorine or anhydrous hydrochloric acid gases, it is entirely possible to select another type that will out-perform the premium turbine oils. Good results have been recorded, using oil containing alkaline type additives.

Oils from turbomachinery should be tested periodically to determine their suitability for continued use; however, visual inspection of the oil can be useful in detecting contaminated oil when the appearance and odor is changed by the contaminant.

A sample of the oil should be withdrawn from the system and several of its characteristics measured in the laboratory. The usual tests conducted to determine the condition of the used oil include: 1) viscosity, 2) pH and neutralization number, and 3) precipitation. The test results will indicate changes from the original specifications and, depending on how extensive these changes are, the oil can or cannot be ue used in the machines.

Lubricant Contamination

Contamination of oil in a turbomachine is one of the major problems faced by all maintenance crew. Contamination is a continuous problem, but the levels of contamination are what we must be concerned with.

The metal particles from wear and rust particles from reservoir and oil piping corrosion can lead to premature equipment failure and oil deterioration.

Water contamination is a constant threat. The sources of water are many – atmospheric condensation, steam leaks, oil coolers and reservoir leaks. Rusting of machine parts and the effects of rust particles in the oil system is the major product of water in oil. In addition, it forms an emulsion and combined with other impurities, such as wear metal and rust particles, acts as catalysts promoting oil oxidation.

The filter elements should be 5 micron and must be water resistant, have high flow rate capability with low pressure drop, possess high dirt retention capacity and be rupture resistant. The clean pressure drop should not exceed 2 psig at 100°F. The elements must have a minimum collapse differential of 50 psig. Pleated paper elements are preferred provided they meet these requirements. Usually the pleated paper element will yield the 2 psig clean drop when used in a filter that was sized to use depth type elements. This is due to the larger surface area of the pleated element, over twice the area of a conventional stacked disc type or other depth type elements.

Cleanup of Lubricant System

Serious mechanical damage to turbomachinery can result from operating with dirty oil systems. It is essential that the oil system be thoroughly cleaned up prior to the initial startup of a new machine and after each overhaul of existing machines.

Preliminary steps for the initial startup and after the overhaul are similar except for the reservoir and oil requirements on the machine after an overhaul. For the overhauled machine, the oil would be drained and tested for condition. If there is no water or metal changes, the oil may be used again.

Inspect the reservoir interior for rust and other deposits. Remove any rust with scrapers and wire brushes and wash down the interior with a detergent solution and flush with clean water. Dry the interior by blowing the surfaces with dry air and use a vacuum cleaner for liquids to remove trapped water.

Install new 5 micron pleated paper elements in the filters. Connect steam piping to the water side of the oil coolers for heating the oil during the flush. Remove orifice and install jumpers at the bearings, coupling, controls, governor, and other critical parts to prevent damage from debris during the flush. Make provisions for 40 mesh telltale screens at each jumper. The conical shaped screen is preferred, but a flat screen is acceptable if the conical screen is not available. Adjust all control valves in the full open position to allow maximum flushing flow. The effectiveness of the flush is dependent to a large extent on high flow velocities through the system to carry the debris into the reservoir and filters. It may be necessary to sectionalize the system to obtain maximum velocities by alternately blocking off branch lines during the flush.

Fill the reservoir with new or clean used oil. Begin the flush without telltale screens by running the pump or pumps to provide the highest possible flow rate.

Heat the oil to 160°F to thermally exercise the pipe. Tap the piping to dislodge debris, especially along the horizontal sections. Flush through one complete temperature cycle, then shut down and install the telltale screens and flush for an additional 30 minutes. Remove screens and check for amount and type of debris. Repeat the preceding procedure until the screens are clean after two consecutive inspections. Observe the pressure drop across the filters during the flushing operation. Do not allow the pressure drop to exceed 20 psig. When the system is considered clean, empty the oil reservoir and clean out all debris by washing with a detergent solution followed by a fresh water rinse. Dry interior by blowing with dry air and vacuum free water. Replace filter elements. Remove jumpers and replace orifices. Return controls to their normal settings. Refill the oil reservoir with the same oil used in the flush if lab tests indicate it is satisfactory; otherwise, refill with new oil.

This procedure will allow the fastest possible cleanup of the oil system, mainly due to the high flow velocities obtained during the flush. The objective is to carry the debris into the reservoir and filters and the turbulence from the high flows along with the thermal and mechanical exercising of the piping are the main factors necessary for a fast and effective system cleanup.

GEARS

Gears are one of the most used couplings between the driving and driven components. Improper gear selection can cause major problems. Gears often form the weakest link in the whole chain, mainly due to the fact that this is the only item that is required to operate with metal parts in close contact. The performance of a gear is dependent on many factors, some of these are:

1. Pressure Angle

Decisions regarding the pressure angle need to be made early during design. The length of time of action and contact ratio are directly dependent on this factor. The noise generated is dependent inversely on the contact ratio. Values selected for pressure angle range from 17–22°.

2. Helix Angle

The helix angle ranges from 5 to 20° for the single helical gear and from 20 to 45° for the double helical gear. The single helical gears are more accurate and less prone to coupling thrust. They are also less expensive although they require expensive thrust bearings and the heat load of the thrust bearings makes them less efficient. The double helical gear are easier to manufacture, but cutting the gear teeth is more expensive. They are more efficient and do not require expensive thrust bearings.

Gears are also available in varying hardness. Medium range gears are not too sensitive to operational errors and wear slightly before failure. They are also not as susceptible to scouring as hard gears.

The most common used bearings for gears are journal and tilt pad bearings. Thrust bearings vary from ball bearing to self equalizing tilting pad bearings. The gear housing is usually made of steel. It should be stress relieved before final matching. Housing should be rigid enough to prevent misalignment. Also, sufficient cleaning should be provided to prevent oil choking. Lubrication in gears serves as two fold purpose that of lubrication and cooling. The most recommended lube oil is the AGMA No. 2.

While installing the gear unit, care should be taken to align the unit properly. Misalignment of gears can cause unequal distribution of teeth loads and distortion of gear elements. Before startup, gear face contact should be checked. This should be about 90%. Gears with modified helix angle may pose problems, and in such cases, the manufacturers recommendations are useful.

The system may influence the gear operation. Particularly critical speeds (torsional, lateral and axial) must be accounted for. Coupling lockups pose another problem. In addition, the gear must be able to handle the maximum load expected, and not just the design value.

Gear noise is causedsby a variety of factors. Extra heavy cast iron or double wall housing are useful in reducing noise. Acoustic enclosures are useful in reducing noise. When installing gears, in addition to cold alignment, hot alignment is necessary. High speed gearing must be treated with care or the user is in for a long unhappy association.

CONTROL SYSTEMS

The controls for most compressor trains consist of two major systems: one for the lubrication system, and one for the compressor. For the lubrication system, minimum alarms are: low oil pressure, low oil pressure trip (at some point lower than the alarm point), low oil level in the reservoir, high thrust bearing metal temperature, and high oil temperature. Each pressure and temperature sensing switch should be in a separate housing. The switch type should be single pole, double throw, and furnished as open (deenergized) to alarm and close (energize) to trip. Pressure switches for alarms should be installed, with a "T" connection for pressure gage and bleeder valve to test the alarms. Temperatures should be monitored in the oil piping to and from the coolers, and at the outlet of each radial and thrust bearing. Bearing-metal temperature should also be measured, since problems will show up much faster in the metal temperature than in the oil temperature.

Pressure gages should be provided at the discharge of the pumps, bearing header, control oil line, and seal oil line. Each atmospheric oil drain line should be equipped with steel nondestructive bull's eye flow indicators, positioned for viewing through the sides.

For the compressor, a complete instrumentation package should be provided. In this manner, a full analysis can be made for both vibration and performance characteristics. For vibration analysis, both proximity probes and accelerometers have been suggested. The reason for two different transducers is that proximity probes show only the movement of the shaft at the location of the probe. The probes do not pick up high frequency problems such as blade passing or gear mesh frequencies.

High subharmonic frequencies indicate oil whirl, and second harmonics indicate some misalignment. In some cases, catastrophic failure has occurred even though no indication was given by the proximity probes. When properly used, proximity probes are excellent in locating subharmonic and misalignment problems. Thus, by using a combination of proximity probes and accelerometers, a complete picture of rotor dynamics is obtained.

Surge control instruments on the market are basically static devices. These measure the pressure rise across the compressor, and sense the flowrate from pressure drop across an orifice. The pressure rise and flowrate are then compared with values programmed into the device; and when they exceed these values, the signal is given for a bypass valve to open.

The biggest drawback of this system is that it does not provide for any changes in the molecular weight of the gas or the degradation of the compressor itself. Presently, programs are underway that measure the dynamic change in the pressure head in the boundary layer. When flow reversal is sensed, the bypass valve is opened. This program is still being evaluated.

Until now, monitoring systems have consisted of no more than vibration monitors. With the advent of real time analyzers and minicomputers, a whole new era is opening in monitoring and diagnostic systems. These new systems indicate problem areas give expected life of various components, and schedule maintenance. They correlate both process and mechanical parameters to fully diagnose the compressor train and run it at its most efficient point. These systems use a performance matrix in combination with trending data and real time analysis to properly project and diagnose problems. A good monitoring and diagnostic system can save thousands of dollars in downtime and energy.

REFERENCES

1. Anderson, R. J., Riter, W. K., and Dilding, D. M., "An Investigation of the Effect of Blade Curvature on Centrifugal-Impeller Performance," NACA TN NO. 1313, 1947.
2. Benser, W. A., and Moser, J. J., "An Investigation of Backflow Phenomena in Centrifugal Compressors," NACA Report No. 806, 1945.
3. Campbell, K., and Talbert, J. E., "Some Advantages and Limitations of Centrifugal and Axials Compressors," SAE Trans, Vol. 53, 1945, p. 607.
4. Woodhouse, H., "Inlet Conditions of Centrifugal Compressors for Aircraft Engine Superchargers and Gas Turbines," J. Inst. Aeron. Sc., Vol. 15, 1948, p. 403.
5. Stepanoff, A. J., "Centrifugal and Axial Flow Pumps," John Wiley & Sons, Inc., N.Y., 1968.

6. Wislicensu, G. F., "Fluid Mechanics of Turbomachinery," McGraw-Hill Book Co., Inc., N.Y. 1947.
7. Stodola, A., "Steam and Gas Turbines," McGraw-Hill Book Co., N.Y., 1927.
8. Cheshire, L. J., "Centrifugal Compressors for Aircraft Gas Turbines," Proc. L. Mech. E., Vol. 153, 1965; reprinted by the ASME, 1967, "Lectures on the Development of the British Gas-Turbine Jet Unit."
9. Concorida, C., and Dowell, M. F., "Analytical Design of Centrifugal Air Compressors," J. App. Mechs., Vol. 13, 1946, p. A-271.
10. Stanitz, J. D., "Two Dimensional Compressible Flow in Conical Mixed Flow Compressors," NACA TN No. 1744, 1948.
11. Stanitz, J. D., "Two Dimensional Compressible Flow in Turbomachines with Conic Flow Surfaces," NACA Report No. 935, 1949.
12. Stanitz, J. D., and Ellis, G. O., "Two Dimensional Compressible Flow in Centrifugal Compressors with Straight Blades," NACA Report No. 954, 1950.
13. Ellis, G. D., and Stanitz, J. D., "Two Dimensional Compressible Flow in Centrifugal Compressors with Logarithmic-Spiral Blades, NACA TN No. 2255, 1951.
14. Gregory, A. T., "Elastic Fluid Compressor," United States Patent Office, 3305165, Feb. 21, 1967.
15. Boyce, M. P., and Bale, Y. S., "A New Method for the Calculations of Blade Loadings in a Radial Flow Compressor," ASME Paper No. 71-GT-60.
16. Stanitz, J. D., and Brian, V. D., "A Rapid Approximate Method for Determining Velocity Distribution on Impeller Blades of Centrifugal Compressors," NACA TN No. 2421, 1951.
17. Ellis, G. O., and Stanitz, J. D., "Comparison of Two- and Three-Dimensional Potential-Flow Solutions in a Rotating Impeller Passage," NACA TN No. 2806, 1952.
18. Hamrick, J. T., Ginsberg, A., and Osborn, W. M., "Method of Analysis for Compressible Flow Through Mixed-Flow Centrifugal Impellers of Arbitrary Design," NACA Report No. 1082, 1952.
19. Stanitz, J. D., "Some Theoretical Aerodynamic Invesitgation of Impellers in Radial and Mixed Flow Centrifugal Compressors," ASME Trans. Vol. 74, 1952, p. 472.
20. Balje, O. E., "A Contribution to the Problem of Designing Radial Turbomachines," ASME Trans., Vol. 74, 1952, p. 451.
21. Kearton, W. F., "The Influence of the Number of Impeller Blades on the Pressure Generated in a Centrifugal Compressor and of its General Performance, Proc. I. Mech. E., Vol. 124, 1933, p. 481.
22. Kasai, T., "On the Exit Velocity and Slip Coefficient of Flow at the Outlet of the Centrifugal Pump Impeller," Memoirs of the Faculty of Engineering, Kyushu Imperial University, Japan, Vol. 8, No. 1, 1936.
23. Peck, J. F., "Investigations Concerning Flow Conditions in a Centrifugal Pump, and the Effect of Blade Loading of Head Slip," Proc. I. Mech. E., Vol. 164, No. 1, 1951, p. 1.
24. Sheets, H. E., "The Flow Through Centrifugal Compressors and Pumps," ASME Trans. Vol. 72, 1950, p. 1009.
25. Boyce, M. P. and Bale, Y. S., "Feasibility Study of a Radial Inflow Compressor," ASME Paper No. 72-GT-52.
26. Boyce, M. P., "Design of Compressor Blade Suitable for Transonic Axial-Flow Compressors," ASME Paper No. 76-GT-47.
27. Michel, D. J., Mizisin, J., and Prian, V. D., "Effect of Changing Passage Configuration on Internal-Flow Characteristics of a 48-inch Centrifugal Compressor," NACA TN No. 2706, 1952.
28. Nuell, W. V. D., "The Maximum Delivery Pressure of Single Stage Radial Supercharges for Aircraft Engines," NACA TN No. 949, 1940.

29. Ritter, W. K., and Johnson I. A., "Performance Effect of Fully Shrouding a Centrifugal Supercharger Impeller," NACA ARR E5H23 or E18.
30. Ginsberg, A., Ritter, W. K., and Polasics, J., "Effects of Performance of Changing the Division of Work Between Increase of Angular Velocity and Increase of Radius of Rotation in an Impeller," NACA TN No. 1216, 1947.
31. Ribary, F., "The Friction Losses of Steam Turbine Discs," Brown Boveri Review, Vol. 21, 1934, p. 120.
32. Bullock, R. O., and Finger, H. B., "Surging in Centrifugal and Axial-Flow Compressors," SAE Quar. Trans. Vol. 6, 1952, p. 220.
33. Shepherd, D. G., "Principles of Turbomachinery," The MacMillan Company, N.Y., 1057.
34. Stockman, N. O., and Kramer, J. L., "Method for Design of Pump Impellers Using a High Speed Digital Computer," NASA TND 1562.
35. Stanitz, J. D., and Ellis, G. O., "Two Dimensional Flow on General Surfaces of Revolution in Turbomachines," NACA TN 2654.
36. Wu, C. H., "A General Theory of Three Dimensional Flow in Subsonic and Supersonic Turbomachines of Axial, Radial and Mixed-Flow Type," NACA TN 2604, 1952.
37. Owczarek, J. A., "Fundamentals of Gas Dynamics," International Textbook Company, Pennsylvania, 1968.
38. Yuan, S. W., "Foundations of Fluid Mechanics," Prentice Hall (New Delhi), 1969.
39. Vavra, M. H., "Aerothermodynamics and Flow in Turbomachines," John Wiley & Sons, Inc., N.Y., 1960.
40. Petermann, H., Untersuchungen am Zentsipetalrad fur Kreiselverdichter Forshung, Vol. 17, 1952.
41. Watabe, K., "Experiments on Fluid Friction of Rotating Disc with Blades," Japan Society of Mechanical Engineers Bulletin, Vol. 5, No. 17.
42. Watabe, K., JSME Bulletin, Vol. 7, Nov. 26, 1966.
43. Watabe, K., "Effects of Clearances and Grooves on Fluid Friction of Rotation Discs," JSME Bulletin, Vol. 8, No. 29, Feb. 1965, p. 55.
44. Schmieden, C., "Uber den Widerstand einer in einer Flussigkeit Rotierenden Scheibe," ZAMM 8, 1928, p. 460.
45. Schultz-Grunow, F., "Der Eirbungswiderstand Rotierender Scheiben in Gehansen" ZAMM 15, 1935, p. 191.
46. Bammert, K., and Rautenberg, M., "On the Energy Transfer in Centrifugal Compressors," ASME Paper No. 74-GT-121.
47. Boyce, M. P., "A Practical Three-Dimensional Flow Visualization Approach to the Complex Flow Characteristics in a Centrifugal Impeller," ASME Paper No. 66-GT-83.
48. York, R. E., and Woodard, H. S., "Supersonic Compressor Cascades – An Analysis of the Entrance Region Flow Field Containing Detached Shock Waves," ASME Paper No. 75-GT-33.
49. Delamey, R. A., and Kavanagh, P., "Transonic Flow Analysis in Axial Flow Turbomachinery Cascades by a Time-Dependent Method of Characteristics," ASME Paper No. 75-GT-8.
50. Eckardt, D., "Instantaneous Measurements in the Jet-Wake Discharge Flow of a Centrifugal Compressor Impeller, ASME Paper No. 74-GT-90.
51. Klassen, H. A., "Effect of Inducer Inlet and Diffuser Throat Areas on Performance of a Low Pressure Ratio Sweptback Centrifugal Compressor," NASA TM X-3148, Lewis Research Center, Jan. 1975.
52. Rodgers, C., and Sapiro, L., "Design Considerations for High Pressure Ratio Centrifugal Compressors," ASME Paper No. 72-GT-91.
53. Bhinder, F. S., and Ingham, D. R., "The Effect of Inducer Shape on the Performance of High Pressure Ratio Centrifugal Compressors," ASME Paper No. 74-GT-122.

54. Sapiro, L., "Preliminary Staging Selection for Gas Turbine Driven Centrifugal Gas Compressors," ASME Paper No. 73-GT-31.
55. Mikolajczak, A. A., Weingold, H. D., Nikkanen, J. P., "Flow Through Cascades of Slotted Compressor Blades," *Journal of Engineering for Power,* ASME Trans., Jan. 1970, p. 57.
56. Rodgers, C., "Typical Performance Characteristics of Gas Turbine Radial Compressors," *Journal of Engineering for Power,* ASME Trans. Apr. 1964, p. 161.
57. Stahler, A. F., "Transonic Flow Problems in Centrifugal Compressors," SAE Reprint 268C, Jan. 1961.
58. Rodgers, C., "Influence of Impeller and Diffuser Characteristics and Matching on Radial Compressor Performance," SAE Reprint 268B, Jan. 1961.
59. Kramer, J. J., Osborn, W. M. and Hamrick, J. T., "Design and Test of Mixed-Flow and Centrifugal Impellers," *Journal of Engineering for Power,* ASME Trans., Series A., Vol. 82, 1960, p. 127.
60. Pfleiderer, C., "Kreisel pumpen," Springer, F155.
61. Mechanical Engineers Handbook, "Centrifugal and Axial Fans," 6th Edition, p. 14–66 to 14–79.
62. Balje, O. E., "Loss and Flow Path Studies on Centrifugal Compressors – Part I & II," ASME Paper No. 70-GT-12-A and 70-GT-12.B.
63. Eckert, B., "Axial and Radialkompressoren," Springer, 1953.
64. Wiesner, F. J., Jr., "Practical Stage Performance Correlations for Centrifugal Compresors," ASME Paper No. 60-Hyd-17.
65. Ferguson, T. B., "The Centrifugal Compressor Stage," Butterworth and Co., Ltd., London, 1963.
66. Balje, O. E., "A Study of Reynolds Number Effects in Turbomachinery," *Journal of Engineering for Power* ASME Trans. Vol. 86, Series A, 1964, p. 227.
67. Schlichting, H., "Boundary Layer Theory," McGraw-Hill Book Co., N.Y., 1968.
68. Moody, L. F., "Friction Factors," ASME Trans. Nov. 1944, p. 672.
69. Balje, O. E., "A Study on Design Criteria and Matching of Turbomachines Part 13," *Journal of Engineering for Power,* ASME Trans., Vol. 84, Series A, 1962, p. 103.
70. Lazarkiewicz, S., and Troskolanski, A. T., "Impeller Pumps," Pergamon Press, 1965.
71. Kovats, A., "Design and Performance of Centrifugal and Axial Flow Pumps and Compressors," The Macmillan Co., 1964.
72. Stepanoff, A. J., "Turboblowers," Wiley, 1955.
73. Batman, J., "Design and Development of a Family of Natural Gas Compressors for a 3000 hp Gas Turbine Engine," ASME Paper No. 72-GT-10.
74. Speer, I. E., "Design and Development of a Broad Range High Efficiency Centrifugal Compressor for a Small Gas Turbine Compressor Unit, ASME Paper No. 52-SA-14.
75. Flaherty, R., "A Method for Estimating Turbulent Boundary Layers and Heat Transfer in an Arbitrary Pressure Gradient," United Aircraft Research Laboratories Report UAR-G51, Aug. 1968.
76. Reshotko, E., and Tucker, M., "Approximate Calculation of the Compressible Turbulent Boundary Layer with Heat Transfer and Arbitrary Pressure Gradient," NACA Tn 4154, 1957.
77. Schlichting, H., "Application of Boundary-Layer Theory in Turbomachinery," ASME Trans. Vol. 81, 1959, p. 543.
78. Boyce, M. P., and Bale, Y. S., "Duffusion Loss in a Mixed Flow Compressor, Paper No. 729061, Intersociety Energy Conversion Compressor Impeller," NRCC ME-220, Ottawa, Jul. 1966.
79. Fowler, H. S., "An Investigation of the Flow Processes in a Centrifugal Compressor Impeller," NRCC ME-220, Ottawa, Jul. 1966.

80. Boyce, M. P., Schiller, R. N., and Desai, A. R., "Study of Casing Treatment Effects in Axial Flow Compressors," ASME Paper No. 74-GT-89.
81. Boyce, M. P., and Desai, A. R., "Clearance Loss in a Centrifugal Impeller," Paper No. 739126. Proc. of the 8th Intersociety Energy Conversion Engineering Conference, Pennsylvania Aug. 1973, p. 638.
82. Boyce, M. P., and Desai, A. R., "A Theoretical Analysis of Non-Isentropic Flow of a Compressible Viscous Gas in Narrow Passages," ASME Trans., Vol. 16, Number 2, Apr. 1973, p. 132.
83. Fowler, H. S., "Experiments on the Flow Processes in Simple Rotating Channels," NRCC ME-229, Ottawa, Jan. 1969.
84. Senoo, Y., and Nakase, Y., "An Analysis of Flow Through a Mixed Flow Impeller," ASME Paper No. 71-GT-2.
85. Senoo, Y., and Nakase, Y., "A Blade Theory of an Impeller with an Arbitrary Surface of Revolution," ASME Paper No. 71-GT-17.
86. Katsanis, T., "Use of Arbitrary Quasi-Orthogonals for Calculating Flow Distribution in the Meridional Plane of a Turbomachine," NASA TND-2546, 1964.
87. Prian, V. D. and Michel, D. J., "An Analysis of Flow in Rotating Passage of Large Radial Inlet Centrifugal Compressor at Top Speed of 700 Feet Persecond," NACA TN No. 2584, 1951.
88. Betz, A., "Introduction to the Theory of Flow Machines," Pergamon Press, 1966.
89. Sovran, G., and Klomp, E. D., "Fluid Mechanics of Internal Flow," Elsevier Publishing Co., N.Y. 1967.
90. Kuethe, A. M. and Schetzer, J. D., "Foundations of Aerodynamics," John Wiley & Sons, N.Y. 1950.
91. Curle, N., and Skan, S. W., "Approximate Method for Predicting Separation Properties of Laminar Boundary Layers," Aero Wuart, G, 257, Aug. 1957.
92. Hansen, A. G. (Editor), "Symposium on Fully Separated Flows," ASME Fluids Energy. Divn. Conf., Philadelphia, Pa., May 18–20, 1964.
93. Hayes, W. D., "The Three Dimensional Boundary Layer" NAVORD Report No. 1313, Washington, D.C., 1051.
94. Shouman, A. R. and Anderson, J. R., "The Use of Compressor Inlet Prewhirl for the Control of Small Gas Turbines" *Journal of Engineering for Power,* Trans. ASME Ser. A., Vol. 86, 1964, pp. 136–140.
95. Dean, R. C., "The Fluid Dynamic Design of Advanced Centrifugal Compressors," Von Karman Institute, Lecture Notes No. 50, 1972.
96. Harold Lown and Frank J. Wiesner, Jr., "Prediction of Chocking Flow in Centrifugal Impeller," Transaction of ASME *Journal of Basic Engineering,* March, 1959.
97. Coppage, J. E., et al, "Study of Supersonic Radial Compressors for Refrigeration and Pressurization Systems," WADC Technical Report 55-257, Astia Document No. AD 110467, 1956.
98. Senoo, Y., et al, "Viscous Effects on Slip Factor of Centrifugal Blowers," ASME publication 73-GT-56, 1973.
99. *Sawyer's Gas Turbine Engineering Handbook* – Vol. I, II, III.
100. Thomson, W. T., Mechanical Vibrations, Second Edition. Prentice-Hall, Inc., Englewood Cliffs, N.J., 1961.
101. Gunter, E. J., Jr. "Rotor Bearing Stability," Proceedings of the First Turbomachinery Symposium, Texas A&M Univ. College Station, Texas, October, 1972.
102. Haag, A. C., "The Influences of Oil Film Journal Bearings on the Stability of Rotating Machines," J. of Appl. Mech. Trans. ASME Vol. 68, p. 211, 1946.
103. Ehrich, F. F., "Identification and Avoidence of Instabilities and Self-Excited Vibrations in Rotating Machinery," Adopted from ASME Paper 72-DE-21. Lynn, Mass.: General Electric Co., Aircraft Engine Group, Group Engineering Division, May 11, 1972.

104. Alford, J. S., "Protecting Turbomachinery from Self-Excited Rotor Whirl," Journal of Engineering for Power, ASME Transactions, Oct. 1965, pp. 333–344.

105..Newkirk, B. L., "Shaft Whipping" General Electric Review, Vol. 27, p. 169, 1924.

106. Housmann, J. G., Turbomachinery Specifications, Proc. First Turbomachinery Symp., pp. 77–78, Texas A&M Univ. College Station, Texas 1972.

107. Davis, H. M., Centrifugal Compressor Operation and Maintenance, Proc. First Turbomachinery Symp. pp. 10–25, Texas A&M University, College Station, Texas 1972.

108. Wilcock, D. E. S. and Booser, E. R., "Bearing Design and Application, McGraw-Hill, New York, 1961.

109. Gunter, E. J., "Dynamic Stability of Rotor Bearing Systems, NASA Sp-113, 1966.

110. Herbage, B. S., High Speed Journal and Thrust Bearing Design, Proc. First Turbomachinery Symp., pp. 55–61, Texas A&M Univ. College Station, Texas, 1972.

111. Boyce, M. P., and Hanawa, D. A., Development of Techniques for Monitoring Turbomachinery, Proc. Gas Turbine Operations and Maintenance Symp., Edmonton, Canada, 1974.

112. Boyce, M. P. and Hanawa, D. A., Parametric Study of a Gas Turbine, *J. Eng. Power,* Dec. 1975.

113. Clapp, A. M., Fundamentals of Lubrications Relating to Operation and Maintenance of Turbomachinery, Proc. First Turbomachinery Symp. pp. 67–74, Texas A&M Univ. College Station, Texas, 1972.

114. Cameron, J. A. and Danowski, F. M. Some Metallurgical Considerations in Centrifugal Compressors, Proc. Second Turbomachinery Symp., pp. 116–128, Texas A&M Univ. College Station, Texas, 1973.

115. Jackson, C. J., Cold and Hot Alignment Techniques for Turbomachinery, Proc. Second Turbomachinery Symp. pp. 1–7 Texas A&M University, College Station, Texas 1973.

116. Lesiecki, G., Evaluation of Liquid Film Seals: Associated Systems and Process Considerations Proc. Sixth Turbomachinery Symp. pp. 145–148, Texas A&M Univ. College Station, Texas 1977.

117. Lewis, R. A., Mechanical Contact Shaft Seal, Proc. Sixth Turbomachinery Sump., pp. 149–151, Texas A&M University, College Station, Texas, 1977.

118. Boyce, M. P., Morgan, E. and White, G. Simulation of Rotor Dynamics of High Speed Rotating Machinery, Proc. First Int. Conf. Centrifugal Compressor Technol., pp. E6-32, Madras, India, 1978.

119. Boyce, M. P., How to Achieve Online Availability of Centrifugal Compressors, Chemical Engineering, pp. 115–117, June 5, 1978.

CHAPTER 4

THE LIQUID RING VACUUM PUMP

A. M. DAVE
Nash Engineering Co.
Norwalk, CT

INTRODUCTION

Vacuum technology is an essential part in most major industries like chemical processing, filtering, power generating, pulp & paper, textile production, and many others. In today's market, vacuum producing equipment includes liquid ring, reciprocating, rotary lobe or piston and sliding vane pumps, centrifugal blowers, and steam jet ejectors.

The *liquid ring pump* consists of a single rotating element generally called "rotor." Pumping is accomplished by means of a rotating rotor chamber (bucket) entering and leaving a solid ring of liquid. The liquid ring pump is not new; it has been used successfully since the turn of the century. This chapter will provide general information on the operating principle, selection, installation, operation and maintenance of the liquid ring pump.

Liquid ring pumps are manufactured in conical, flat sided, and cylindrical design. The conical, flat sided, or cylindrical describes the shape of stationary port. In a conical design, ports are located on a conical surface which fits into matching tapered bore of the rotor. The unassembled view of a conical type is shown in Figure 4.1. Similarly in a flat sided design, the port plate fits close and parallel to rotor end shrouds, which have matching ports between the blades as illustrated in Figure 4.2. The flat sided design of a liquid ring pump is usually less expensive to manufacture than the conical design; but the flexible port design in a conical type exhibits a low internal pressure loss and capability of handling extra liquid. Liquid ring pumps are available in single or double acting design, as illustrated in Figure 4.1 and 4.3. Single acting is less expensive and usually used for vacuum pumps. The double acting balances hydraulic forces across the pump and thus has advantages as a compressor. The double acting completes two inlet and discharge strokes during each revolution. All liquid ring pumps have essentially the same principle of operation, regardless of their design.

PRINCIPLE OF OPERATION

The pump consists of a circular casing and a rotor with a series of blades projecting from a hollow cylindrical hub through which a shaft is pressed.

Rotor blades are shrouded at the sides to form chambers (buckets) such as (A).

Figure 4.1. Single acting conical design liquid ring pump (Photo courtesy of Nash Engineering Company).

Figure 4.2.

Rotor chambers contain a liquid compressant, usually water (4). As a result of centrifugal force, water follows the contour of the casing as it rotates with the rotor.

Starting at "A," the rotor chamber is full of water. As the rotor moves clockwise, water recedes until the chamber is empty (5). As the rotor continues further, the converging casing forces water back into the rotor chamber until it is again full at (6). This cycle takes place once each revolution in a single acting pump. The

Figure 4.3. Unassembled view of double acting liquid ring compressor.

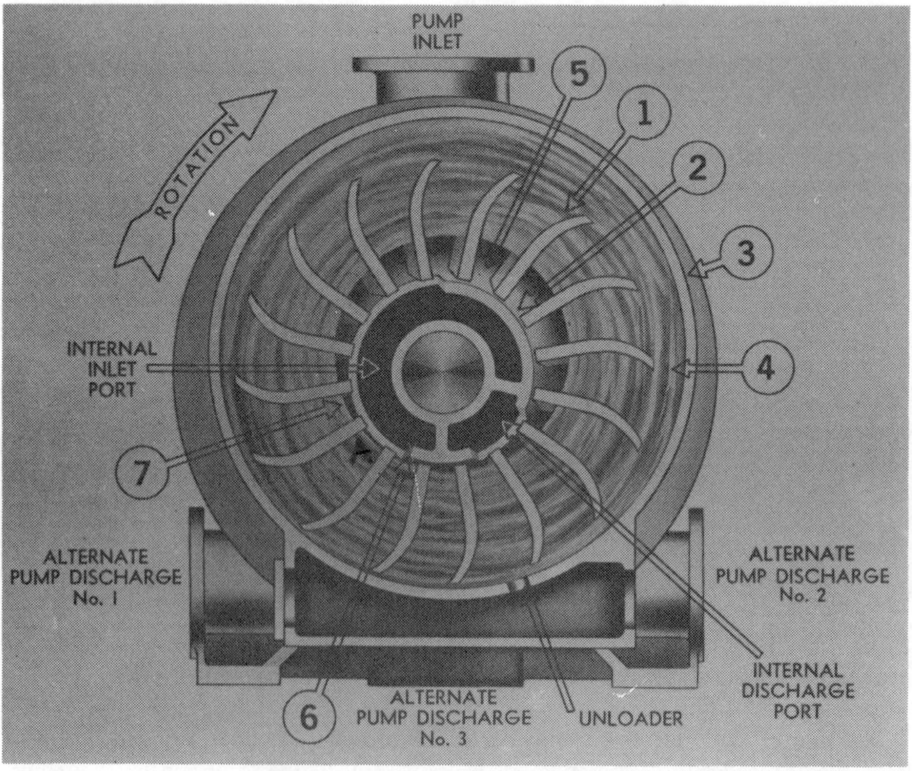

Figure 4.4. The liquid ring principle.

operation is similar in a double acting as shown in Figure 4.5. It completes two inlet and discharge strokes during each revolution of the rotor and, therefore, it includes two internal inlet and discharge ports.

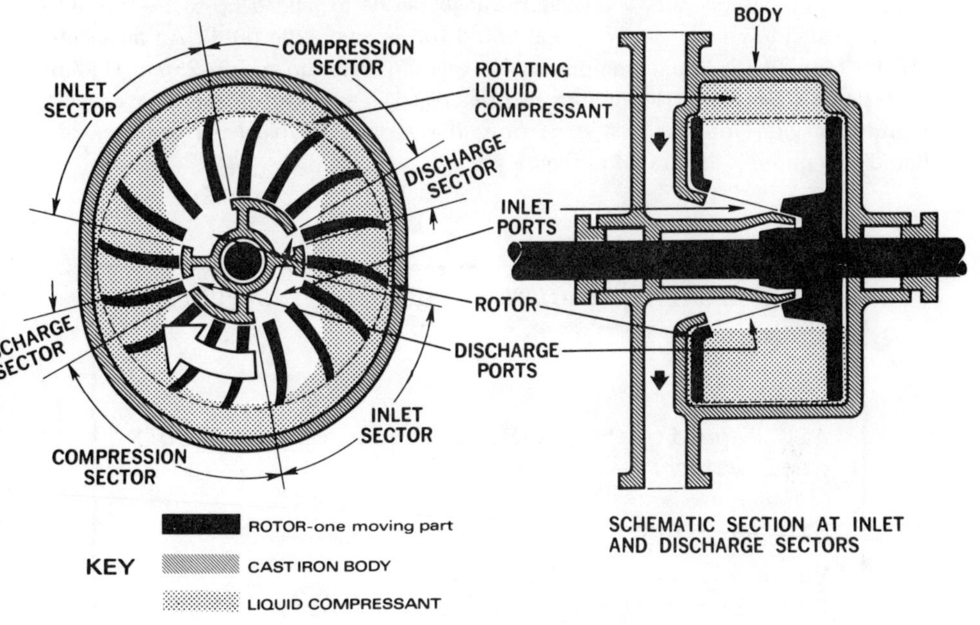

Figure 4.5. Typical cross section of double acting compressor.

In a liquid ring pump, the compression is a direct result of the energy imparted to the liquid ring as it is thrown out centrifugally by the rotor blades. This energy is a function of the rotor velocity (RPM) and the specific gravity of the liquid compressant. A portion of this imparted energy is recovered as it compresses the gas in the discharge cycle. The unrecovered energy is lost in pumping of the seal liquid and fluid friction loses. If energy in seal liquid is not adequate to meet the compression ratio, the liquid ring velocity slows down prior to the discharge port opening, and the pump loses its solid liquid piston (ring) in the rotor bucket. This creates instability in the pump due to collapse of liquid ring. This is known as "stall," and is usually accompanied by an increase in power and noise and reduction in pump capacity.

Liquid ring pumps can be operated as a vacuum pump or as a compressor. In fact, vacuum pump is actually a compressor, compressing from a sub-atmospheric condition to atmospheric pressure.

CAPABILITY OF LIQUID RING PUMP

Generalized capabilities of the liquid ring pump is indicated in Table 4.1. Liquid ring pumps are available in capacities up to 14,000 CFM (396 M^3/min) as a vacuum pump or as a compressor. The vacuum range can be expanded up to 5–10mm Hg Abs by using low vapor-pressure seal liquid (oil or hydraulic fluid). An air ejector can be staged with liquid ring pumps for vacuum levels up to 20–25mm Hg Abs. Two single stage liquid ring pumps can be staged in series when there are capacity limitations with integral two stage unit, if necessary. Typical performance of a liquid ring pump is illustrated in Figure 4.6.

Table 4.1

APPROXIMATE CAPACITY OF LIQUID RING PUMPS

Air capacity 50% R.H. Std. Barometer & 60°F seal water.

Single Stage

Vacuum up to 2.0" Hg Abs.
Capacity up to 11,500 ACFM (326m^3/min)

Two Stage

Vacuum up to 0.78" Hg Abs.
Capacity up to 11,500 ACFM (326m^3/min)

Vacuum range can be extended by use of low vapor pressure liquid (oil or hydraulic fluid). Also an air ejector can be staged with L.R. pump.

Compressor

Single Acting - Pressure up to 35 psig (2.5 Kg/cm^2g)
Capacity up to 14,000 CFM (396m^3/min)

Double Acting - Pressure up to 125 psig (8.8 Kg/cm^2g)
Capacity up to 300 CFM (8.5m^3/min)

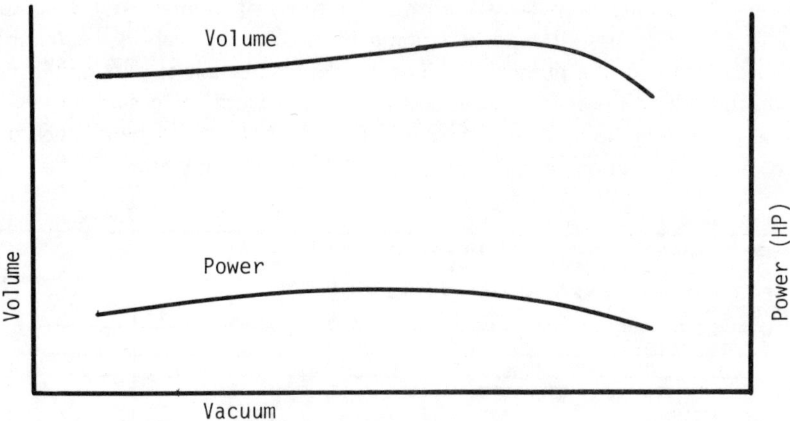

Figure 4.6. Typical performance curve of liquid ring pump.

OWNER BENEFITS

Liquid ring pumps have only one moving part (rotor) and pumping is accomplished without pistons, valves, sliding vanes, and without any metallic contact between rotating and stationary element. In addition, they provide a continual source of vacuum or pressure without pulsation.

Liquid ring pumps provide significant advantages over other vacuum producing equipment when wet, contaminated, corrosive, and explosive gases are handled. Proper selection of a vacuum pump system can reduce environmental pollution and energy consumption. Following is an explanation of some of the owner benefits with liquid ring pumps:

1. Liquid Compressor Cools the Gas – The gases being handled in a liquid ring pump are cooled while the heat of compression is absorbed by the liquid compressant and discharged into a gas liquid separator. The separator removes liquid from the gas. It is obvious that the cooling of a liquid ring compressor is direct rather than through the walls of the casing as in the other types of compressors. Because of intimate contact of gas and liquid, the final discharge temperature can be held close to the temperature of the inlet compressant liquid and, thus, liquid ring pumps eliminate the need for after-coolers that may be required with other types of compressors.

2. Wet Gas with Possible Liquid Carryover – The liquid ring pump can handle saturated vapor gas mixtures with liquid carryover with no adverse effect on pump performance, while this type of service can be detrimental to other types of vacuum pumps.

3. Saturated Vapors – Saturated vapors are condensed in a liquid ring pump when liquid compressant is cooler than the saturated vapor due to the compression.

Thus, the liquid ring pump acts as a condenser and eliminates the need of a pre-condenser which may be necessary with other types of pumps. As the liquid ring cools the incoming gas, it reduces volume by condensing vapors, increasing net pumping capacity at the pump inlet. The condensate becomes part of the sealant liquid during the compression cycle, and is removed from the gas-liquid separating device with sealing fluid. A typical curve (Figure 4.7) shows percentage gain in capacity at a fixed vapor. temperature and 15°C (60°F) seal water.

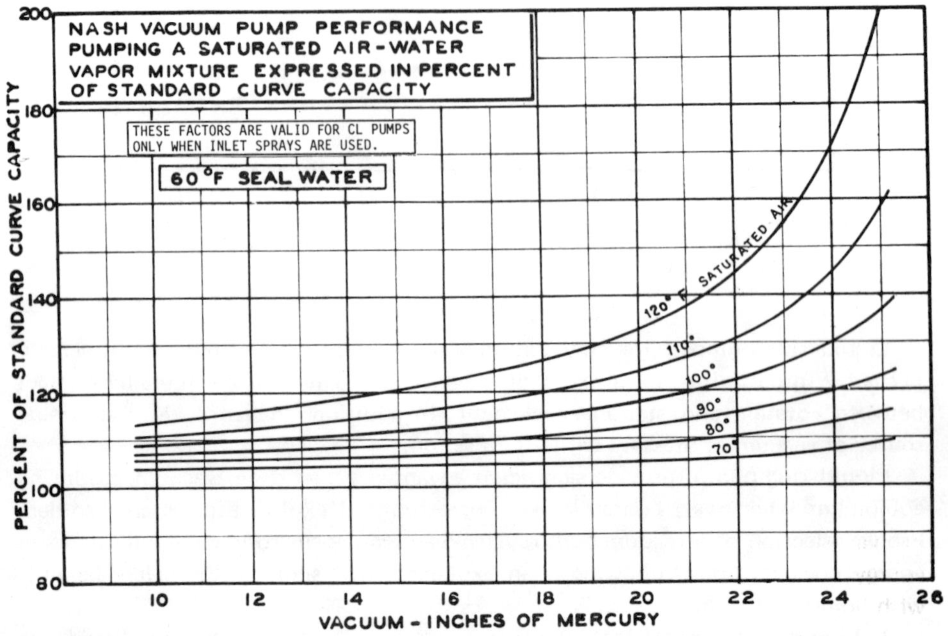

Figure 4.7.

The analogy of saturated gas handled in a liquid ring pump is indicated for understanding purposes.

Figure 4.8 (a) illustrates a saturated mixture before the pump with a very small partial gas volume and a large partial volume of water vapor. As this gas enters the pump, hot saturated vapor immediately condenses with a cooler seal liquid and reduces the vapor volume as shown in Figure 4.8 (b). The capacity of the pump to handle gas has now been increased as shown in Figure 4.8 (b).

4. Compressant Liquid Scrubs The Gas – In a liquid ring pump there is intimate mixing between the gas and liquid. The liquid scrubs the gas removing solid particulates down to micron sizes. Many solids, like pulp or textile fiber, can be passed through the liquid ring pump without any damage. However, abrasive solids in a gas stream can shorten pump life and should be removed ahead of the pump. Some designs are available with patented purge lines to remove abrasives that are trapped in the pump.

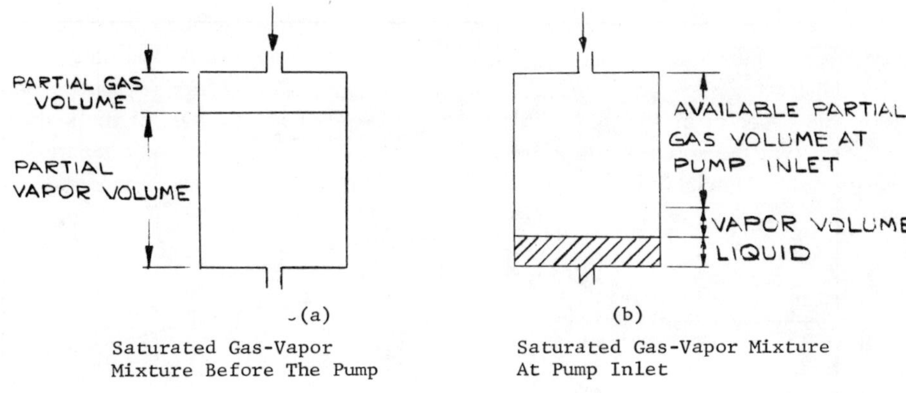

(a)
Saturated Gas-Vapor Mixture Before The Pump

(b)
Saturated Gas-Vapor Mixture At Pump Inlet

Figure 4.8.

5. Corrosive Gas Can Be Handled – Corrosive gas can be handled when proper compressant liquid is selected. One must select a liquid which will neutralize the corrosive gas, or select a liquid which is neutral to the corrosive gas. For example, dry chlorine gas can be handled with 93–98% sulfuric acid.

6. Explosive Gases – Explosive gases can be handled very safely in a liquid ring pump as gas remains cool and moist during compression. Chances of sparking are greatly reduced because of the compressant liquid, and there is no metal-to-metal contact in the pump. For example, water sealed pumps are used on acetylene service because acetylene is less hazardous in presence of moisture.

7. Product Recovery – When saturated vapors are handled, the vapor can be condensed in the pump, and this condensate can be used as the compressant liquid. The liquid ring pump acts as a condenser and recovers the product for reuse. For example, to handle nitrogen and hexane vapors, hexane liquid is used as compressant and liquid ring pump recovers the product, and also reduces discharge of pollutant to atmosphere. This cannot be accomplished with other pumps without an upstream condenser.

8. Save Energy – The optimum utilization of energy and effective pollution control have become top priorities of many process plants. The low operating efficiency and wastefulness of steam ejectors coupled with the accelerating cost of fuel to produce steam, have prompted many users of steam ejectors to re-evaluate their existing steam jet and compare operating costs with liquid ring pumps.

Generally, liquid ring pumps are anywhere from 2 to 10 times more efficient than steam ejectors. The overall thermal efficiency[1], a ratio of theoretical work to purchased energy, is shown in Figure 4.9, and indicates liquid ring pumps have a higher efficiency than the steam jets. Liquid ring pumps are widely accepted in many chemical plants over the more efficient type blower or mechanical type pump when the previously described advantages are compared.

Operating cost comparison between steam jet and vacuum pump can be made by

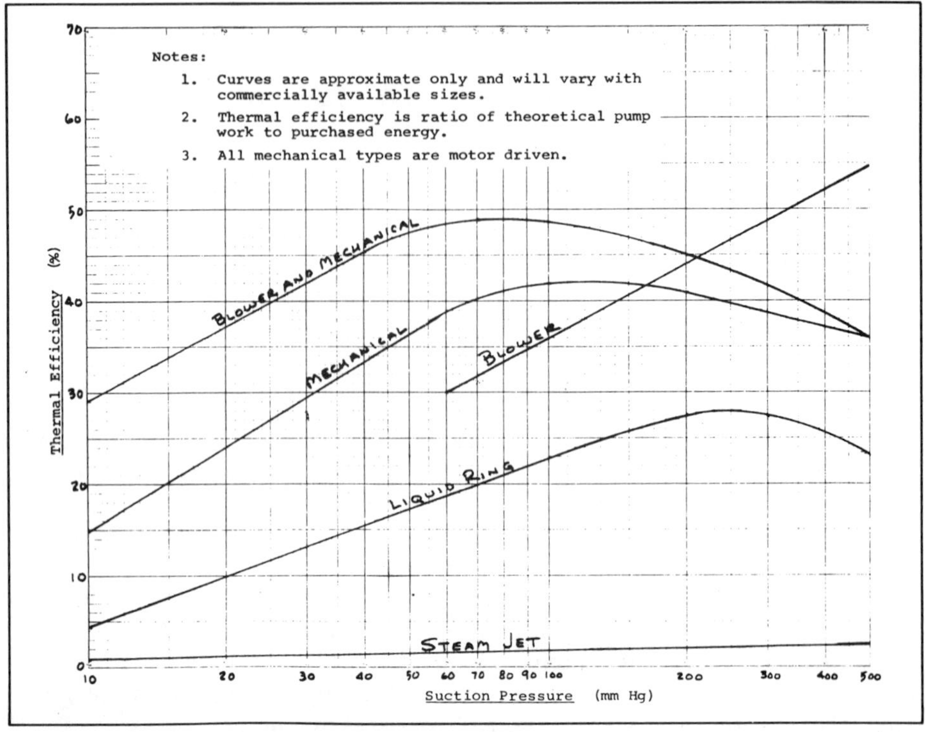

Figure 4.9. Overall thermal efficiency of vacuum pumps.

determining the required cost to produce steam vs kilowatt usage of an equivalent vacuum pump. Table 4.2 illustrates the comparison of induced heat load between vacuum pump and jet. Table 4.3 relates operating cost per hour of steam jet vs electrical cost of an equivalent vacuum pump. Table 4.4 shows an average cost comparison for steam at $3.0/1000 lbs and electricity at 2.5c/KWH. It shows the operating cost of a vacuum pump is approximately one-fifth that of steam jet ejectors. Generally, the pay-back period with an installation of a liquid ring vacuum pump can result between 3–20 months.

9. Reduce Pollution – In applications where steam jet ejectors are used to provide vacuum, often it may contaminate motive steam by mixing with a polluted process steam. Gas is heated and this may result in thermal pollution. The necessity of treating contaminated condensate makes liquid ring pumps an attractive replacement. The use of liquid ring vacuum pumps in many applications can reduce pollution and save the cost of waste water treatment by closed-loop recirculation of compressant liquid.

Table 4.2, 4.3 and 4.4 Cost Comparison of Steam Jet Ejector vs Liquid Ring Pump

HERE'S HOW OPERATING COSTS COMPARE

Table 4.2. Steam Jet Ejector vs. Mechanical Vacuum Pump

Vacuum level required	20-in.Hg	24-in.Hg	26-in.Hg	28-in.Hg
Steam jet ejectors:				
1000 psi steam in lb/hr	900	1,650	3,300	2,500
Heat load in Btu/hr . .	924×10^3	$1{,}690 \times 10^3$	$3{,}380 \times 10^3$	$2{,}570 \times 10^3$
Vacuum Pump:				
ACFM required at vacuum	326	543	815	1,630
Motor bhp required. . .	20	36	45	101
Kilowatts required. . .	16.6	29.9	37.2	83.5
Heat load in Btu/hr . .	50×10^3	93×10^3	114×10^3	255×10^3

How Steam and Electrical Costs Relate

Table 4.3. Cost per Hour to Remove 500 lb/hr Dry Air

Steam jet ejectors: Cost per 1,000 lb	20-in.Hg	24-in.Hg	26-in.Hg	28-in.Hg
$0.20	$0.18	$0.33	$0.66	$0.50
0.40	0.36	0.66	1.32	1.00
0.80	0.72	1.32	2.64	2.00
1.00	0.90	1.65	3.30	2.50
1.20	1.08	1.98	3.96	3.00
1.40	1.26	2.31	4.62	3.50
1.60	1.44	2.64	5.28	4.00
2.00	1.80	3.30	6.60	5.00
2.40	2.16	3.96	7.92	6.00
3.00	2.70	4.95	9.89	7.50
4.00	3.60	6.60	13.20	10.00
Vacuum pump: Cost per kWh				
1.0c	$0.17	$0.30	$0.37	$0.82
1.5c	0.26	0.45	0.56	1.23
2.0c	0.34	0.60	0.74	1.64
2.5c	0.43	0.75	0.93	2.05
3.0c	0.51	0.90	1.11	2.46

And How You Can Save With a Vacuum Pump

Table 4.4. Costs with Steam @ $3/1,000 lbs and Electricity @ 2.5c/kWh

	20-in.Hg	24-in.Hg	26-in.Hg	28-in.Hg
Steam jet ejectors	$2.70	$4.95	$9.89	$7.50
Vacuum pump	0.43	0.75	0.93	2.05

Disadvantages

Liquid ring pumps have pressure limitation up to 125 psig (8.8 kg/cm^2g). In most instances they consume more energy than blower and reciprocating type pumps, but less than the steam ejector. A portion of the energy consumed by a liquid ring pump is recovered in the compression cycle. The rest of the energy is consumed in pumping the seal liquid and overcoming fluid friction. Also, it requires a constant supply of seal liquid, either on once-thru or recirculated seal basis. Liquid ring pumps are considered less efficient than the blower or mechanical pump, but the superior gas handling capabilities of liquid ring pumps greatly compensate for the relatively lower efficiency.

OPERATION OF LIQUID RING PUMP

1. Vs. Steam Jet – The typical performance of a two stage condensing jet vs a liquid ring two stage is illustrated in Figure 4.10. It indicates significant capacity differences at vacuum below design point or 24″ Hg Vac. The liquid ring pump is a constant volume machine while a jet is designed on a constant weight basis. As a result, the problem in excess leakage can reduce jet capacity significantly. Liquid ring pump performance increases at lower vacuum. Liquid ring pumps show noticeable capacity difference below 24″ Hg Vac. The increased capacity at lower vacuum cuts down evacuation time by 2/3 as compared to a steam jet. In an emergency condition, the time difference can be of value in either start-up or regaining lost vacuum. The following advantages are summarized:

1. Automation. Liquid ring vacuum pump systems can be easily automated compared to special control valve arrangements required for steam jets.
2. Simplified Piping. The liquid ring pump does not require high pessure insulated piping.
3. Energy Conservation
4. Reduced water pollution.
5. Low installation cost.
6. Low noise pollution.
7. Better gas handling capability. Saturated vapors can be handled without precondenser.

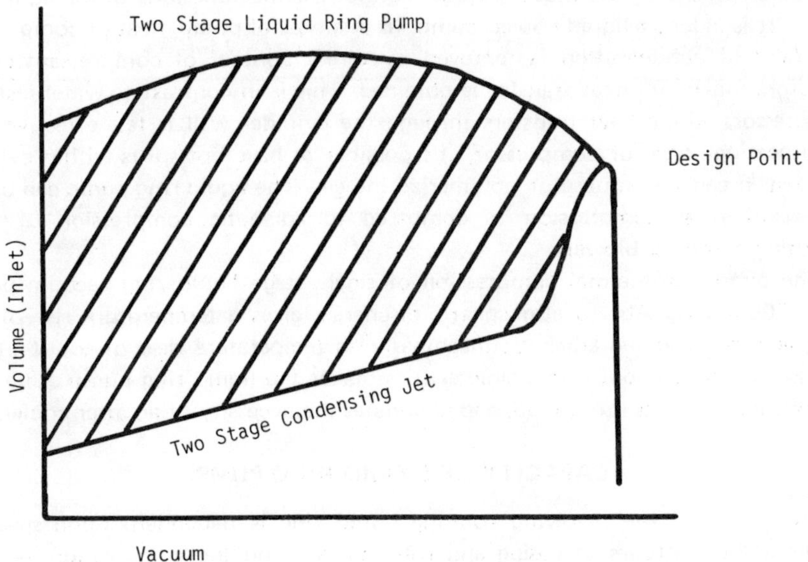

Figure 4.10. Typical performance of a two stage liquid ring pump vs two stage condensing steam jet.

The liquid ring pump can be used for backing up steam jet ejector and still provide substantial savings when compared with a multi-stage steam ejector system.

Reliability of equipment is quite important to plant personnel. The liquid ring pump has been used successfully for years in condenser exhausting in power plants where reliability is of most importance.

2. Vs. Blower and Mechanical Pumps – Liquid ring pumps have pressure limitation of 125 psig (8.8 kg/cm^2) and it is limited by the energy of liquid ring. Blowers and mechanical pumps are capable of achieving higher compression ratios and they are considered more efficient than liquid ring pumps.

The liquid ring pump is non-pulsating, which eliminates vibration problems and the need for mandatory receiver and heavy foundations. Since there is only one moving element, rotating without metallic contact, there is nothing to wear, realign, or to be adjusted as may be required with other types. As discussed earlier, a liquid ring pump does not need an after-cooler. In addition, superior gas handling capabilities of liquid ring pump greatly compensate for lower efficiency.

THERMODYNAMIC CHARACTERISTICS OF LIQUID RING PUMP & RELATED GAS LAWS

In a liquid ring pump, the compressant liquid and incoming gas or vapors undergo intimate mixing. Gas reactions can be performed, if required, by appropriate selection of liquid.

Cooling is one of the most obvious thermodynamic functions of the liquid ring pump. It is indeed a liquid cooled pump. In this type of pump, heat of compression and heat of condensation is removed by direct contact of compressant liquid, therefore maximum heat transfer is obtained. This is in contrast to water jacketed compressors where heat transfers through the cylinder wall is less effective. In a reciprocating type of compressor, it's possible to have hot spots within cylinder wall which can decompose or polymerize the gas. The liquid ring pump can obtain near isothermal compression as compared to adiabatic compression of reciprocating pumps or blowers.

The almost isothermal compression of single stage liquid ring vacuum pump, from 200mm Hg Abs to atmospheric discharge, gives a temperature rise of 3°C compared to uncooled adiabatic machines with temperature rises of about 111°C. The isothermal compression characteristics makes the liquid ring pump quite ideal where cool air or gas is required, and eliminates the necessity of an after-cooler.

CAPACITY OF LIQUID RING PUMP

The shape of the revolving rotating liquid ring is dependent upon speed of rotation, the contours of casing and rotor blades, and the compression work. It does not describe a pattern that can be clearly defined or measured, therefore it is emphasized that the liquid ring pump cannot be related similarly to positive displacement type reciprocating or rotary pumps. Generally, liquid ring pumps are tested for actual delivered capacity. Some suppliers test all pumps; others test none or a statistical sample of production. Standard performance curves published by manufacturers are based on air (50% R.H.) at 760mm Hg Abs barometer, and 15°C water. Any variation on these conditions will change the pump performance. The maximum vacuum that can be obtained by the liquid ring pump is limited by the vapor pressure of compressant liquid.

Capacity for some vendors is subject to a 5% tolerance, while for others as high as 10%; and horsepower varies from 0% to 10% depending upon manufacturer. This is an important consideration in evaluating published curve.

PERFORMANCE

The following conditions play a significant role in pump selection and have definite effects on a liquid ring pump's performance:

1. Operating Pressure — Operating pressure determines whether single stage, two stage, low vapor pressure sealed, or water sealed pumps are required. The operating pressure is effected by physical properties of liquid compressant (vapor pressure, sp. gr., sp. ht. and viscosity).

2. Gas Composition — The types of gases and vapors with its composition determines if pump is required to handle any condensible vapors. During condensation of vapors, the latent heat of vaporization is given off to the compressant liquid, raising its temperature and affecting pump performance.

3. Dry Gas – When dry gas enters the rotor bucket of a liquid ring pump, some of the seal liquid vaporizes and occupies some space, reducing ideal pumping capacity. If compressant water entering the pump is greater than 60°F (15°C), actual capacity is less than published curve. Table 4.5 illustrates general seal water correction factors with respect to various seal temperature for a vacuum pump handling dry air only.

Table 4.5. Typical Liquid Ring Vacuum Pumps

Temp. °F	INCHES HG VACUUM										
	SINGLE STAGE								TWO STAGE		
	5	10	15	20	22	24	25	26	27	28	28-1/2
40	.987	.984	.979	.968	.959	.945	.934	.917	.883	.816	.741
50	.993	.991	.988	.989	.976	.967	.956	.950	.929	.885	.832
60	1.00	1.00	1.00	1.00	1.00	1.00	1.00	1.00	1.00	1.00	1.00
70	1.01	1.01	1.02	1.03	1.03	1.05	1.06	1.08	1.11	1.21	1.38
80	1.02	1.03	1.04	1.07	1.08	1.12	1.15	1.20	1.32	1.70	2.79
90	1.04	1.05	1.08	1.12	1.16	1.23	1.30	1.41	1.74	3.62	
100	1.07	1.09	1.12	1.20	1.27	1.41	1.55	1.85	3.00		
110	1.11	1.14	1.19	1.33	1.46	1.76	2.10	3.06			
120	1.15	1.20	1.29	1.54	1.80	2.54	3.81				

Correction Factor

NOTE: This data applies to Vacuum Pumps selected for handling relatively dry air, that is, up to a maximum of 50% relative humidity. These factors do not apply for pumps handling saturated air and vapor mixtures. (For Reference Only)

EXAMPLE: Wanted, a vacuum pump to operate at 22″ vac with a capacity of 1000 cfm at 22″ vac. Seal water available at 100°F, barometer 30″ mercury. Correction factor from table is 1.27. Select pump from standard Nash curves for capacity of 1270 cfm at 22″ vacuum.

4. Wet or Saturated Gas – Liquid ring pumps sealed with water act as a heat sink capable of cooling and condensing vapors. If seal water in the pump is cooler than the incoming vapors, pumping capacities will be greater than standard published curve. As the vapors come in contact with cool seal water, it reduces volume by condensing vapors before entering the rotor chamber, increasing pumping capacity. A typical performance curve (Figure 4.11) shows the gain in capacity when saturated air-water vapor mixture is handled. The net pumping capacity of a liquid ring pump increases with greater temperature difference between vapor and a seal water.

5. Liquid Compressant Temperature – High seal temperature not only causes the incoming gas to expand, but also raises the compressant liquid vapor pressure, thus decreasing pumping capacity. Conversely, lower temperature reduces volume of incoming gas to allow more weight flow for a given pump. The higher the seal liquid temperature, the more vaporization occurs and occupies more space in the bucket, reducing pumping capacity.

6. Solubility of incoming gases into liquid compressant. The gas solubility affects the pump selection as some of the dissolved gases in the liquid ring flashes back at the inlet of the pump, occupying some space in rotor bucket, reducing its net handling capacity. Soluble gases, like sulfur dioxide, carbon dioxide, and some others, have been successfully handling in a liquid ring pump with proper consideration of capacity loss.

7. Effect of Altitude on Pump Performance – The capacity of the liquid ring pump remains constant for a pressure ratio unless the inlet pressure is above atmospheric. The operating pressure of a vacuum pump is limited by exerting barometer. It is absolutely necessary to convert operating pressures given at altitude to the reference sea level conditions using Eq (1), then compare required capacity with standard pump performance.

$$\frac{(P1)}{(P2)} \text{ at sea level} = \frac{(P1)}{(P2)} \text{ at altitude} \qquad \text{Eq-1}$$

where

P1 (at sea level) = Ref. absolute pressure
P2 (at sea level) = Sea level barometer 760mm
P1 (at altitude) = Ref. Abs. Press. at altitude
P2 (at altitude) = Barometer at altitude

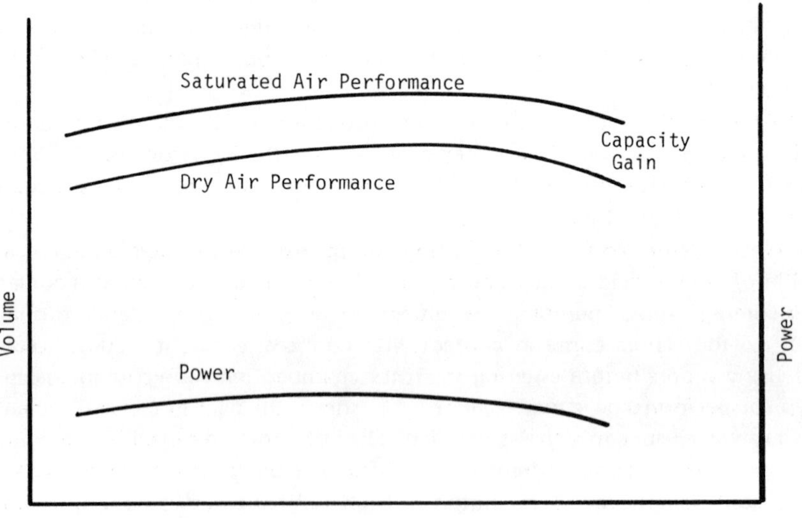

Figure 4.11. Increase in capacity of liquid ring pump handling saturated mixture.

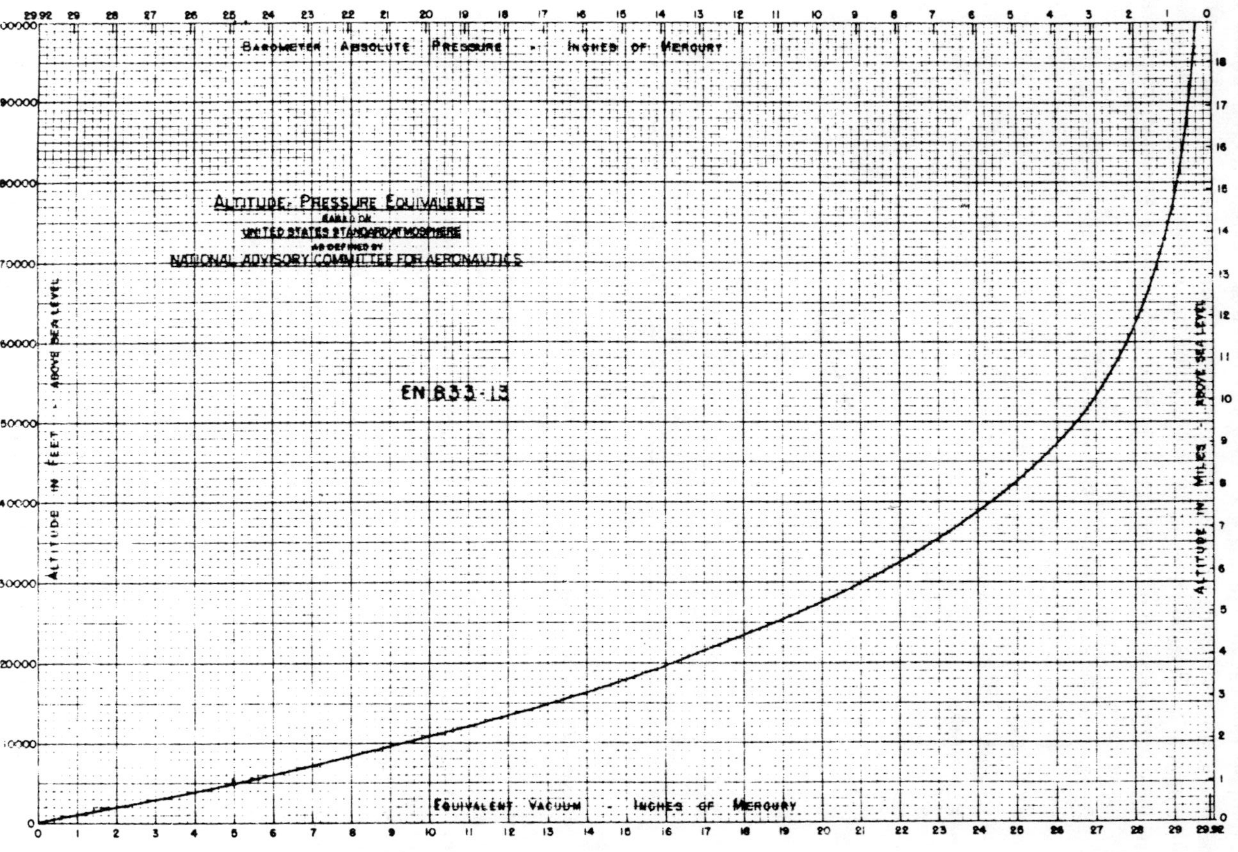

Figure 4.12. Relation of altitude to barometric pressure or vacuum.

Typical barometric pressure vs altitude is shown in Figure 4.12.

The liquid ring vacuum pump operating at altitude shows reduction in power as it operates against a reduced pressure ratio (discharge pressure/inlet pressure), due to lower discharge pressure.

The typical characteristics of reduction in power is shown in Figure 4.13.

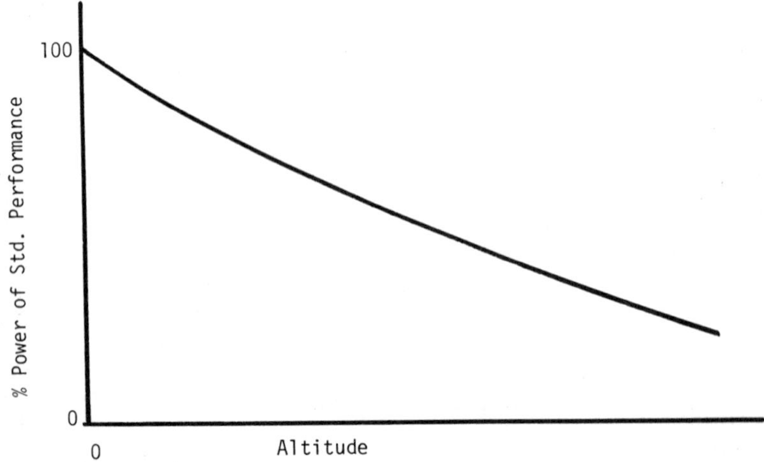

Figure 4.13. Typical reduction in power consumption of a liquid ring vacuum pump operating at the altitude.

In a compressor, the greater compression ratio tends to be off-set by lower weight of gas (less moles). It could be safe to assume power consumption of the liquid ring compressor remains the same at an altitude as at sea level for a given absolute discharge pressure.

The motor selection at the altitude requires special attention for a type of insulation and a service factor, as air is less dense at the altitude and has less ability to cool the motor. Consult motor manufacturer for specific motor selection at the altitude.

PUMP SELECTION

Pump selection can be straight-forward for handling dry air or a water vapor mixture, and quite involved for complex applications. For optimum pump selection, contact an experienced manufacturer. Most liquid ring pump applications will involve conditions other than the standard, as stated on manufacturer's performance curve. The effect of such variations can be determined by reference to basic gas laws which may be stated mathematically (below). For details, refer to a basic thermodynamic book.

1. Combination of Boyles & Charles Law

$$\frac{P1V1}{T1} = \frac{P2V2}{T2} \qquad \text{Eq-1A}$$

Where pressures and temperatures are expressed in absolute terms.

Volume from one set of pressure and temperature conditions to the other can be computed.

2. Ideal Gas Law

$$PV = mRT$$

$$\text{or} \qquad \text{Eq-2}$$

$$V = \frac{mRT}{P}$$

Volume can be computed from given mass flow.

3. Dalton's Law

$$P = P1 + P2 + P3 + \cdot\cdot$$

$$\text{if} \qquad \text{Eq-3}$$

$$V = V1 = V2 = V3$$

Applying Dalton's Law to Ideal gas mixture

$$Vt = \frac{ma\ Ra \times T}{Pa} \qquad \text{Eq-4}$$

Based on Dalton's Law, the amount of free dry air in a saturated air or water vapor mixture is shown in Figure 4.14. This provides quick determination of dry air in saturated air-water vapor mixtures.

4. Roulte's Law

$$Pa = P_A \times m_{Fa} \qquad \text{Eq-5}$$

where

Pa = Partial pressure of a constituent
P_A = Vapor pressure of A
m_{Fa} = Mole fraction

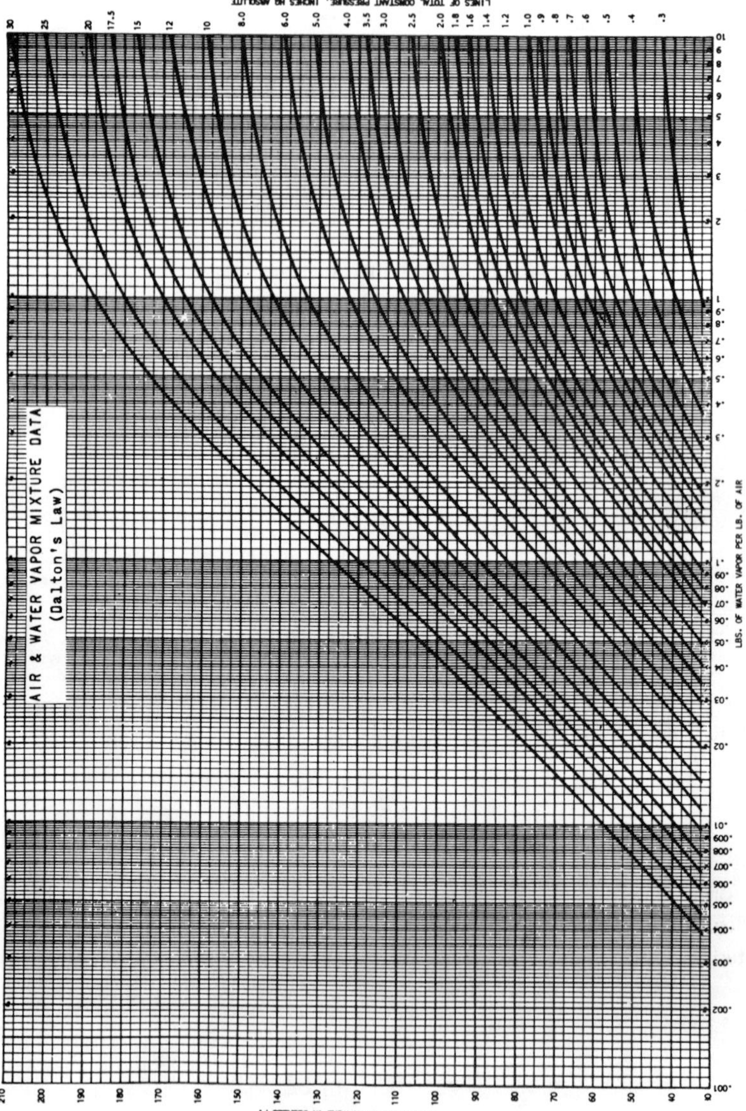

Figure 4.14. Free dry air in saturated air-vapor mixtures.

Calculation of Vapor Pressure Effect For:

a. Miscible Seal Liquid – when two or more liquids are miscible (example: methanol and water), the total vapor pressure exerted by the liquid compressant (seal liquid) follows Roulte's Law, and mathematically stated in Eq-5.

$$PVt = P_A \times mFa + P_B \times mF_b + \cdots \qquad \text{Eq-6}$$

b. Immiscible Seal Liquid – when two or more liquids are immiscible (example: benzene and water), the total vapor pressure exerted by the liquid compressant is an algebric sum of the individual vapor pressure as stated in Eq-7.

$$PVt = P_A + P_B + P_C + \cdots \qquad \text{Eq-7}$$

Volume required at the pump inlet is computed, considering vapor pressure effect at the pump operating temperature and assuming gas mixture follows Ideal behavior.

Temperature Rise Across The Liquid Ring Pump

In a liquid ring pump, ideally assumed, all heat of compression and heat of condensation is transmitted to the liquid compressant due to intimate mixing of gas and liquid. Temperature rise can be computed by simplified heat-mass transfer calculations. Typical temperature rise across the pump, without any condensation, is between 10–15°F (6–8°C). Temperature rise across the pump can be lowered by increasing liquid compressant flow.

The curve presented in Figure 4.14 provides the data to determine the amount of free dry air in saturated air-vapor mixtures following Dalton's Law.

As an example: Given 30 lbs of saturated mixture at 29 in. vacuum and 60°F; how much free dry air is contained in the mixture? Referring to curve below read across the line corresponding to 60°F mixture temperature until it intersects the curve line corresponding to one inch absolute vacuum (equivalent to 29 in. gauge vacuum). Reading down from the curve line we have 0.68 lbs of water vapor per lb of air. That is to say, an air-vapor mixture at 60°F and one inch mercury absolute vacuum will consist of one pound of air and .68 lb of water vapor; therefore in 30 lb of saturated mixture under these conditions of temperature and pressure there will be 30/1.68 = 17.9 lb of free dry air.

COMPRESSANT SELECTION

Water is the most common compressant or seal liquid used due to its availability and suitable properties. However, other liquid compressants can be used with the following desirable characteristics.

1. Low Vapor Pressure — The vapor pressure of a liquid compressant increases with increasing temperature. An increase of the vapor pressure reduces available bucket space and thus reduces net pumping capacity. If the vapor pressure of the compressant is lower than water at operating pressure and temperature, it increases pumping capacity. Capacity gain using oil as the seal liquid is illustrated in Figure 4.15. For operating pressure lower than 20mm Hg A (0.78″ Hg Abs), oil or hydraulic fluid can be used as the compressant. If the liquid phase of vapor condenses in the liquid ring pump, proper vapor pressure correction for miscible or immiscible liquids must be made.

2. High Specific Heat — Generally it is assumed that all energy (power) required to operate a liquid ring pump is ultimately transmitted to the liquid compressant in the form of heat. The temperature rise across the liquid ring pump increases with decreasing specific heat of the compressant. The greater the temperature rise, the greater the vapor pressure of liquid compressant, and a corresponding reduction in capacity. Therefore, it is preferable to select compressant liquid with higher specific heat, if possible.

3. Specific Gravity — The liquid ring energy is a function of rotor velocity and a specific gravity of the liquid compressant. In a vacuum pump, when the specific gravity of the seal liquid is greater than 1.0, the standard curve capacity remains unchanged. However, this consumes more power than shown in the standard curve. When specific gravity is less than 1.0, capacity and power remains unchanged, but pump speed must be increased to provide adequate energy to the less dense compressant. In a compressor, the performance changes with a specific gravity of the

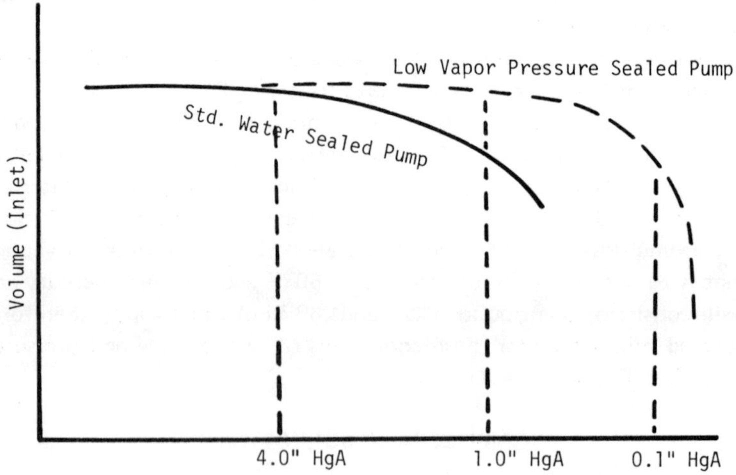

Figure 4.15. Capacity gain of a liquid ring vacuum pump sealed with low vapor pressure liquid compressant.

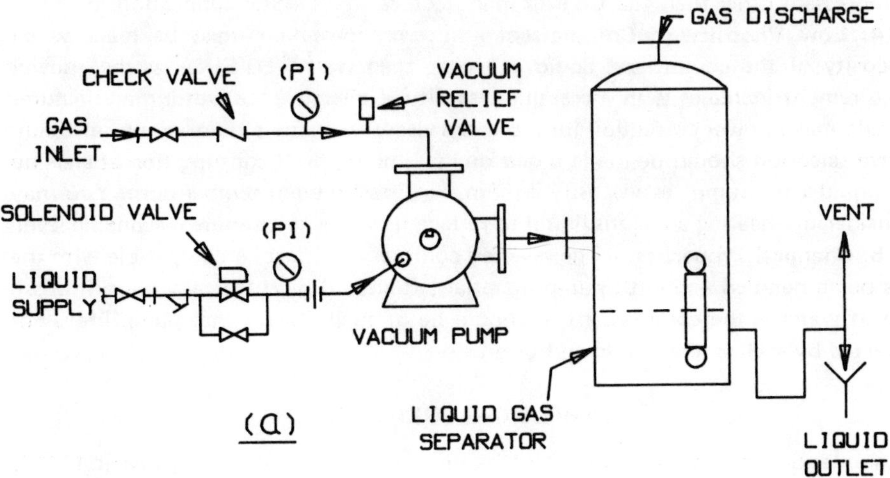

Figure 4.15a. Vacuum pump once through seal.

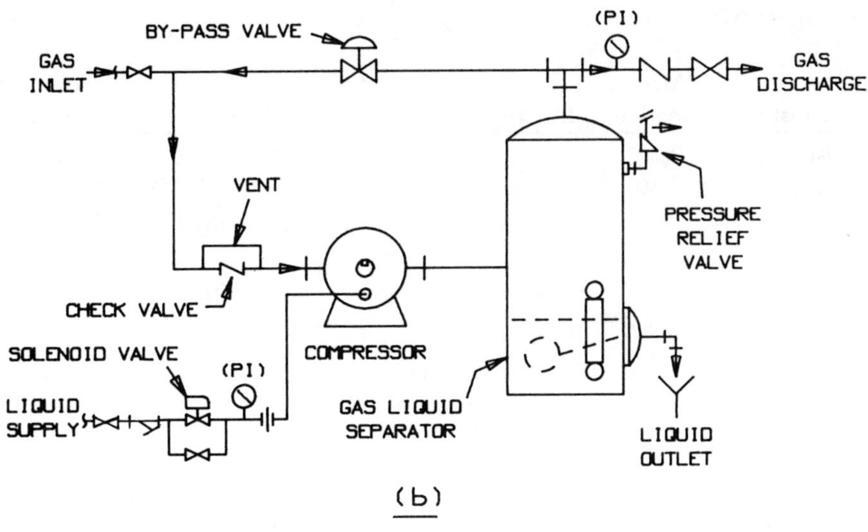

Figure 4.15b. Compressor once through seal.

compressant other than 1.0. Consult manufacturer for specific application.

4. Low Viscosity of Compressant – Power correction must be made when viscosity of the comprssant liquid is higher than water (50 SUS), as the power requirement increases with increasing viscosity. Generally, the motor manufacturer should make power correction for increased viscosity at the time of pump selection. Drive selection should be made based on maximum power consumption at start-up of liquid ring pump, as viscosity is higher at low ambient temperature. You may consider pre-heating a viscous liquid to reduce its viscosity to some reasonable level.

5. Chemical Properties – The selected compressant must be compatible with the gas being handled and with pump materials. It should be chemically inert, if possible. If water is the compressant, it should be of such quality that pump life is not affected by scaling, corrosion, and/or erosion.

SAMPLE CALCULATIONS

Example: Select pump to handle 700 lbs/hr of gas mixture at 25″ Hg Vac at 110°F. The gas analysis by weight % is as follows:

Air – 60%, CO_2 – 40%
Available seal water at 85°F,
discharge pressure 29.92″ Hg Abs.

Solution:

1. Calculate Ave M Wt of the gas mixture

Const.	Mass Fraction	Lbs/Hr	M.Wt. (#/#mole)	Moles/Hr	Mole Fraction
Air	0.60	420	29	14.48(420/29)	0.69
CO_2	0.40	280	44	6.36(280/44)	0.31
				Total 20.84	

$$\text{Mole Fraction} = \text{Moles/Total Moles}$$

$$\text{Average M.Wt.} = \text{M.Wt Air} \times mF_{Air} + \text{M.Wt } CO_2 \times mF_{CO_2}$$

$$= 29 \times 0.69 + 44 \times 0.31$$

$$33.44\#/\#\text{mole of mixture}$$

2. $$\left.\begin{array}{l}\text{Volume ahead of}\\ \text{liquid ring pump}\end{array}\right) = \frac{700\#/\text{Hr}}{60} \times \frac{379}{\text{AV.M.Wt}} \times \frac{29.92}{(29.92-25)} \times \frac{460+110}{520}$$

$$= 882 \text{ ACFM}$$

Based on this volume, make preliminary pump selection, then calculate temperature rise across the pump.

3. Assume temperature rise across the pump and verify actual temperature rise is within ± 1°F. For this example assume temperature rise of 6°F (ΔT).

$$\begin{matrix}\text{Operating pump temp)} \\ \text{(T effective)} \qquad)\end{matrix} = 85°F \text{ (seal water)} + 6°F\ (\Delta T) = 91°F$$

$$\text{Vapor pressure of water at } 91°F = 1.47''\ \text{HgA}$$

$$\text{Calculate partial gas pressure} = P_{Total} = 4.92''\ \text{HgAB}\ (29.92\text{-}25)$$

$$P_{VapH_2O} = 1.47''\ \text{HgAB}$$

$$@\ \text{Teff}$$

$$P_{gas\ press} = 3.45$$

Actual volume required at pump inlet is equal to:

$$V = \frac{700}{60} \times \frac{379}{33.44} \times \frac{29.92}{3.45} \times \frac{400 + 91\ \text{Teff}}{520} = 1216\ \text{ACFM}$$

Actual pump selection should be capable of handling 1216 ACFM at 25″ Hg Vacuum.

NOTE: For simplicity, solubility of CO_2 is not considered. In an actual selection, solubility loss must be considered.

SAMPLE INQUIRY

It is important that the application inquiry forwarded to the manufacturer is properly filled with all required information. Sometimes a customer may not be aware of information required for proper pump and system selection, and lack of this information often leads to time-consuming communications and results in delay of a quotation. A typical inquiry sheet is shown on the Inquiry Information Sheet.

SEAL SYSTEMS

A liquid ring pump requires continuous supply of the liquid compressant to maintain its ring. It is important that the flow requirement, as specified by the manufacturer, is met. If the flow requirements are reduced significantly, it affects pump performance and even creates instability of the pump due to collapse of the liquid ring. Typically there are three types of seal systems used for a liquid ring pump and are explained as follows:

1. Once-thru-seal system — may be employed where compressant liquid (water) supply is abundent and there is no danger of liquid contamination. Once-thru-seal arrangement is illustrated in Figure 4.15. The compressant liquid is supplied from the main source, passed through the pump into a separator, and discharged to a drain.

INQUIRY INFORMATION SHEET

Engr. ____________

Inquiry No. ____________ Customer ____________

Date Rec'd. ____________ Due ____________ Complete ____________

GAS HANDLED

Air ________ Other ____________

Mol.Wt. ________ Sp. Gravity ________ Temp.________ °F/°C.

CAPACITY REQUIRED

ACFM (measured at inlet)________ Lbs./Hr________SCFM (Corrected to 14.7 psia and 60°F) ____________ Other ________ Saturated (%) ________ Dry ________

OPERATING CONDITIONS

Continuous ________ Intermittent ________ Cycles/Hr ____________

Altitude (Ft./Meters) ____________ Ambient Temp (°F/°C) ____________

Inlet Pressure ″Hg Abs ________ ″Hg Vac ________ Psia ________Psig ________

Other ________ Min. ________ Max. ________ Inlet Temp. ____________

Discharge Pressure (Psia) ____________Psig ________ Other ____________

Min. ____________ Max. ____________

SEAL OR COOLING LIQUID

Water ________ Other ____________ Temp ________ °F/°C

Available at ________ Psig. Once thru ____________ Recirc ____________

Partial Recirc ____________ Vapor Pressure ________Psia/″Hg Abs at ________°F/°C

________ Psia/″Hg Abs at ______°F/°C ______ Psia/″Hg Abs at ________ °F/°C

Specific Gravity ____________Specific Heat ____________ Mol. Wt. ____________

Viscosity ____________________Temperature (at ten temperatures)

ELECTRICAL

Available power supply, Volts ____________ Phase ____________ Hz. ________

Control Circuit Voltage ____________________ .

MOTORS

Enclosure ODP ________ TEFC ________ Severe Duty ______ Explosion Proof ______

Service Factor ____________ Insulation ____________ List Special Features________

____ .

CONTROL

Furnish for wall mount ________ Mount ________ Mount & Wire ____________

Describe fully ____

SPECIAL REQUIREMENTS

Material of Const: All Bronze ________ All Iron ______ Bronze Fitted ________

Stainless ________ Std. Cast Iron ________ Other ____________

Mech Seals (Single) ____________ (double) ______ Type ________ Tank Mount ______

Piped Package ____________ Accessories Required ____________

____

List on reverse side all attachments, supplemental information, and comments.

2. Partial recirculation system — may be employed where liquid compressant is in short supply or desired to reduce liquid consumption. Some of the liquid compressant is recirculated through a specifically sized orifice without use of heat exchanger. This may result in a 10–67% saving of liquid compressant, depending upon the temperature of non-recirculated seal and operating vacuum. Partial recirculated seal arrangement is shown in Figure 4.16. Generally it operates at higher temperature than once-thru-seal and may affect pump performance.

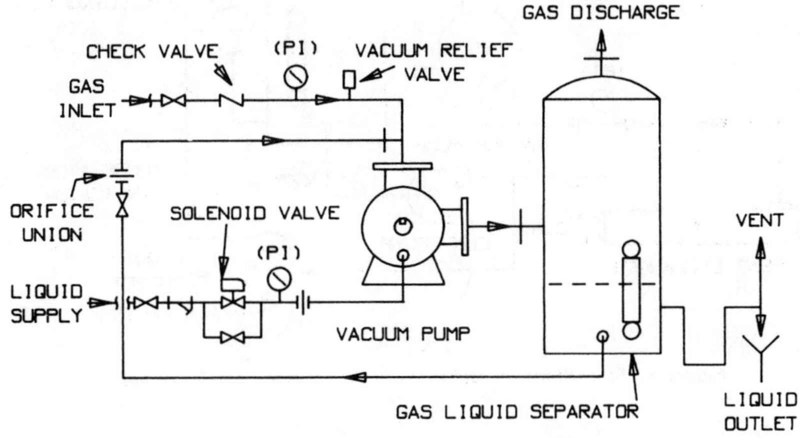

Figure 4.16. Partial recirc. seal for liquid saving.

3. Full recirculation seal system — this type of arrangement is employed where liquid compressant is in short supply and/or the cost of the liquid compressant (other than water) necessitates the recirculated seal system. It is often required to reduce liquid contamination or pollution and reduce cost of waste treatment in certain applications. Typical recirculation systems with and without recirculation pumps are illustrated in Figure 4.17. The liquid compressant is cooled through the heat exchanger where heat of compression and condensation is removed and then recirculated back into the pump. Installation of a compressor and vacuum pump with inlet seal sometimes do not require a recirculation pump. However, a vacuum system without a recirculating pump requires a careful piping arrangement to lower pressure drops.

Tank Mounted Unit — Vacuum Pump

This arrangement requires no external heat exchanger or circulating pump. Cooling of seal liquid is accomplished by natural heat transfer through the wall of the tank, piping, and pump. Typical piping arrangement is shown in Figure 4.18.

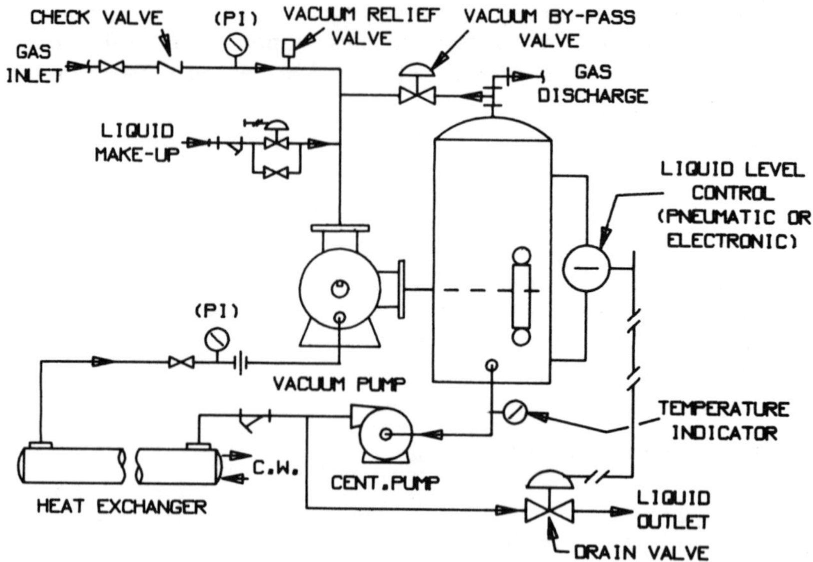

Figure 4.17a. Full recirc. seal vacuum pump with centrifugal pump.

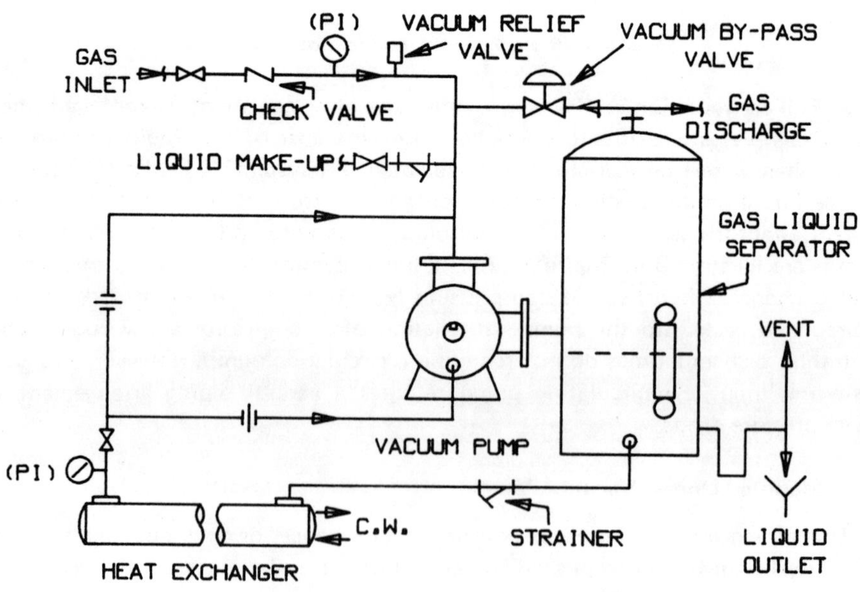

Figure 4.17b. Vacuum pump without recirc. pump.

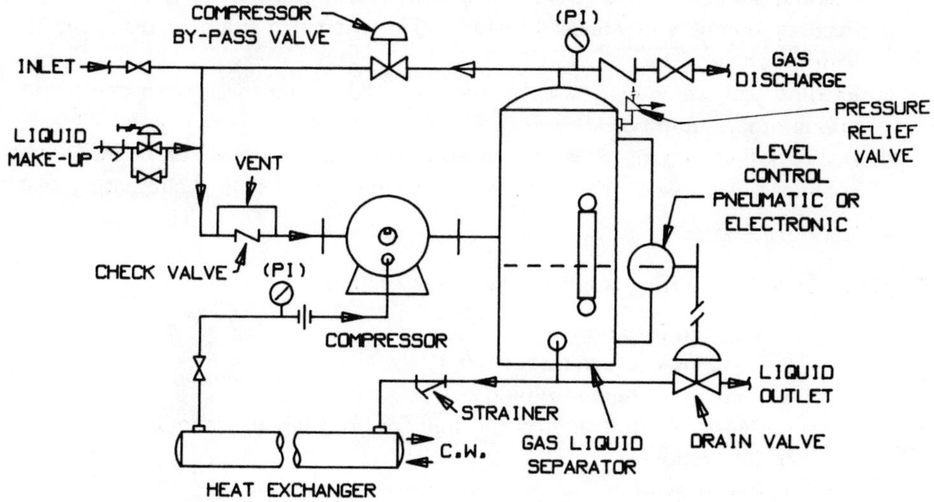

Figure 4.17c. Compressor (generally recirc. pump not req'd).

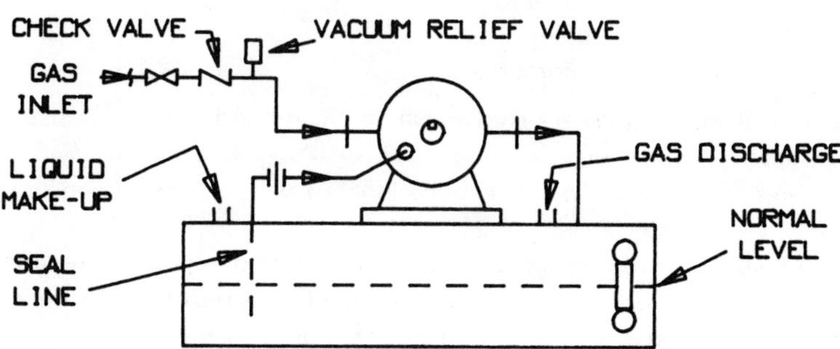

Figure 4.18. Tank mounted vacuum pump.

GENERAL NOTES

1) Vacuum and pressure relief valves suggested for pump and system protection
2) Liquid level controls often recommended to protect pump from operating dry or overflooding
3) By-pass control valve is necessary when regulation of system pressure required.

Water Conservation

Variations on recirculated seal systems are numerous. Seal water may be conserved by using pre-used water or by re-using water after pump discharge. For example:

1. Feeding pump with fresh water and discharging to another system.
2. Feeding pump with pre-used water and discharge to drain or discharge to another system.
3. Feeding seal water in a cascade manner, feeding discharge from one pump into another pump.
4. Recirculating cooling tower water from a pump to tower, or recirculating cooling tower water through heat exchanger for cooling circulating seal liquid.

Typical List of Liquid Ring Applications Are as Follows:

1. Vacuum filtration
2. Condenser exhauster
3. Centrifugal pump priming
4. Central vacuum and compressair for hospitals and laboratories
5. Deaeration
6. Recarbonation
7. Vapor recovery
8. Process gas compressor
9. Vinyl-chloride monomer or butadiene recovery
10. Chlorine compressing
11. Tail gas compressor
12. Instrument air compressor

Some of the applications are shown schematically and briefly discussed on the following pages.

The reference chart indicates applications of liquid ring pumps in various industries.

FILTRATION: In this arrangement, the liquid ring pump evacuates the filtrate receiver.The constant, non-pulsating vacuum it maintains results in the deposit of an even, uniform filter cake. A small compressor supplies air for cake blowing. (Figure 4.19a)

SOLVENT RECOVERY: Vacuum tumble drying draws solvent vapors off solids. In the vacuum pump, solvent vapors come in contact with cooled solvent liquid. The pump itself serves as a partial condenser. Non-condensibles are separate out mechanically, and the recovered solvent is stored for re-use. (Figure 4.19b)

MOISTURE EXTRACTION: Textile fabric containing mosture comes out of a wash box and passes over a slotted tube. The vacuum pump draws moisture, along with a considerable amount of air, from the extractor so as to maintain a continuous vacuum in the system. Any textile fibers that come through in the mixture create no problem. They can be disposed of when the separator is cleaned out periodically. (Figure 4.19c)

CONDENSER EXHAUSTING: Maintaining vacuum depends on removal of air

INDUSTRIES \ APPLICATIONS	aeration-agitation	blending	blowing	breathing air	chlorine gas	container washing	filtering	folding	gas boosting	head box	instrument air	laboratory	liquid transfer	sampling	solvent recovery	waste gas
beverage	✓					✓			✓		✓					
bottling						✓										
brick & clay							✓		✓		✓					
building													✓			
chemical				✓	✓		✓		✓		✓	✓			✓	✓
electrical									✓							
film mfg.			✓													
food	✓	✓				✓	✓				✓	✓			✓	
hospital											✓	✓				
lime & cement														✓		
mining				✓					✓		✓					✓
paint & varnish			✓								✓					
petroleum							✓				✓	✓				
plastics									✓			✓			✓	
power											✓					✓
pulp & paper							✓	✓	✓	✓	✓					
rubber			✓								✓					
sewage plant	✓						✓		✓		✓	✓	✓			
steel & iron	✓			✓					✓		✓					
textile													✓			
water supply	✓						✓		✓		✓	✓				

Reference Chart.

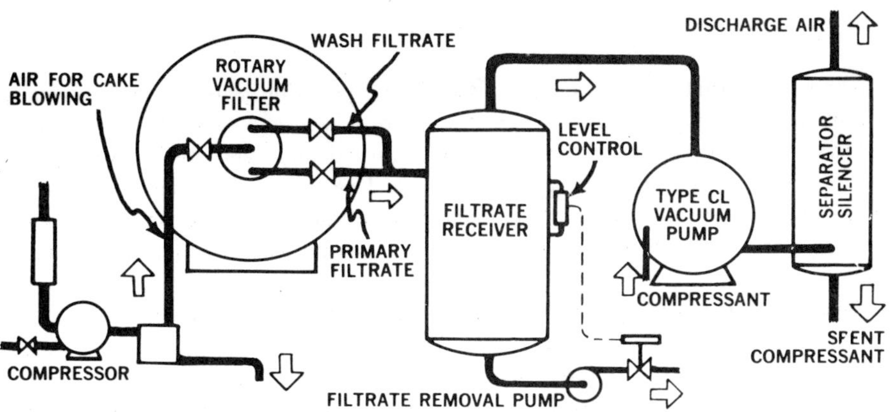

Figure 4.19a.

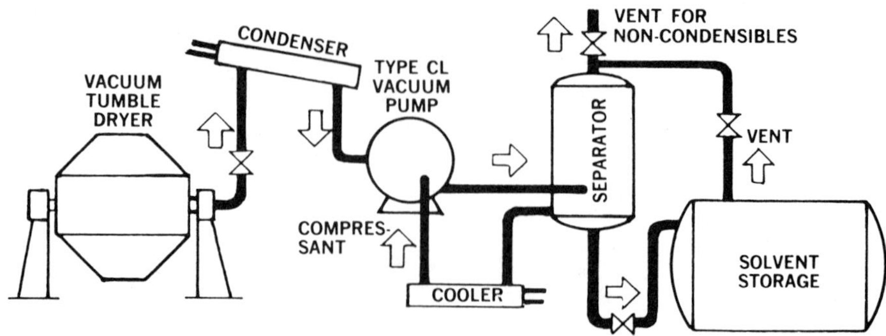

Figure 4.19b.

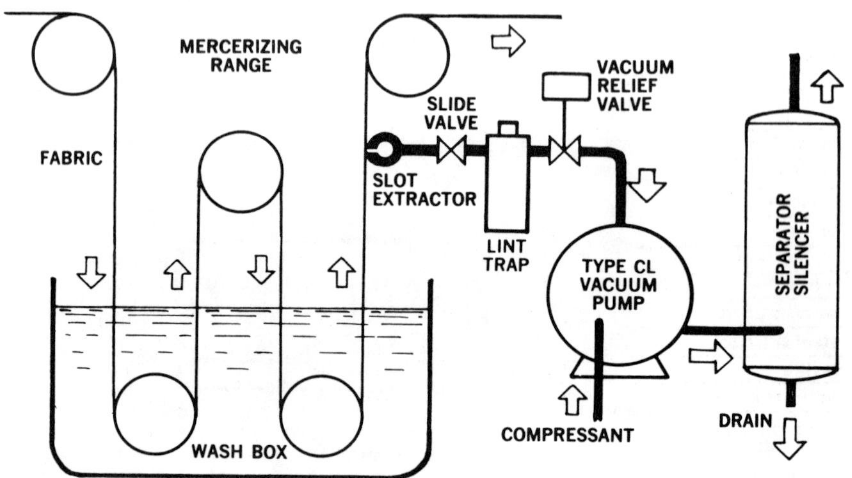

Figure 4.19c.

and other gases that cannot be condensed. The liquid ring vacuum pump normally operates in series with: an air ejector at condenser vacuums above 27″ of mercury. At lower vacuums, the ejector is by-passed automatically. and the vacuum pump works directly from the condenser at increased capacity. Automatic changeover between hog and hold operation is a major benefit that the Nash exhausting system offers as compared with steam ejectors which must be adjusted manually. (Figure 4.19d)

PRIMING: A centrifugal pump located above its suction level is kept full of liquid and ready to pump on startup. Instead of depending on a foot valve, a vacuum is drawn in the pump casing. The priming system, served by a vacuum pump, is connected to each centrifugal pump through a priming valve. It is a float

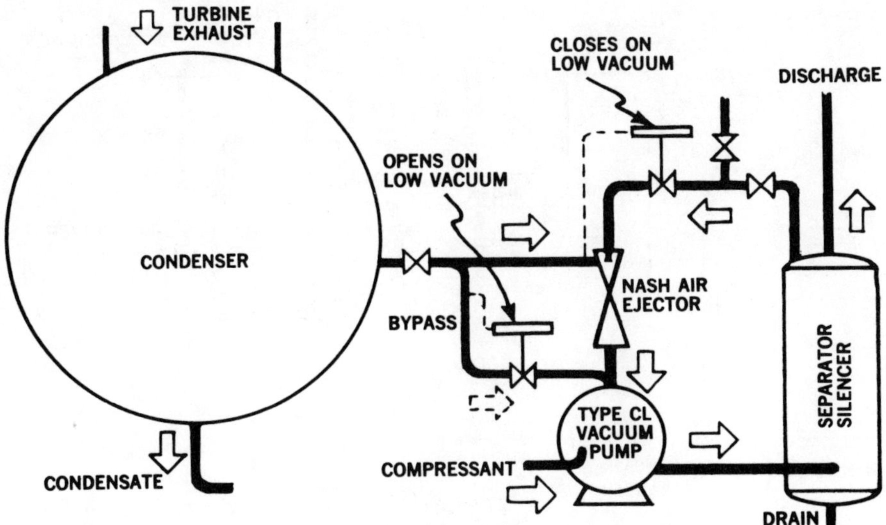

Figure 4.19d.

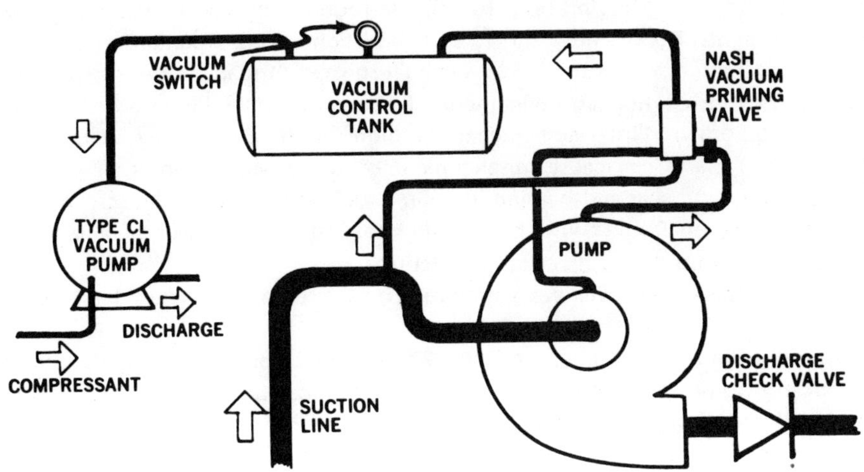

Figure 4.19e

valve – closes when liquid rises in it. Several connections may be needed to evacuate air from all high points in the centrifugals pump's suction eye and volute. The priming valve is tapped at three places for those connections. (Figure 4.19e)

In chlorine production, vacuum pump removes by-product hydrogen, and concentrated sulfuric acid is used as the liquid compressant in the chlorine gas compressor.

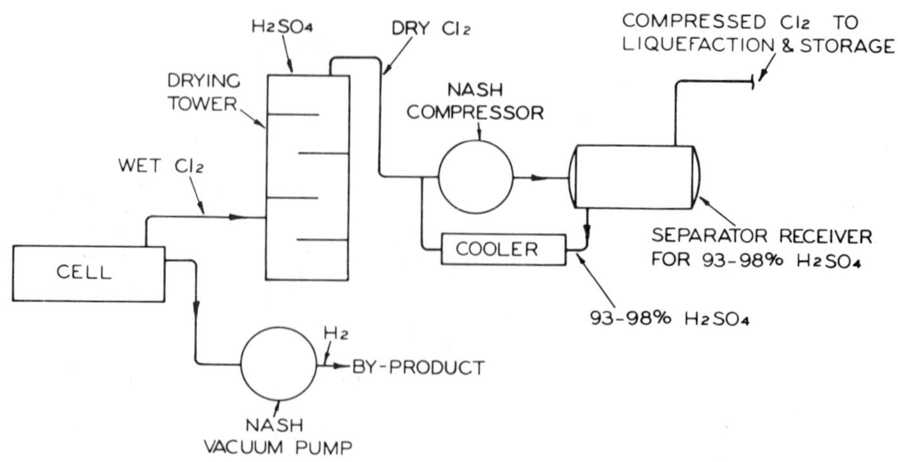

Figure 4.19f

INSTALLATION

Liquid ring pumps have only one moving part (rotor) and for all practical purposes most of them are vibration free. Single foundations can be designed to support the weight of equipment and to suit local conditions. Foundation layout and general installation instructions are usually furnished in manufacturer's instruction manual. The installation and alignment of the drive is similar to other balanced, rotating equipment. Inlet and discharge piping for the pump must be free of any strain, otherwise it can cause misalignment of the drive, rub internal parts, and reduce mechanical seal and bearing life. On a new installation, inlet screens, clean-outs, and dirt pockets should be installed to prevent carryover of any harmful foreign matter into the pump at initial start-up.

Piping Arrangement. Well designed piping must provide for the unique characteristics of the liquid ring pump.

Start-up. Pump casing must be ¼ to ½ full of seal liquid. If underfilled, there is insufficient liquid to mate a ring, and unit will be unstable or will seize. Overfilling leads to flooding and broken shaft or rotor blades.

Some designs are vertical ("12 o'clock") discharge. These usually require an auxiliary casing drain to prevent flooded starts.

Shutdown. A vacuum pump with inlet check valve will suck liquid back into pump casing. A tight check valve will hold liquid in upper part of casing and give the effect of a flooded start-up. Small vacuum breakers are available for installation on pump casing to counteract this undesirable condition.

Operating. Piping between pump discharge and the liquid-vapor separating device is critical. This should be short, direct, and without excessive elevation. For pump designs with horizontal discharge, the piping is usually 0–12″ above shaft centerline. Designs with vertical (12 o'clock) discharge should likewise minimize discharge elevation as discharge pipe floods with liquid on pump shutdown. Refer to Figure 4.20.

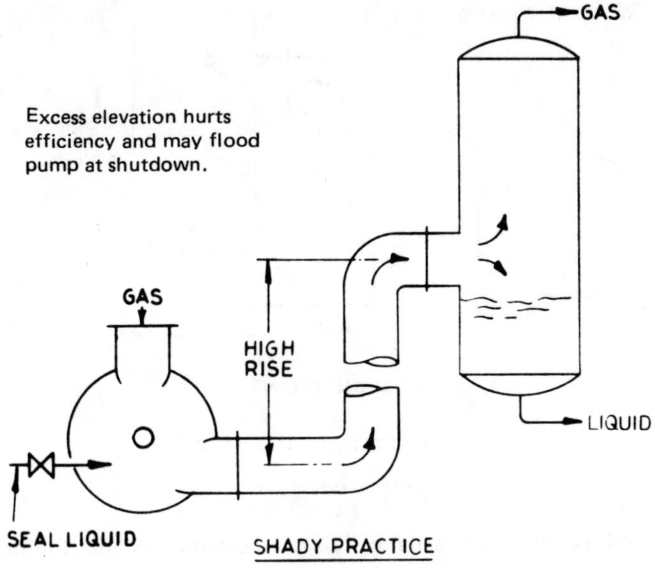

Figure 4.20a.

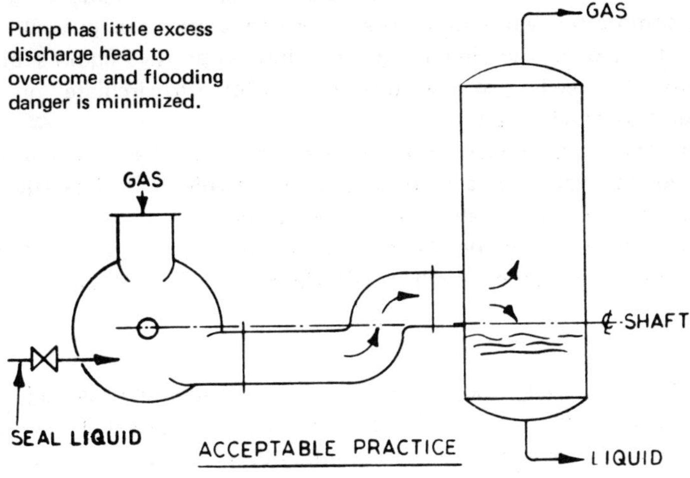

Figure 4.20b.

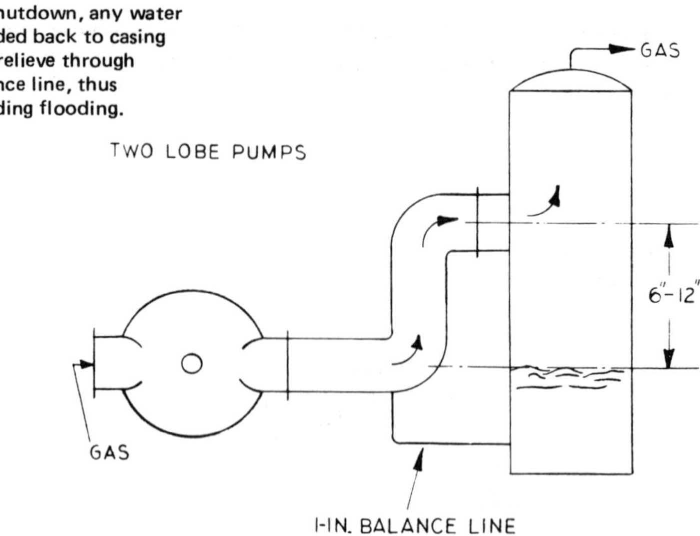

Figure 4.20c.

Water Piping. This is critical in partial or full recirculation systems. Piping should be located to avoid traps and maximize free flow from liquid-vapor separator back to vacuum pump. Piping shoudl always be located below pump centerline.

Liquid ring pumps require 1 to 5 1/min per bhp seal liquid. Piping should be designed to handle this with minimum pressure drop. In a recirculating system, a liquid ring compressor will require no auxiliary circulating pump, while a vacuum pump should use one. The circulating pump insures positive liquid supply to the vacuum pump. It likewise provides added driving force to circulate liquid through clogged strainers or sealed-up piping.

Controls. These are similar to those required with other mechanical vacuum pumps, except for water line controls and level controls. It is imperative that unit receive a constant supply of water at proper flow rate.

To achieve this, recirculating systems use a variety of level controls, flow alarms, and high temperature shutdowns when unit will be running unattended.

Piping Errors

Traps in Suction Piping. (See Figure 4.21). The condensate can collect in a low spot as shown in bottom of "U." The pressure drop increases and pump operation gets eratic. This can eventually pull a slug of liquids and, in extreme cases, results in internal pump damage. A proper liquid drainage line must be provided.

Long Unbroken Vertical Drops. (See Figure 4.22). If slugs of liquid carryover,

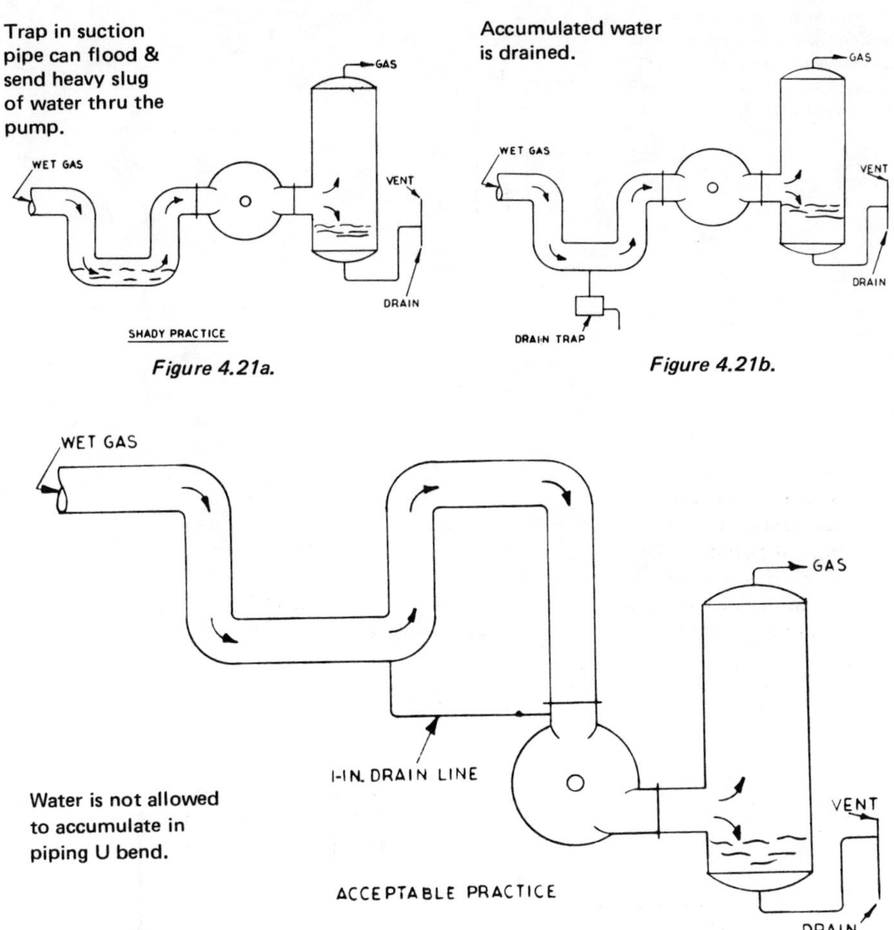

Figure 4.21a.

Figure 4.21b.

Figure 4.21c.

they can attain high kinetic energy in falling down the pipe, hitting pump internals with considerable force. This can crack or break the rotor blades. Where vertical drops cannot be avoided, soften blows from slug carryover by making it change direction. Also dirt pockets should be provided to catch foreign matters.

Uncleanable Start-up Screens. (See Figure 4.23). The standard type screens are just flat wire mesh on pumps with top inlets; the removal of screens by sliding out between flanges would result in a pump concentrated amount of debris. In such installations, provision for cleanout or conical shape screen should be considered.

Inlet Check Valve. (See Figure 4.24). When a liquid ring compressor operating at high discharge pressure is shut off, the gas in the separator expands back towards the inlet. This carries a slug of liquid from within the compressor casing to the

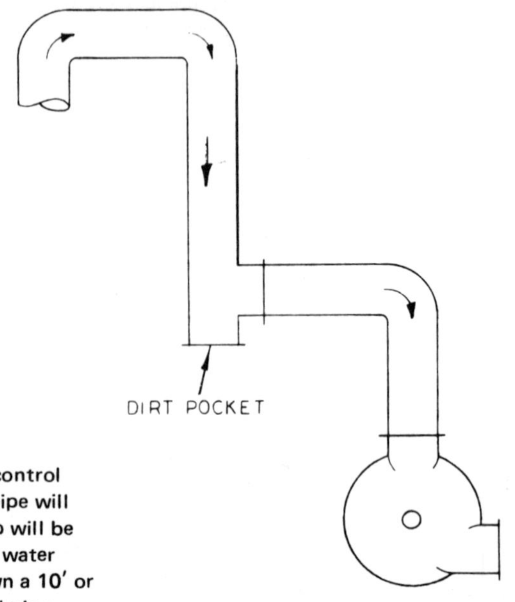

Figure 4.22a.

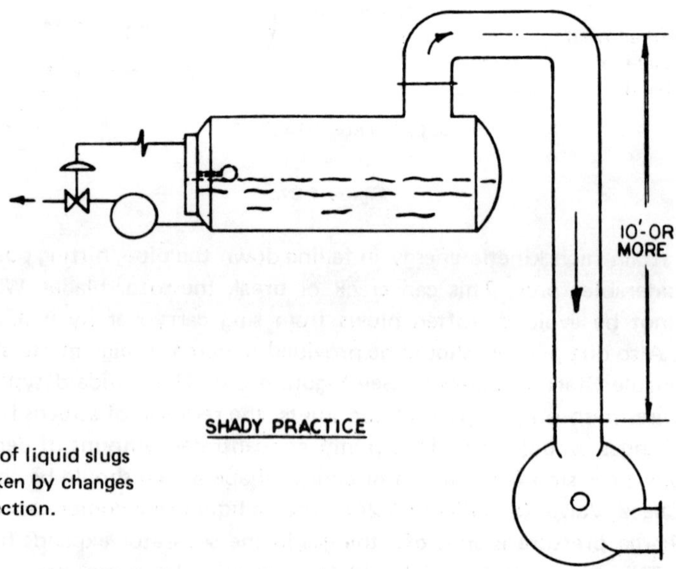

Figure 4.22b.

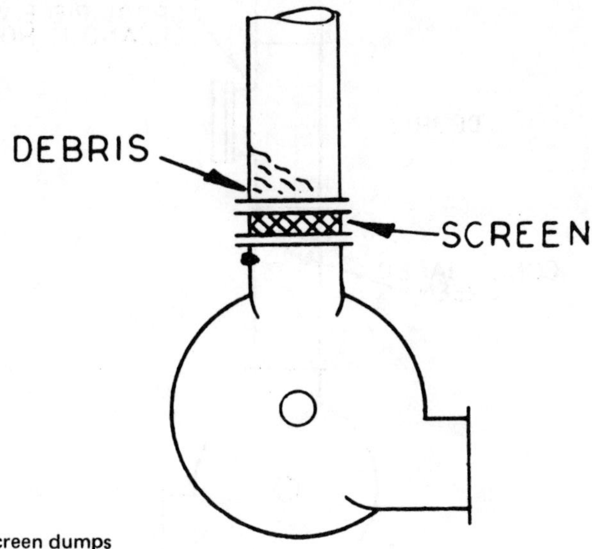

Removal of screen dumps debris directly into pump.

SHADY PRACTICE

Figure 4.23a.

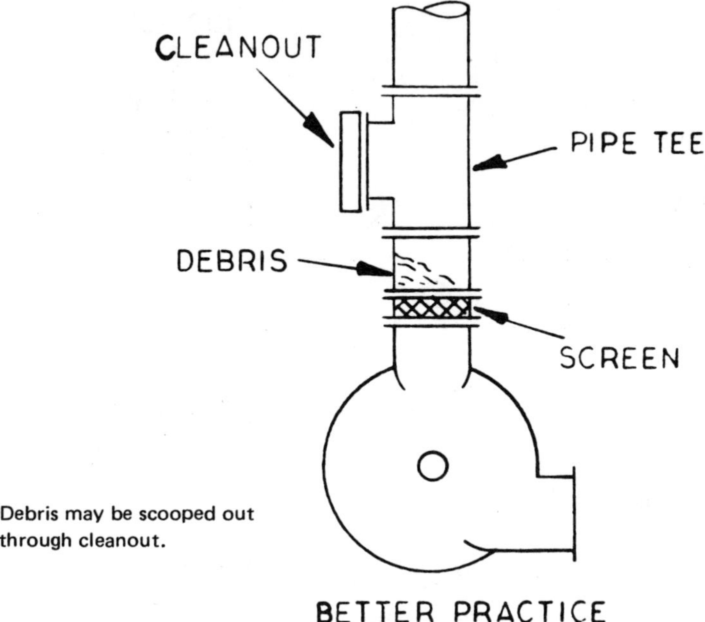

Debris may be scooped out through cleanout.

BETTER PRACTICE

Figure 4.23b.

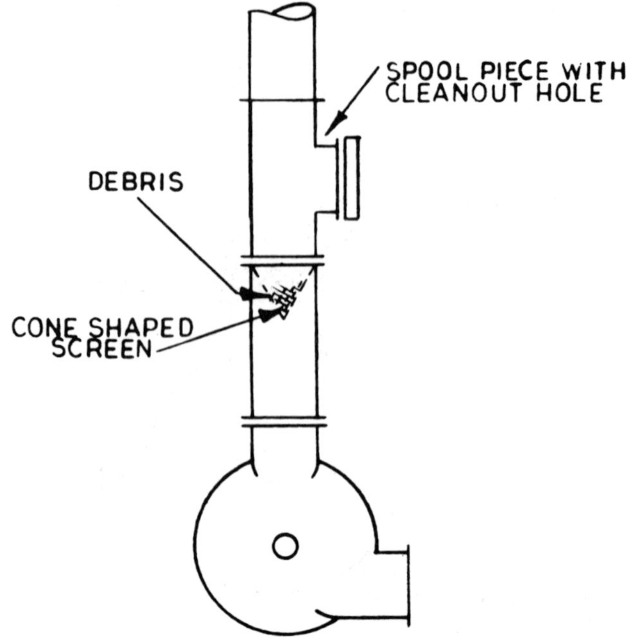

Figure 4.23c.

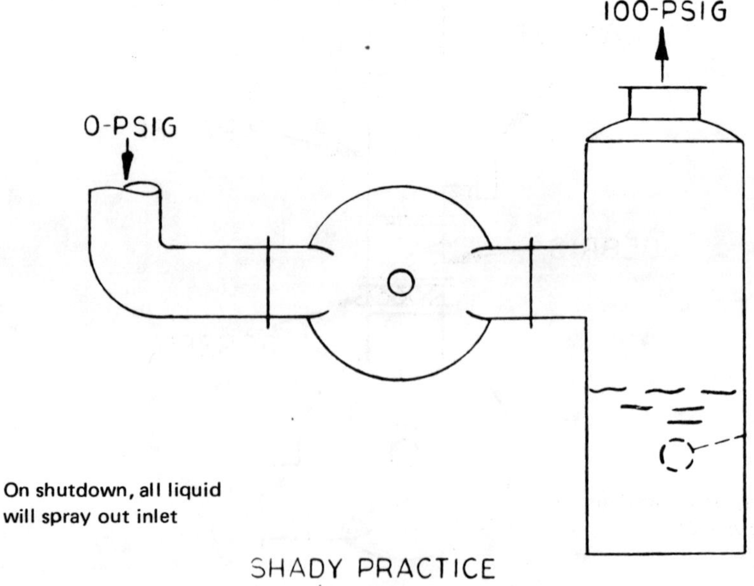

Figure 4.24a.

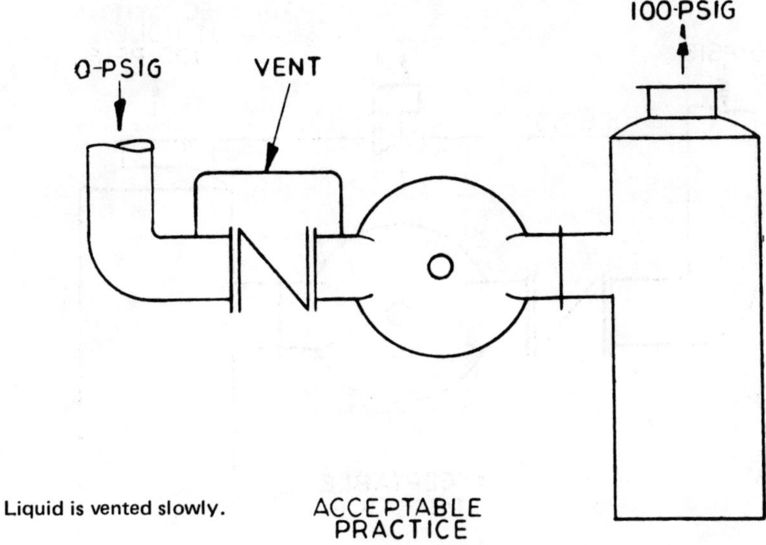

Figure 4.24b.

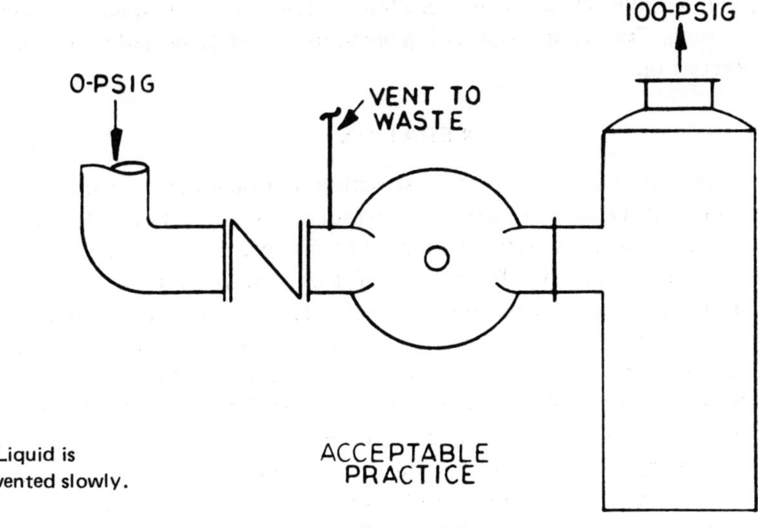

Figure 4.24c.

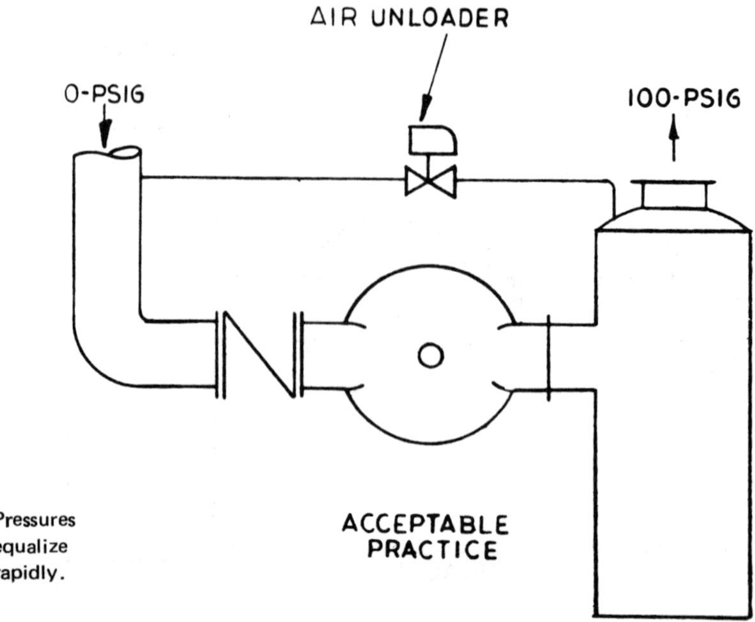

Figure 4.24d.

suction piping. An inlet check valve stops liquid blow-back and liquid settles in the compressor casing. The compressor casing pressure should be vented to atmosphere prior to next start-up.

OPERATION

Prior to pump start-up, review the instruction and operation manual generally supplied by the manufacturer. Check all piping for correct installation. Be sure inlet screens are installed at initial start-up to protect pump from possible carryover of bolts, nuts, welding shots, etc. The screen must be removed after a few weeks of operation, otherwise it would corrode and possibly fall inside the pump. Before starting, turn pump shaft one full rotation to check for possible binding. At initial start-up, it is necessary to add compressant liquid to fill half the pump casing and the complete seal system. NEVER OPERATE A LIQUID RING PUMP DRY. If the pump is started dry, serious damage to the pump will result. Bump the driver to verify correct rotation of the pump. Start the pump. Check for stability and temperature of the pump casing for possible overheating. If the pump is overheated, shut down until source of trouble is located.

It is essential *not* to operate liquid ring pump with blocked-off suction. It can result in cavitation. Always start up pump at atmospheric pressure.

In certain types of liquid ring compressors at start-up, gas should be by-passed from discharge of separator to inlet of pump for a few minutes, to help smooth formation of liquid ring.

MAINTENANCE

In a liquid ring pump there is only one moving element, rotating without metallic contact. There are no rubbing surfaces to wear so maintenance is minimum. If necessary, clearances can be adjusted easily by removal or addition of gaskets.

The only preventive maintenance required on liquid ring pumps is proper lubrication of bearings. The bearings are either grease or oil lubricated. The grease lubricated bearings are recommended to be hand-packed rather than use of grease gun. Over-greasing can shorten bearing life. Oil lubricated bearings require periodic inspection of oil level.

Stuffing boxes should be checked once a year or when leak develops.

ACKNOWLEDGMENT

I would like to thank The Nash Engineering Company for their cooperation in preparation of this chapter.

REFERENCES

1. Monroe, E. S., "Energy Conservation and Vacuum Pump," Chemical Engineering Progress, October 1975, p. 71.
2. Hied, P. T., "Save Energy With Mechanical Vacuum Pump," Modern Power and Engineering, October 1977.
3. Dadik, W. B., "Hook-up and Controls," Internal report of Nash Engineering Company.

CHAPTER 5

JET EJECTORS AND CONDENSERS AS VACUUM PRODUCING DEVICES

ROGER G. MOTE
AMETEK
Schutte & Koerting Div.
Cornwells Heights, PA

INTRODUCTION

Rapid, or high-capacity vacuum pumping of gases or vapors to 10^{-3} Torr is generally performed most economically by a jet-ejector vacuum system. Mechanically, it is the simplest of all present day types of vacuum pumps or compressors with the consequent benefit of minimum maintenance demands. Jet vacuum systems have the advantage of very low initial cost as well as low installation cost. Jets contain no moving parts and therefore have no lubrication or oil problems. They do not require extremely close tolerances, and they can be made from practically any corrosion-resistant material. In applications which require unusual materials of construction, glass for highly corrosive process vapors as one example, a jet-ejector may be the only available type or unit. Those fabrication materials lacking sufficient strength of their own, can be used as liners with steel backing.

A jet vacuum system frequently has higher utility (steam/air and water) requirements than conventional mechanical type vacuum pumps. Also, in the case of direct contact condensing systems, the motive gas and condensing water contact the process vapors with the result that the effluent may require special treatment if the suction vapors should contain any objectionable contaminants. The use of surface condensers does not eliminate the problem but greatly reduces the volume of effluent to be treated. For applications having great fluctuations of suction load at a given suction pressure, the use of parallel units (called dual or multiple element) are advantageous in keeping utility consumption to a minimum.

A jet ejector (Figure 5.1) is a rigid, hollow, thermodynamic device, which through controlled expansion in the motive nozzles, produces a jet of motive gas moving at supersonic speed, thus converting the static energy of the motive gas pressure to kinetic energy. Introduced circumferentially around this jet, is the relatively slow moving stream of suction vapor. This centrally located, high velocity jet of motive gas picks up and mixes with the slower suction stream as it passes through the converging section of the venturi diffuser. It then enters the throat section of the diffuser, completely mixed, at the critical (sonic) velocity of the mixture. The mixture from the throat passes through the diverging section of the

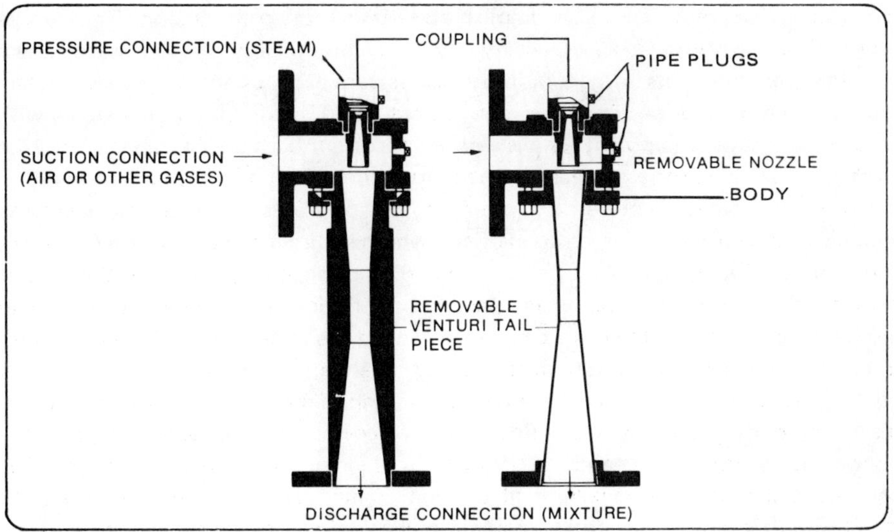

Figure 5.1. Typical jet ejectors

venturi diffuser, and as the cross sectional area increases the velocity decreases, converting its high kinetic energy into static pressure. The discharge area is designed to recover some 90–97% of the kinetic energy in the throat. The pressure in the throat corresponds to the critical value of the discharge pressure of the jet. During this conversion of pressure energy at the inlet to the motive nozzle, to velocity energy in the throat, to pressure energy at the discharge, work is performed in the form of entrainment, mixing and compression so that the discharge pressure is always intermediate to the motive and suction pressures.

Critical jets are those that involve a compression ratio greater than or equal to the critical pressure ratio of the suction vapor. For steam (water vapor) this ratio is 1.81 and for air the value is quite close to the same. That is, if the suction pressure is less than 0.55 times the discharge pressure, then the jet is a critical jet. If the compression ratio is less than this critical pressure ratio, then the jet is a non-critical jet.

In the same sense, the motive gas flow through the nozzle may be critical or non-critical. However, this flow does not determine whether the jet is called critical or non-critical. Usually, this motive gas flow is critical and must be if the jet itself is critical.

With critical jets, those designed for compression ratios of 4 to 10, the efficiency is greatly impaired if the motive gas pressure is not at least seven times the discharge pressure; efficiency is greatly improved when this same ratio is eleven or greater.

With extremely low compressions, the above ratios are not advantageous. Critical jets can be designed for lower motive pressures but with reduced compression ratios more stages are required to accomplish the same total compression. This can be economically advantageous, especially in the recycling of low pressure waste steam to reduce energy costs in generating supply steam. The quantity of waste steam required may be several times the requirement of 10.55 kg/cm^2 supply steam with general effects on equipment. Larger jets are necessary because the capacity of each stage is lower while the decrease in maximum compression available, especially in the last two stages, requires a greater number of stages. Further, the increased volume of steam means larger condensers, whether direct contact or surface type, and more cooling water. Nonetheless, depending upon the application, the volume of waste steam available, and an analysis of costs, critical jets designed for the lower motive pressure of waste steam can be not only a cost effective equipment acquisition but one which contributes greater savings as energy costs inevitably rise.

A jet ejector is composed of three basic components – body, nozzle, and diffuser. The body is essentially a hollow vessel, whose prime purpose is to support the nozzle, the suction port, and the diffuser in their proper physical positions. The conventional motive gas nozzle is of the converging, diverging type and is used to measure and expand the motive gas to a predetermined pressure, volume and velocity. The inlet of the converging section must be well rounded to obtain a high and dependable orifice coefficient. The orifice is sized to give the correct quantity of steam usually calculated at critical flow. The diverging section allows the motive gas to expand and aims it centrally through the diffuser inlet into the diffuser throat. The diffuser is a converging, diverging venturi with a constant diameter throat section between the two conical sections. Figure 5.2 shows the relationship of these basic components and their corresponding pressure, velocity, and work parameters.

For further reference to standard nomenclature, configurations, fundamentals, etc. refer to HEI (1) (Heat Exchange Institute) "Standard for Steam Jet Ejectors" and ASME (2) (American Society of Mechanical Engineers) standards. Several excellent technical papers to supplement certain sections of this text are noted in References 6 to 13.

VACUUM

Vacuum is simply described as a pressure less than the local atmospheric pressure. For selecting vacuum producing equipment, however, vacuum is quantified as pressure related to a perfect vacuum or absolute zero pressure. Since ambient or atmospheric pressure decreases measurably with increasing altitude, care must be taken when translating from vacuum to absolute pressures. Normally, barometric pressures are measured in millimeters of mercury, inches of mercury, or pounds per square inch absolute pressures. In the past few years, the term "Torr" has become popular. Torr is described as 1/760th of a standard atmosphere. Since there are 759.95 millimeters of mercury in one standard atmosphere, a Torr and one millimeter of mercury are for all intents and purposes identical. In the case of very slight

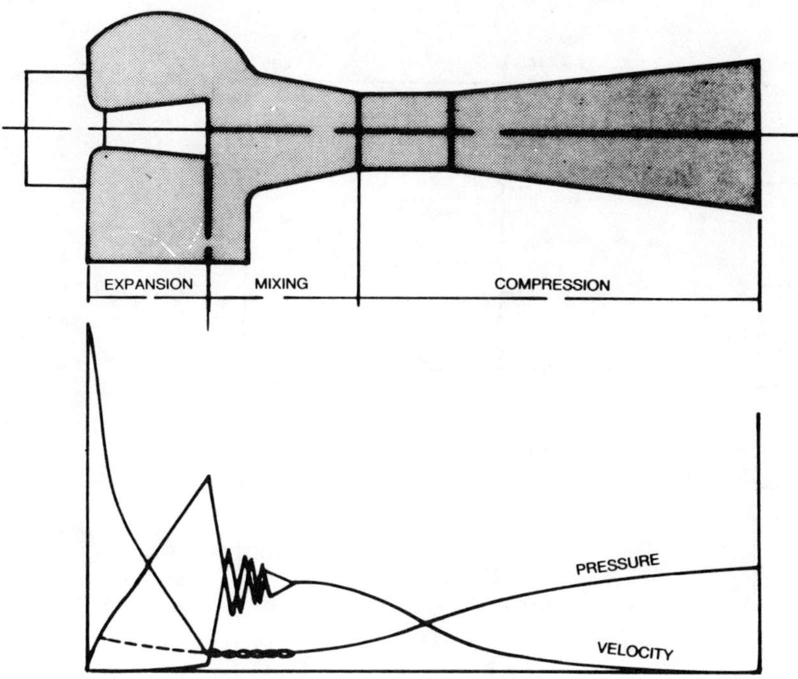

Figure 5.2. Plot of the jet ejector process.

vacuums, of about one inch of mercury or less, it is customary to state the vacuum in "inches of water vacuum or draft."

The three standard methods of measurement are, of course, equivalent with proper transposition constants. Figure 5.3 shows that a standard atmosphere can be represented by 760 Torr (or mm of mercury), 29.92 inches of mercury (both at 0°C) or 14.696 pounds per square inch absolute. Equivalent pressures are represented by horizontal intersects on Figure 5.3.

An example of the importance of utilizing either the absolute pressure or referencing the local barometric pressure to absolute units when designing a vacuum system can be illustrated as follows:

> Two companies, one in Philadelphia and one in Denver, want equipment to produce 635 mm of mercury vacuum. The mm mercury barometric pressure in Philadelphia is about 762 mm mercury, so that company requires equipment which will produce a 137 mm mercury absolute pressure. However, in Denver where local barometric pressures are 635 mm of mercury, the company has unknowingly requested zero suction pressure which is technically unobtainable.

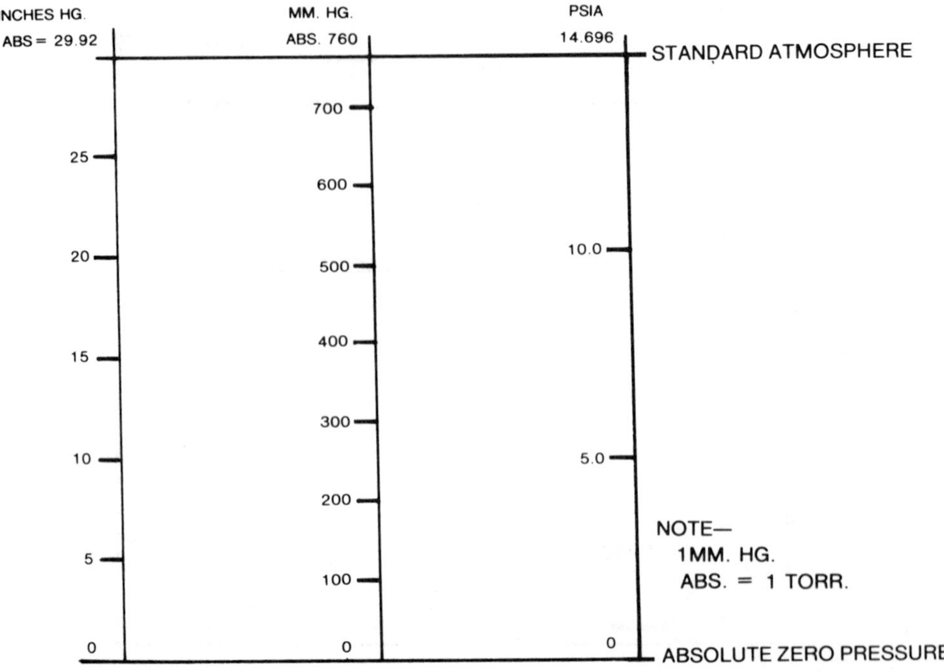

Figure 5.3 Vacuum relationships.

SUCTION LOADS

In addition to those parallel pressure measurement systems already described, flow rates are often specified in varying ways. The relation between these several approaches is summarized as follows:

Capacity loads are specified in either weight or volumetric rates (i.e. weight or volume per unit of time). For this text weight rates in kilogram per hour (kg/hr) will be followed. Other commonly used weight rates are pounds per hour (PPH), pounds per second (lbs/sec), tons per day, and so on. Simple conversion constants are available for these various rates from many published sources.

Volumetric rates for incompressible fluids are very similar to weight flow rates in that conversions from volumetric to weight flow rates are simple and direct. However, conversion for compressible fluids may often depend on actual operating pressure and temperatures. Very often the chemical and allied industries refer to SCFM (standard cubic feet per minute). Gases measured in SCFM means gas at standard conditions and pressures (70°F and 14.696 PSIA). If the rate is listed as ACFM (actual cubic feet per minute) the load being considered is at the actual flowing pressure and temperature. Note that a volume flow rate such as CFM (cubic feet per minute) is definitely incomplete: it must be further amplified by stating "actual" or "standard" conditions; a similar amplification must be made for temperature if other than the 21°C which is standard in the jet vacuum industry.

The conversion from either ACFM or SCFM, cubic meters per second, barrels per day, gallons per minute, liters per hour, and so on to kilograms per hour can also be obtained from published conversion charts. Though gas mixtures require a slightly more complex calculation to determine volumetric to weight rate conversion these too can be obtained from many published conversion charts.

The basis for selecting proper jet vacuum equipment is determined by the required suction pressure in Torr, and suction loads in weight flow rates of kilograms per hour of *dry air equivalent* (DAE) at 21°C. The method of converting weight flow rate in kg/hr to DAE at 21°C involves corrections for temperature if other than 21°C, molecular weight if other than air, and water vapor inclusion if suction load is saturated. Figures 5.4, 5.5 and 5.6 show "Temperature Entrainment Ratio Curve," "Molecular Weight Entraniment Ratio Curve," and "Air and Water Vapor Mixture Data (Dalton's Law)." These curves are extracted from HEI Standards for Steam Jet Ejectors, Third Edition.

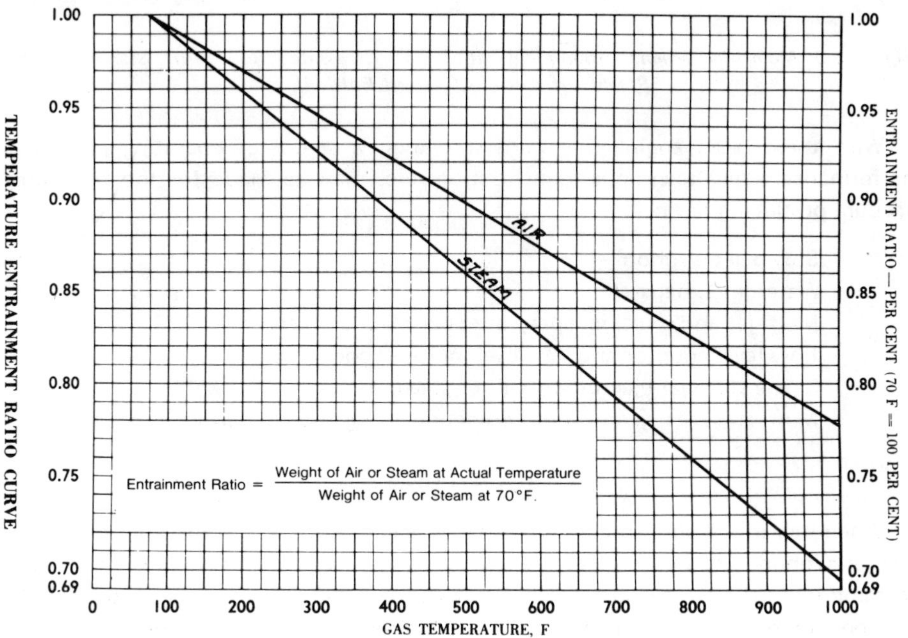

Figure 5.4. Temperature entrainment ratio curve (Reprinted from HEI "Standards for Steam Jet Ejectors," Third Edition)

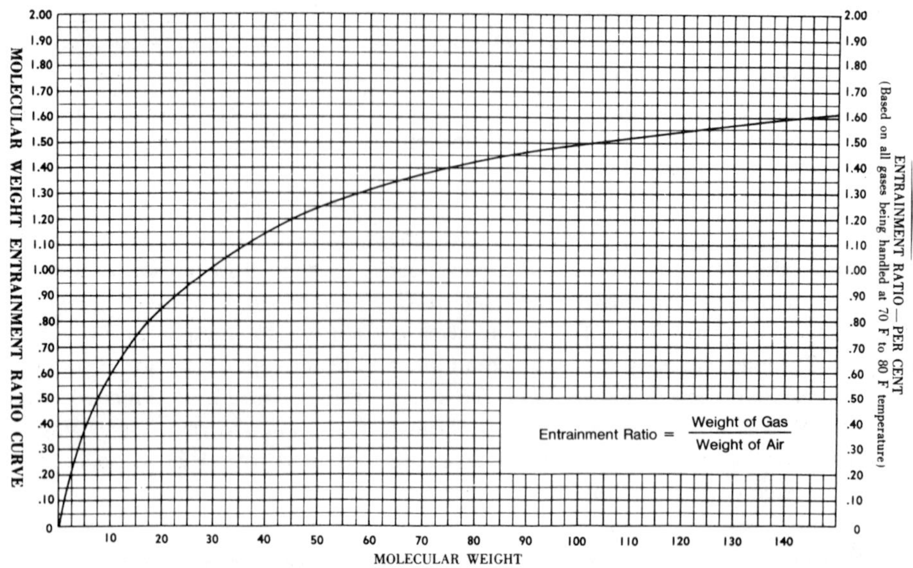

Figure 5.5. Molecular weight entrainment ratio curve (Reprinted from HEI "Standards for Steam Jet Ejectors," Third Edition).

With these basic engineering understandings, a review of operating principle, performance sizing technique, construction, etc. may be made for the following vacuum producing devices:

Steam jet syphon
Water jet exhauster
Multi-jet and multi-jet spray direct contact condenser
Low level eductor and low level multi-jet condenser
Single stage jet-ejector with air or steam motive
Multi-stage jet ejector, both non-condensing and condensing

JET SYPHON

When low initial cost and low available motive pressure are of major importance, the jet syphon offers considerable advantages. This type unit is commonly utilized in sampling, priming, evacuating, and exhausting applications and operates well if the possibility of flooding exists. It is, however, uneconomical for continuous operation because of high motive utility requirements.

The jet syphon (Fig. 5.7) is simple in construction with only two components, body and nozzle. It can also be supplied in any castable material and is normally stocked in cast iron, bronze, and stainless steel.

Performance ranges, suction capacities, and evacuation curves are shown in

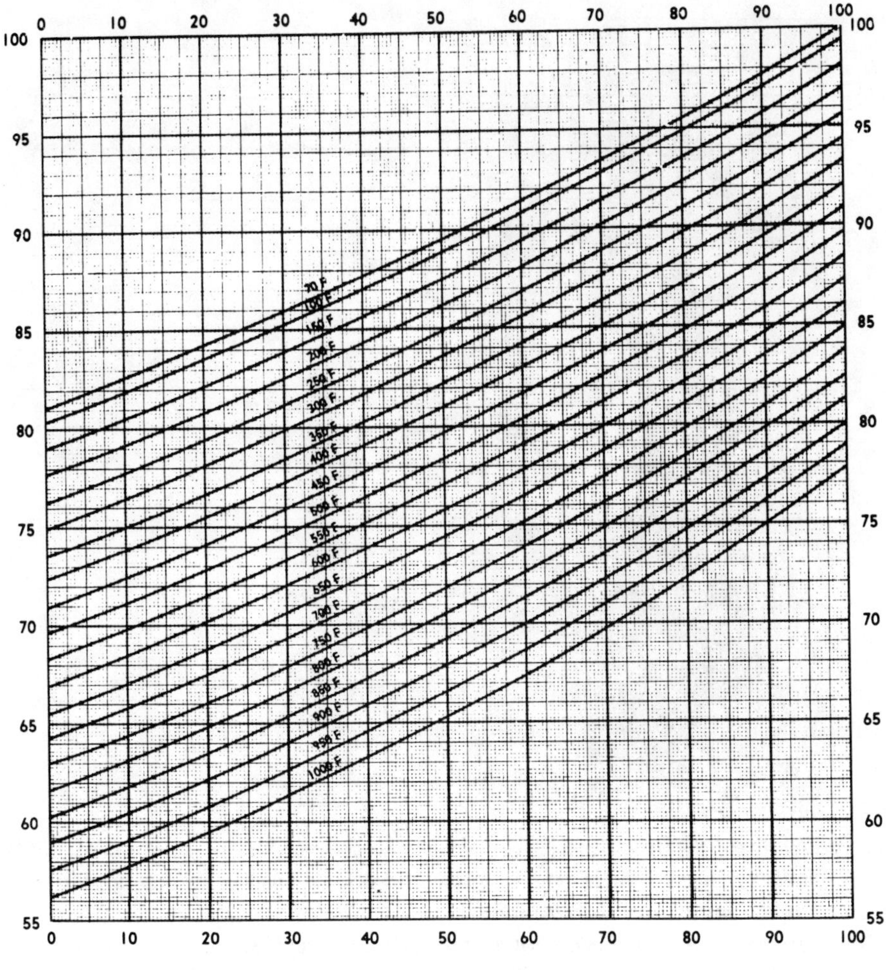

Figure 5.6. Entrainment ratio curves for air-stream mixtures.

Figures 5.8, 5.9 and 5.10 for two common nozzle sizes. These curves are based on steam motive. Suction capacities utilizing air as the motive stream can be approximated by using one half of the listed capacities for steam motive.

WATER JET EXHAUSTER

The water jet exhauster operates on the jet principle utilizing water as the motive force. Because of its condensing capabilities it is ideal for handling mixtures of condensibles and small quantities of non-condensible gases. Jet exhausters are

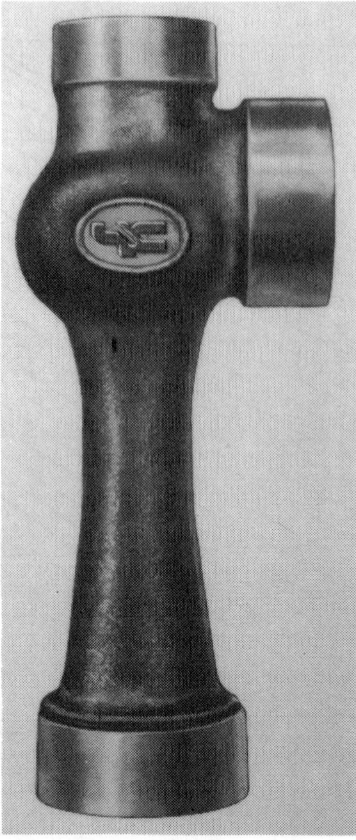

Figure 5.7. Typical jet syphon.

used widely throughout industry to evacuate closed vessels, to prime centrifugal pumps, to pump air and gases and to assist in producing vacua in such processes as filtration, percolation, and evaporation.

There are basically two types of water jet exhausters, single and multi-nozzle types. Single nozzle exhausters will shut-off within 12.7 torr of the water vapor pressure while multi-nozzle type units will shut off within 3.8 Torr of the water vapor pressure provided the water is absolutely free of air. All water jet exhausters require the discharge end to be sealed to prevent air back flow.

Capacities range from 0.5 kg/hr to 50 kg/hr DAE in small laboratory and single nozzle units to 50 kg/hr to approximately 500 kg/hr in the larger multi-nozzle type units. Figure 5.11 illustrates a multi-nozzle water jet exhauster. Figures 5.12 and 5.13 show sizing chart and required water consumption for the unit depicted in Figure 5.11. Standard materials of construction are normally cast iron, bronze, steel, and stainless steel although they can be supplied in Monel, Everdur, Haveg, lead, fiberglass, and glass.

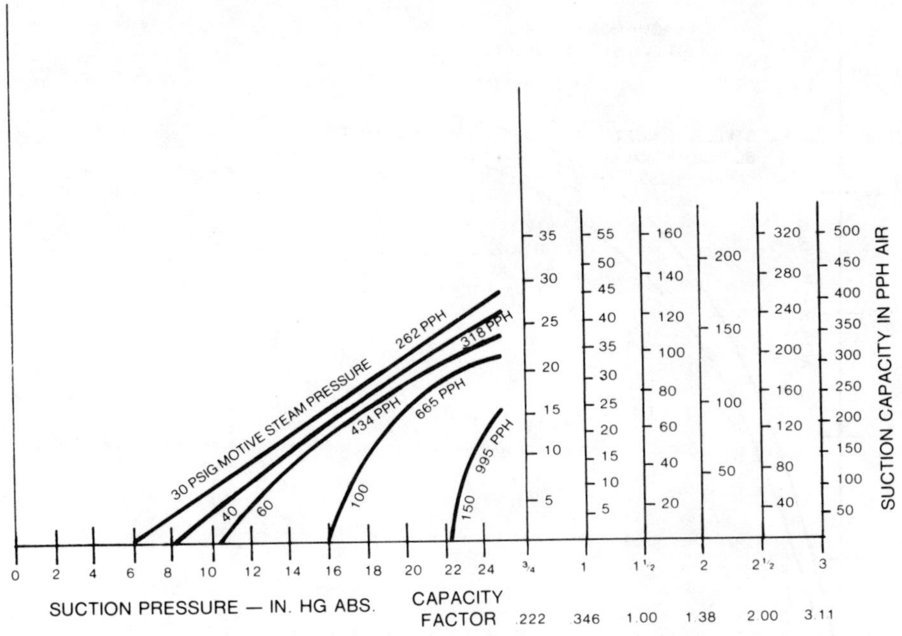

Figure 5.8. Air handling curve for jet syphon with style 60 nozzle.

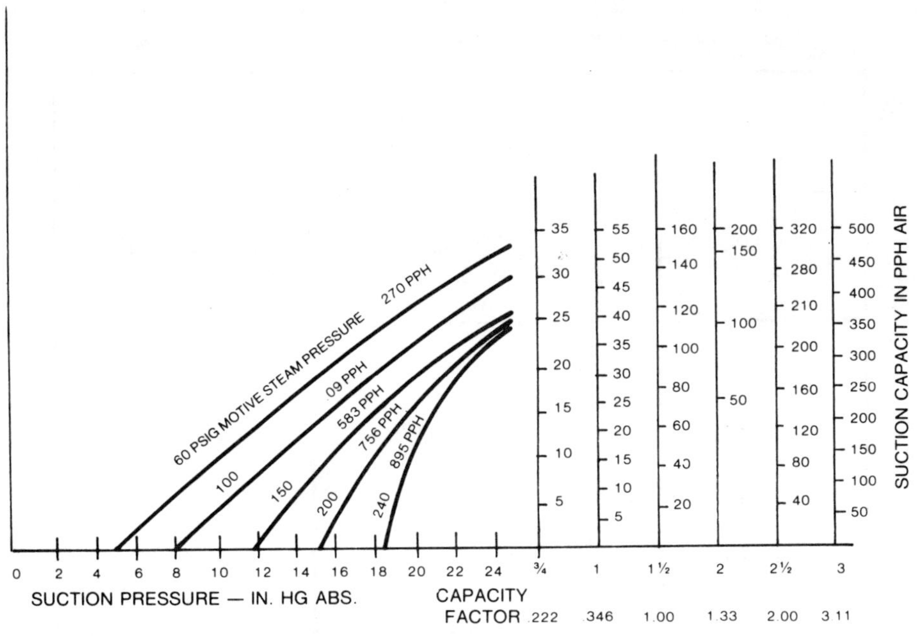

Figure 5.9. Air handling curve for jet syphon with style 115 nozzle.

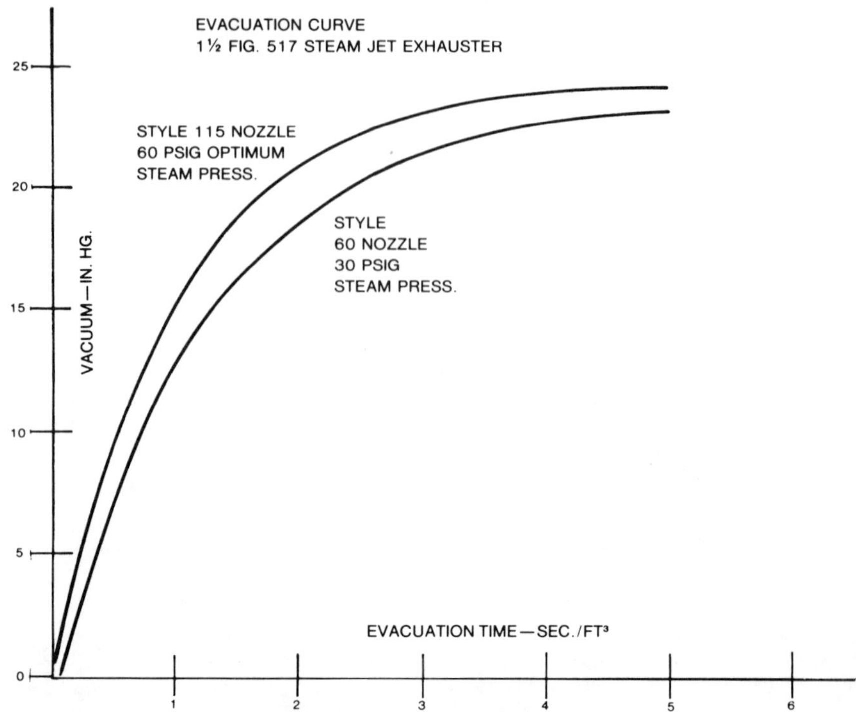

Figure 5.10. Evacuation curve for jet syphon with style 60 and 115 nozzles.

MOTIVE STEAM CONSUMPTION			
CAP. FACTOR	SIZE	60 NOZZLE @ 30 PSIG.	115 NOZZLE @ 60 PSIG.
.222	3/4"	58	61
.346	1"	91	93
1.00	1½"	262	270
1.38	2"	362	375
2.00	2½"	523	537
3.11	3"	815	840

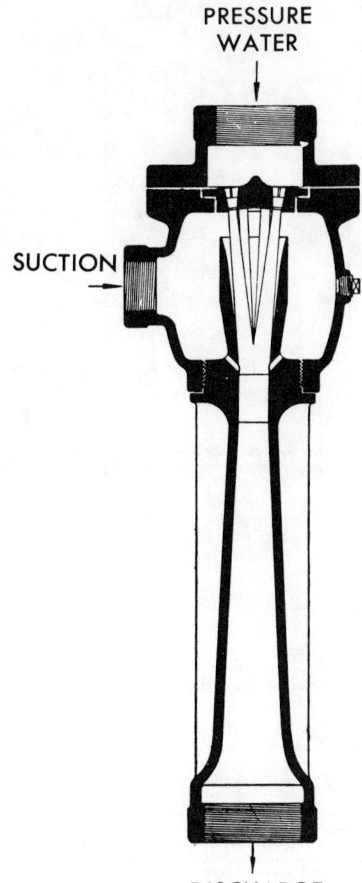

Figure 5.11. Typical multi-nozzle water jet exhauster.

CONVERSION FROM FT. WATER TO IN. HG. FOR PRIMING			
5'	10'	15'	20'
4.4"	8.8"	13.2"	17.6"

DIRECT CONTACT CONDENSERS

Co-Current or Parallel Flow Type Units

The direct contact condenser is employed in a variety of industries as an economical means of removing air, exhaust steam, and other vapors from vacuum

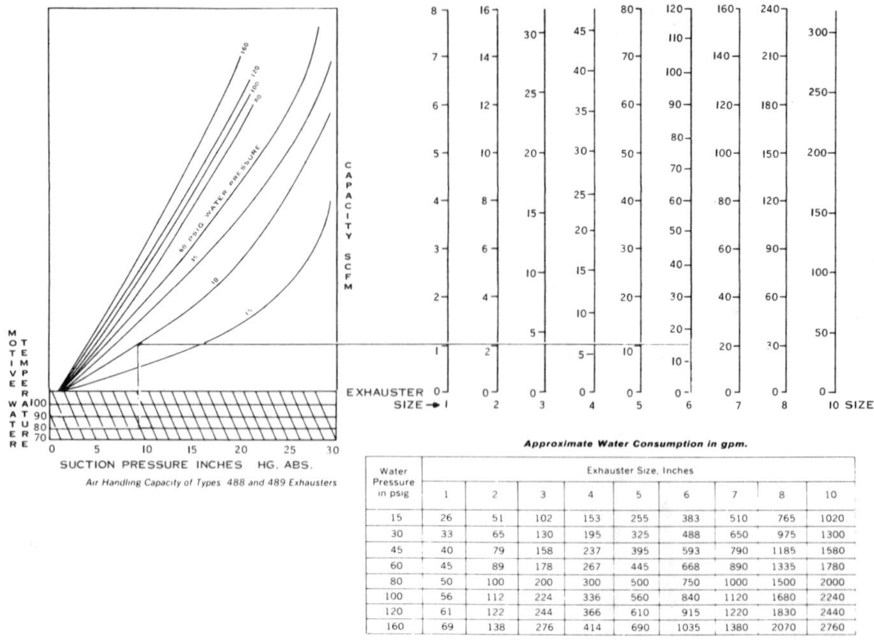

Approximate Water Consumption in gpm.

Water Pressure in psig	Exhauster Size, Inches								
	1	2	3	4	5	6	7	8	10
15	26	51	102	153	255	383	510	765	1020
30	33	65	130	195	325	488	650	975	1300
45	40	79	158	237	395	593	790	1185	1580
60	45	89	178	267	445	668	890	1335	1780
80	50	100	200	300	500	750	1000	1500	2000
100	56	112	224	336	560	840	1120	1680	2240
120	61	122	244	366	610	915	1220	1830	2440
160	69	138	276	414	690	1035	1380	2070	2760

Figure 5.12. Size determination and water consumption table for water jet exhauster.

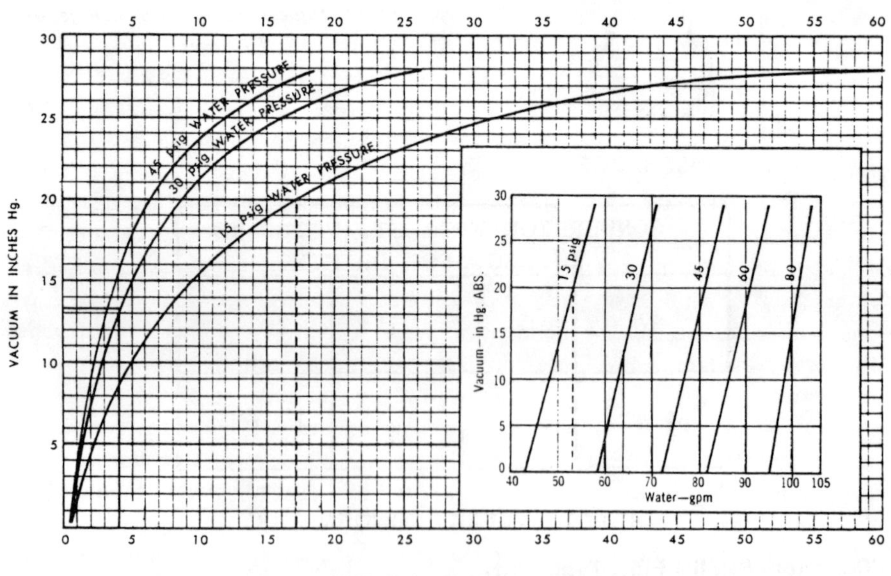

Figure 5.13. Evacuation/priming time for typical water jet exhauster.

equipment. It is found in almost every area of the chemical and process industries which use vacuum stills, calandria pans, multiple-effect evaporators, and vacuum crystalizers. In addition they have been used for decades in the food industry (edible oil, milk, and sugar) as well as the distilling, pulp and paper, and refinery industries.

A principle feature of the direct contact condenser is that injection water may be discharged through a tail pipe by gravity, without requiring a pump. Another advantage is its capability of withstanding flooding in the event of priming or liquid entrainment. In most plants, the vapor exhaust connection of the process vessel under vacuum is located at considerable elevation above ground level. The use of a direct contact condenser at these higher elevations permits shorter exhaust vapor lines which reduces leakage hazards and vapor line pressure drop, thereby reducing initial cost.

There are two basic types of direct contact condensers:

(1) *Co-current,* or *parallel flow,* in which the vapor to be condensed enters at the top of the unit and flows in the same direction as the condensing water, and

(2) *Counter-current,* or *counter flow,* in which the vapor enters near the bottom of the equipment and passes upward against the water flow, with the non-condensables discharging near the top. Injection water is delivered to the condenser in the form of a solid head, jets, sprays, water curtains, or a combination thereof. This section will consider the co-current or parallel flow type units. Counter-current type direct contact condensers will be discussed in the multi-stage jet ejector, condensing system section (82.).

Direct contact condensers of the co-current type can be further categorized as multi-jet, multi-jet spray, or multi-spray type units. These are depicted in Figures 5.14, 5.15 and 5.16 respectively. The denotation of "jet" in the classification indicates that the condenser is capable of handling some non-condensibles, while performing its primary condensing function. The denotation of "spray" in direct contact co-current condensers indicates little non-condensible handling capability.

The performance of any condenser is described by a simple heat balance. The heat added to the system is the quantity of steam being condensed multiplied by the latent heat of vaporization at that pressure and temperature. This must be equal to the heat removed by the condensing water which is the quantity of water multiplied by the temperature rise from inlet to outlet times the specific heat.

It is obvious that the larger the allowable temperature rise of the condensing water, the smaller the amount of water required and the higher the condensate discharge temperature. Under theoretically perfect conditions, a condenser could operate under a vacuum corresponding to its tail and discharge water temperature, but no higher.

Under normal operation this, of course, can never occur. Air entering with the injection water and non-condensibles entering with the vapor load exert a partial pressure. The operating pressure of the condenser then, is the sum of the vapor

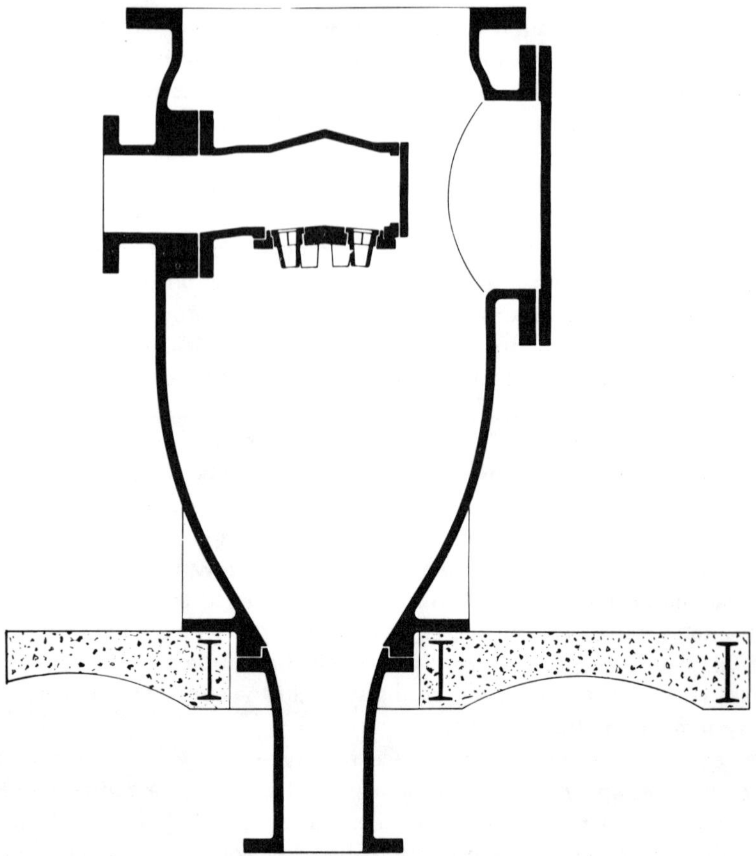

Figure 5.14. Typical multi-jet direct contact condenser.

pressure at the tail temperature plus the partial pressure of the non-condensibles present. The difference between the temperature of the tail water and the temperature corresponding to saturated water vapor at the actual condenser pressure is known as the "terminal difference." The efficiency of a condenser can be measured by the "terminal difference," although it also varies with the percentage of non-condensibles present.

Operation of the multi-jet type and multi-jet spray type direct contact condensers is very similar. Condensing water is delivered into the nozzle case which is designed to operate using a specified quantity of water at stated head pressure (minimum 0.7 kg/cm^2 differential), and a given vacuum in the condenser. The water jets are directed into the tail-piece at the lower end of the body, where they unite to form a single stream. Vapors entering the condenser come into direct contact with the converging water and are condensed. The multi-jet and multi-jet

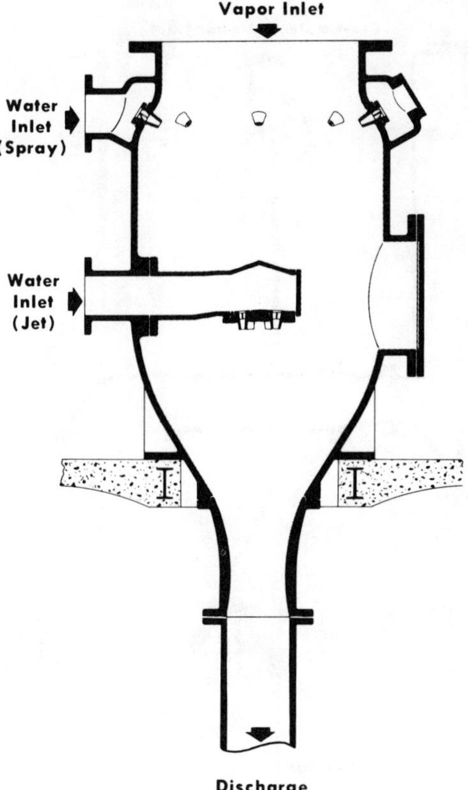

Figure 5.15. Typical multi-jet spray direct contact condenser.

spray type condensers are capable of flushing a certain amount of non-condensibles down the discharge pipe. The quantity is relatively small and varies with the operating pressure of the condenser. Conversely, excess non-condensibles above the quantity which the condenser is capable of handling will cause a rise in operating pressure to a point sufficient to flush the added non-condensibles through to discharge. This increase in operating pressure above the design point is undesirable for several reasons including the possibility of overloading the supporting jet stages or an objectionable pressure rise in the vessel being evacuated. The air handling capabilities of the multi-jet and multi-jet spray type condensers are shown in Figure 5.17. This graph denotes typical capacities over the air present in normal cooling water (which is illustrated in Figure 5.18).

Both type units are capable of terminal differences of 6.1 to 7.2°C. Their condensible vapor handling capabilities and maximum cooling water capacities are shown in Figure 5.19.

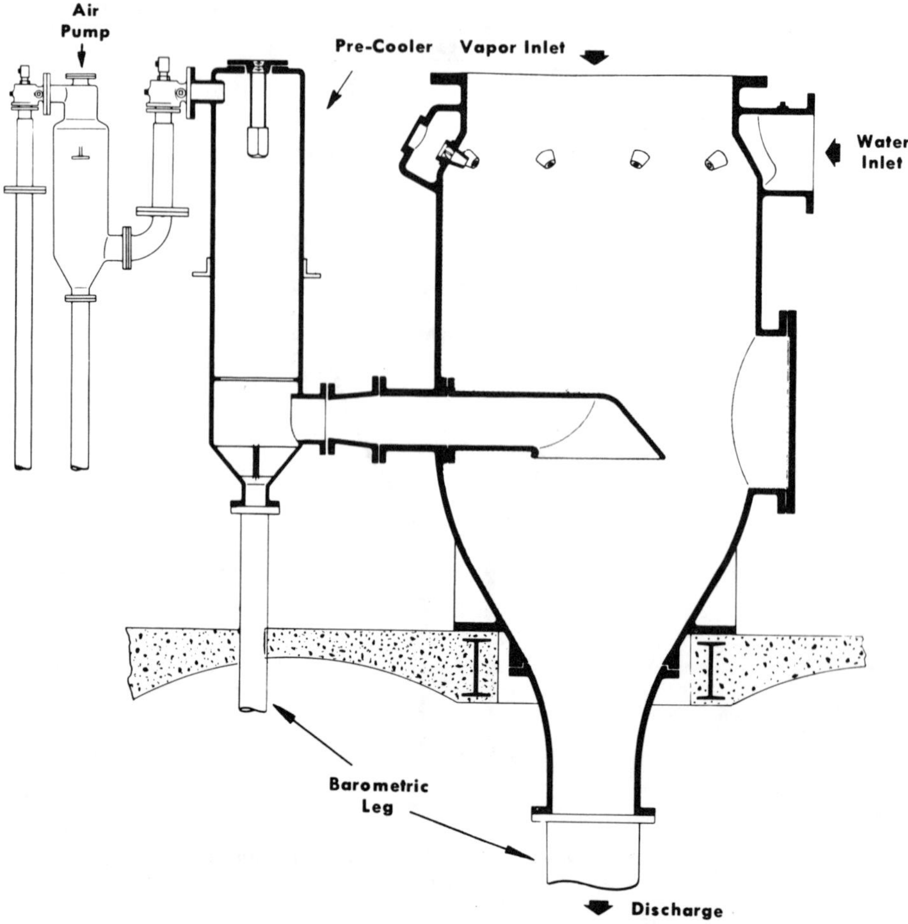

Figure 5.16. Typical multi-spray direct contact condenser with pre-cooler and air pump.

The multi-jet spray unit differs from the multi-jet design in its flexibility of operation. This is apparent from its design as seen in Figures 5.14 and 5.15. For full vapor load the rated water capacity is passed through both the spray and jet nozzles. If the vapor load or water temperature decreases, it is possible to throttle the water to the spray nozzles and ultimately turn them off completely. In the latter case, the condenser is operating in a manner similar to the multi-jet type, but with a minimum of injection water under the given conditions.

The multi-spray direct contact condenser was developed to handle applications involving limited water supply, high water temperatures in relation to vacuum requirement, or when removal of a large volume of non-condensibles is required. This type condenser requires an air pump to draw the non-condensibles off the

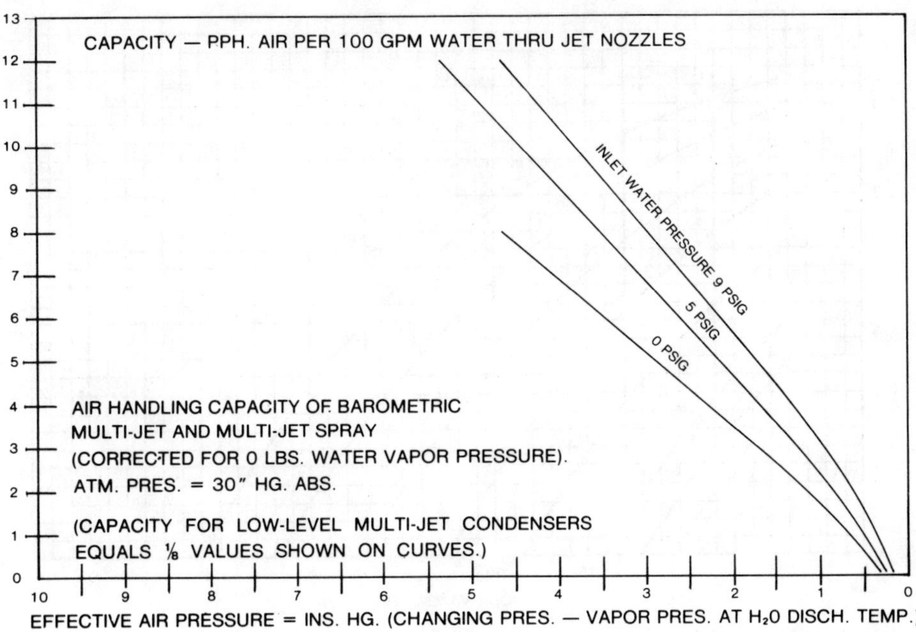

Figure 5.17. Non-condensible (air) handling capacity of multi-jet and multi-jet spray direct contact condensers. (Note: Capacities shown are in addition to air present in cooling water).

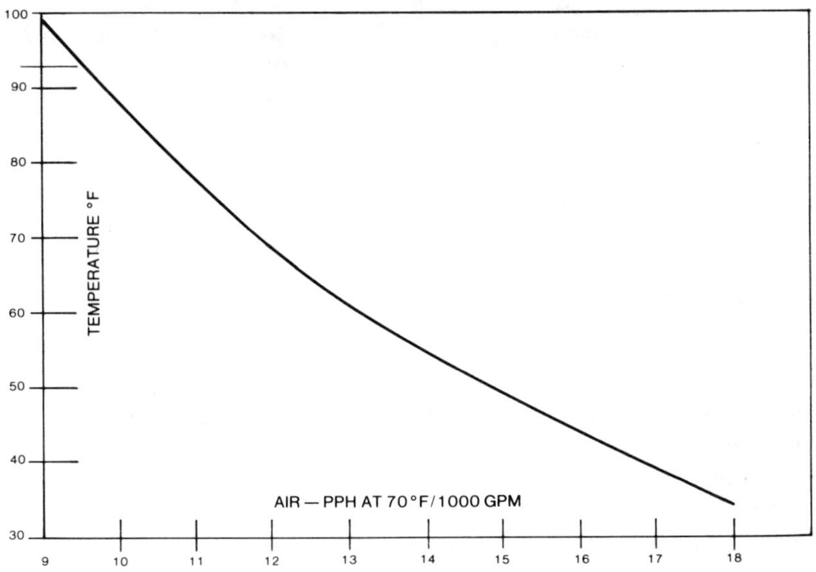

Figure 5.18. Allowance to be made for air present in condensing water.

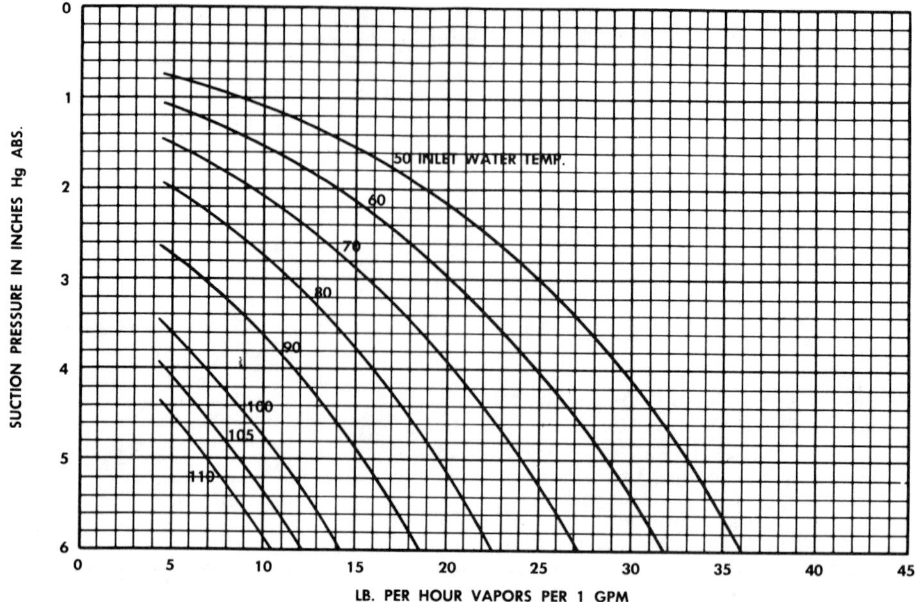

Figure 5.19. Performance curve for typical multi-jet or multi-jet spray direct contact condenser with maximum water capacity chart. (Note: Multi-jet spray unit normally manufactured in size 30 and higher).

Size No.	Maximum Water Capacity gpm.	Size No.	Maximum Water Capacity gpm.
2	50	33	1100
3	85	34	1300
4	130	35	1700
5	210	36	2200
26	375	37	3200
27	425	38	3800
28	475	39	5000
29	550	40	6000
30	625	41	8000
31	750	42	9000
32	950	43	12000

condenser through the pre-cooler. As can be seen in Figure 5.16 the multi-spray condenser has a circular spray nozzle arrangement at the vapor inlet. In this respect it is very similar to a multi-jet spray type unit, however, it does not have a jet spray section. Vapor enters the condenser at the top and mixes thoroughly with the injection water which is delivered through spray nozzles. The downward action of these converging sprays tends to create a suction in addition to their condensing action. The condensed vapors are taken to the hotwell through the barometric leg. The non-condensibles are drawn through an air suction chamber to a small direct-contact counter-current pre-cooler attached to the multi-spray condenser. The pre-cooler lowers the temperature of the air-vapor mixture and condenses as much of the remaining vapor as the water temperature and operating pressure allows. Since the pre-cooler is using its own fresh water and only has to handle a relatively small amount of condensibles, it can reduce the water vapor carry-over to the air pump to a point much lower than is possible with the larger condenser. The spray nozzles are designed for a minimum 0.7 kg/cm^2 throttling pressure with terminal differences of 1.7 to 2.8°C obtainable in a well designed "tight" system. It is also possible to operate a multi-spray direct contact condenser without an air pump for short periods of time at slightly reduced vacua. Performance and maximum water capacities for the multi-spray condenser are shown in Figure 5.20.

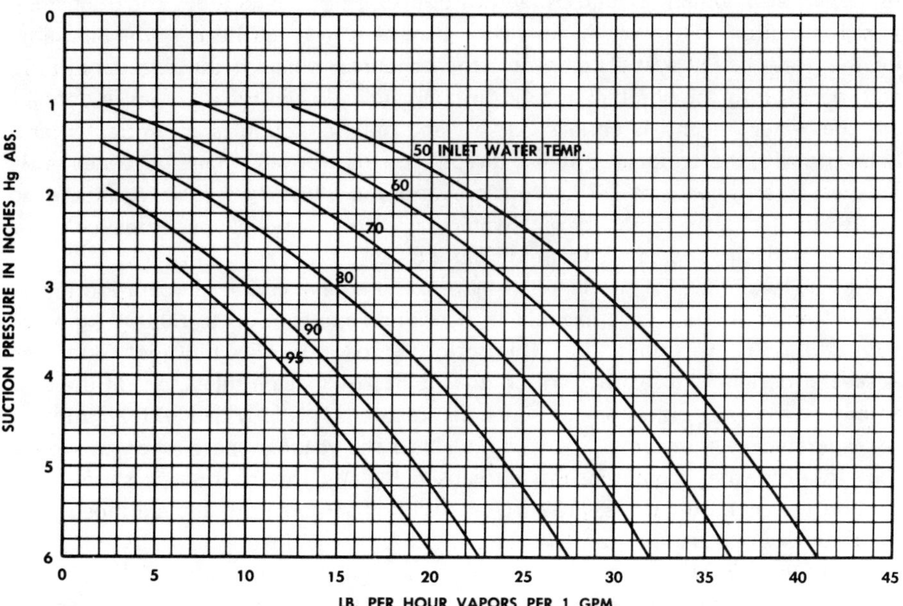

Figure 5.20. Performance curve for typical multi-spray direct contact condenser with maximum water capacity chart.

Size No.	Maximum Water Capacity gpm	Size No.	Maximum Water Capacity gpm
30	625	36	2200
31	750	37	3200
32	950	38	3800
33	1100	39	5000
34	1300	40	6000
35	1700	41	8000
		42	9000

LOW-LEVEL EDUCTOR AND MULTI-JET EDUCTOR TYPE CONDENSERS

The low-level eductor and multi-jet condensers are designed to produce medium to high vacua (735 Torr or higher) depending upon water temperature, for use with small engines or turbines. They employ the energy of water issuing through jet nozzles to create vacuum, condense steam, and handle small amounts of air (approximately 20% of the air handling capability denoted in Figure 5.17). No auxiliary vacuum pump is necessary and the water discharges directly into a hotwell with no requirement of a barometric leg.

The low-level eductor condenser (Figure 5.21) is capable of handling capacities up to 3410 kg/hr of steam at 658 Torr vacuum and 21°C injection temperature. The body of the condenser is a closed cylindrical chamber with a water nozzle in the upper end which extends downward into a combining tube, and discharges through a venturi tail piece. A short pipe (two feet in length) carries the discharge into the hotwell. Exhaust steam enters the condenser through the side inlet in the body, passes into the combining tube through holes in the tube, comes in contact with the water jet and is condensed. Air and non-condensibles are entrained and discharged with the condensed steam directly to the hotwell. Injection water must be delivered at a constant 0.7 kg/cm^2 pressure and a water check valve should be installed in the steam line to pevent water back flow. A typical performance curve and water capacity chart are shown in Figure 5.22.

The low-level multi-jet condenser (Figure 5.23) is very similar in operation to the low-level eductor unit. It is capable of handling capacities up to 69,000 kg/hr of steam at 658 Torr vacuum and 21°C injection temperature. The multi-jet condenser has several design features that are unique. It has two steam inlets, one at the top, and one at the side, either one of which can be used. It is provided with an internal float operated vacuum breaker to admit air to the vacuum space when there is a back flow of water from the hotwell caused by any interruption or shut-off of injection water. Larger units, (132.51/sec and higher), are normally equipped with two vacuum breakers. This unit's performance curve and water capacity chart are shown in Figure 5.24.

A typical arrangement of a low-level multi-jet condenser under a turbine is shown in Figure 5.25. In operation, the turbine is started with atmospheric exhaust. As soon as the injection water pump is primed and running at normal speed, the valve in the pump discharge line is opened, admitting water to the condenser. As the water jet begins to condense steam, the atmospheric relief valve closes automatically and vacuum builds in the condenser.

Injection water pressure should be 0.7 kg/cm^2 or over to create vacuum. After the vacuum is formed, the water nozzles are designed to pass the required flow at specified vacuum with 0.35 kg/cm^2 pressure. Vacuum in the condenser will not build to that required for efficient operation if the injection pump is not properly rated, if friction losses and lifting head have not been calculated properly, or if turbine stuffing boxes or exhaust steam connections are not reasonably tight. Ex-

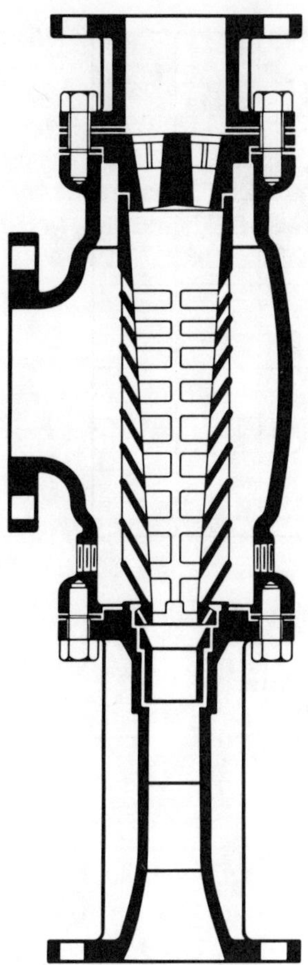

Figure 5.21. Typical low level eductor condenser.

cessive air leakage is indicated by turbulence in the hotwell. The water nozzles should also be protected from foreign matter by using a screen or strainer in the water line between the pump and the condenser.

SINGLE STAGE JET EJECTOR

The general principles of operation for the single stage jet-ejector were discussed in an earlier segment of this chapter. The jet-ejector, it may be recalled, is a thermodynamic device which utilizes the transfer of energy produced by a motive gas to create vacua and compress the suction load to an intermediate pressure. The enthalpy-entropy path of the motive gas (steam in this case) passing through a single stage unit is illustrated in Figure 5.26 by points 0.1-1′-2. The amount of work or energy available for compressing the suction fluid is the difference between the motive fluid-velocity energy $h_0 - h_1$ and the energy required to compress the motive

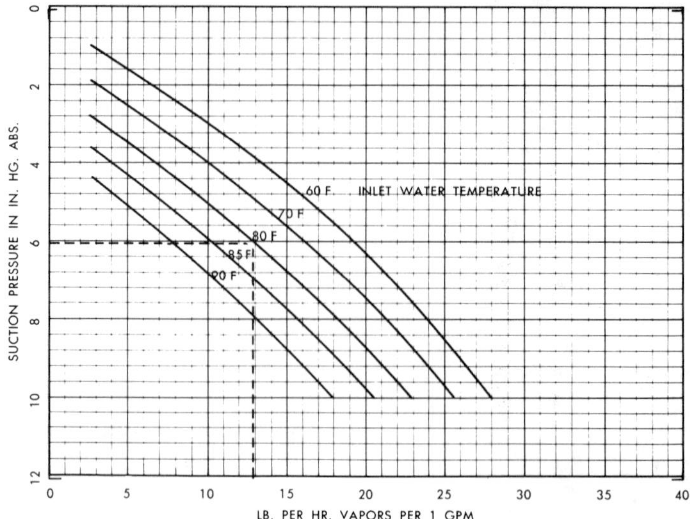

Figure 5.22. Performance curve and capacity chart for low level eductor condenser.

Size in Inches	Maximum Water Capacity in gpm
1½	15
2	26
2½	37
3	52
4	112
6	240
8	450
10	750

fluid back up to the desired discharge pressure, $h_2 - h_1'$. If isentropic expansion and compression were possible, then the ultimate vacuum that could be achieved in any one stage would be absolute zero. Because of inefficiencies in expansion and compression, however, a point is reached in the expansion process of a single stage unit where suction pressure will not go any lower. This occurs at the suction pressure where the actual energy required to compress the motive stream to the discharge pressure, $h_2' - h_1''$, in Figure 5.26, is equal to the actual velocity energy in the motive stream $h_0 - h_1'$. This is the shut-off pressure and zero suction load point of the jet-ejector. This shut-off may be in the range of 1/20 to 1/50 of the absolute discharge pressure.

All jet-ejectors have low thermal efficiencies: a range of 1 to 5 percent. The thermal efficiency expresses the actual work versus the total quantity of heat supplied. This value contains larger percentages of heat or energy that are supplied but are unavailable for use in producing work. The Rankine-cycle efficiency when

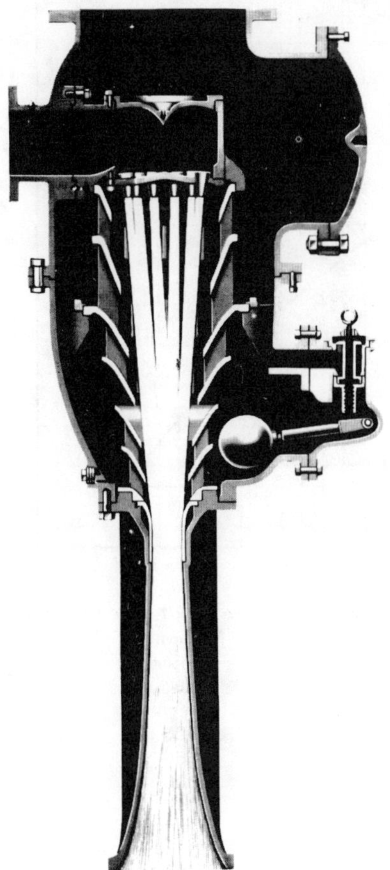

Figure 5.23. Typical low level multi-jet condenser.

used in connection with the thermal efficiency gives a better indication of the performance of the jet-ejector. The Rankine-cycle efficiency defines the maximum amount of energy that is available for use as work, under the ideal isentropic expansion and compression conditions. In the Rankine-cycle process, all of the heat is added at the higher constant supply pressure level and all of the heat remaining after performing work is rejected from the cycle at the lower constant discharge pressure level. This is expressed in the simplest manner by the following well known equation:

$$E_{Rankine} = \frac{Q \text{ supplied} - Q \text{ rejected}}{Q \text{ supplied}}$$

The Rankine-cycle efficiencies for the jet-ejector operating at various discharge pressures and motive pressures are shown in Figure 5.27. These values represent the theoretical maximum that can be obtained or are available for use as work.

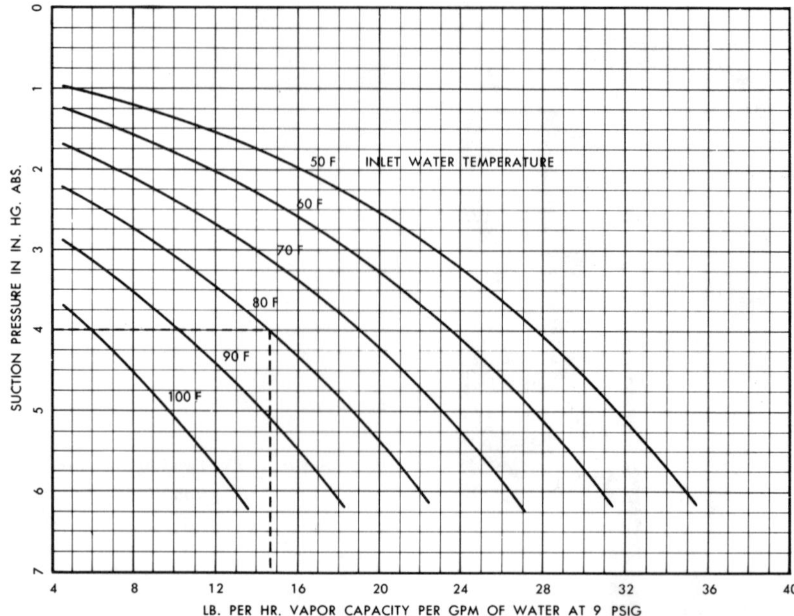

Figure 5.24. Performance curve and capacity chart for low level multi-jet condenser.

Size No.	Maximum Water Capacity In gpm	Size No.	Maximum Water Capacity In gpm
2	50	31	550
3	85	32	650
4	130	33	800
5	210	34	1000
24	170	35	1250
25	200	36	1600
26	250	37	2100
27	300	38	2800
28	350	39	4000
29	400	40	6000
30	450	41	8000

The ratio of the actual work obtained to the theoretical maximum as determined by the Rankine-cycle is a better indication of the operating efficiency of a jet-ejector than thermal efficiency. It can be expressed in several ways:

$$E_{jet} = \frac{E_{Thermal}}{E_{Rankine}} = \frac{\text{actual work}}{\text{Q supplied} - \text{Q rejected}}$$

The general range of attainable efficiencies of a single-stage steam jet-ejector handling air at various compression ratios and operating with steam rates of 182 kg/hr and above are shown in Figure 5.28. One point along a given ejector curve will have

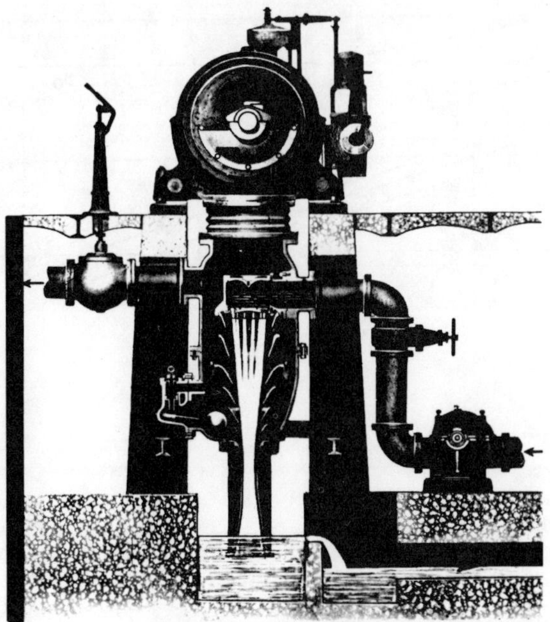

Figure 5.25. Typical arrangement of a low level multi-jet condenser under a turbine.

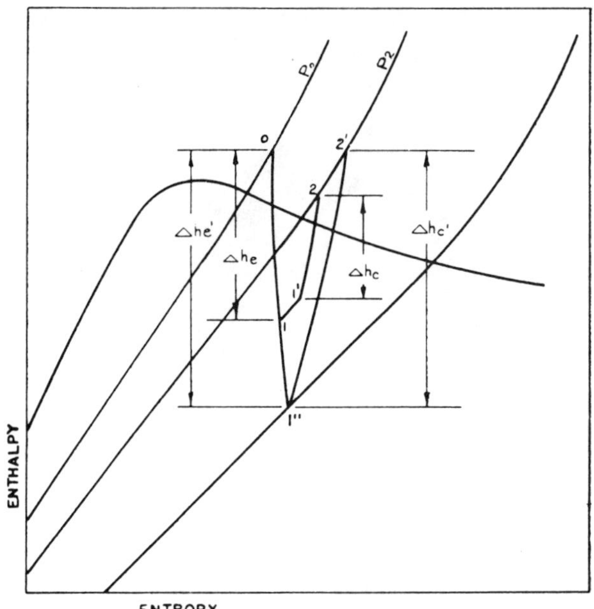

Figure 5.26. Typical thermodynamic cycle for steam jet-vacuum pump operating with a suction load (0-1-2), and at zero suction load (shutoff) (0-1''-2').

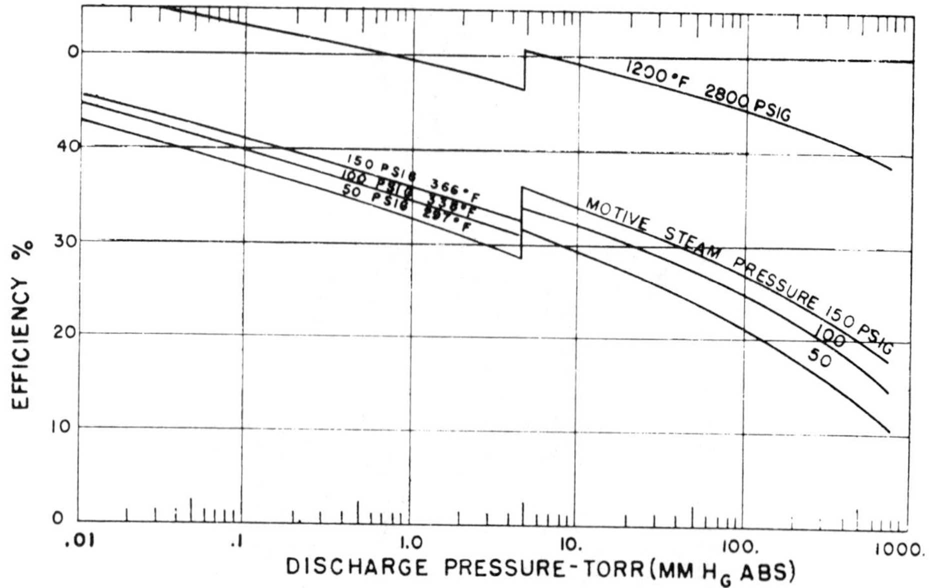

Figure 5.27. Rankine cycle efficiency, steam jet-vacuum pump, single stage.

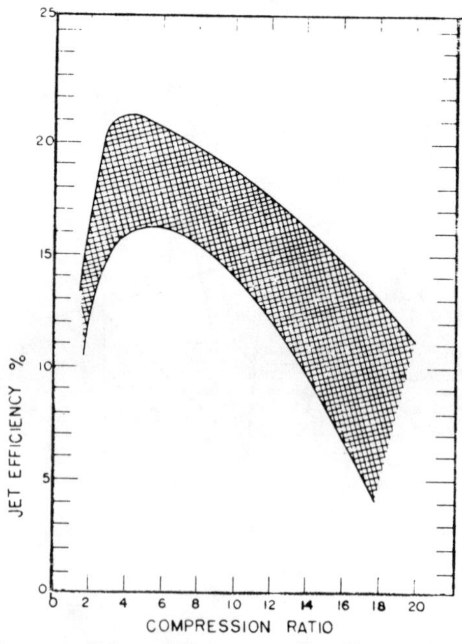

Figure 5.28. Range of attainable jet efficiencies for steam-jet pumping air.

a peak efficiency that reaches the upper curve; other useful pumping points along the same curve will have efficiencies that fall in the cross-sectional region between the upper and lower curves. Because of the mixing action of the motive and suction streams it is extremely difficult to determine the actual or true work of compression performed on the suction fluid. For the purposes of the graph, in Figure 5.28, the compression work performed on a suction fluid of air is assumed to occur at adiabatic conditions. Although the values shown are not thermodynamically precise, the information plotted provides two clear conclusions: (1) A major decrease in efficiency occurs at compression ratios above 10:1, and (2) maximum efficiency occurs in the range of 4:1 compression ratio.

The suction load capacity characteristics for any given size unit can be varied considerably by varying the position of the nozzle with respect to the diffuser and also by using different motive rates of flow. A typical capacity handling curve for a given nozzle position and a given motive rate (steam in this case) is illustrated by the corresponding curves labeled 5 in Figure 5.29. The upper curve in Figure 5.29 is the discharge pressure to which the jet-ejector will compress for a corresponding air capacity-suction pressure point on the lower curve. Corresponding families of Curves 1 to 6 show the various other pumping curves obtainable by putting the nozzle in different positions with respect to the diffuser. Curve 1 is achieved with nozzle location closest to the diffuser, Curve 6 with the nozzle most distant. In general, the curves reveal that as the nozzle position is moved farther away from the diffuser, the air handling capacity at any given suction pressure increases while the discharge pressure and compression ratio decrease. When the nozzle position is moved too far into the diffuser, decreases in capacity occur without any increase in discharge pressure or compression ratio as illustrated by Curves 1 and 2 in Figure 5.29. At the other extreme when the nozzle is too far from the diffuser, the jet becomes unstable at pressures near shut-off, as indicated by Curve 6 in Figure 5.29. For any one set of conditions there would be only one theoretically best nozzle position. The family of curves shown produces an envelope of attainable performance of all nozzle positions at a given steam rate.

By varying the steam rate, the individual performance curve can also be varied. At another steam rate, changing the nozzle position will produce another family of pumping performances similar to, but not the same as, that shown in Figure 5.29. The effect of changing steam rate upon the envelope of the family of nozzle-position curves is shown in Figure 5.30. In general, it has been observed that increasing the steam rate at any given nozzle position increases the compression ratio and also increases the shut-off pressure. As illustrated in Figure 5.30, depending upon the suction pressure, the air handling capacity may decrease or increase. Near shut-off the air-handling decreases as the steam rate increases, whereas at higher suction pressures, the air-handling capacity incrases as the steam rate increases.

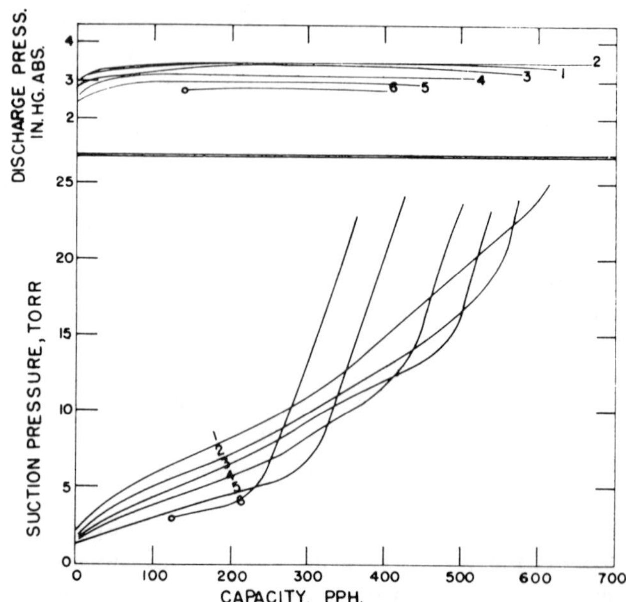

Figure 5.29. Typical effect of nozzle location on stage performance.

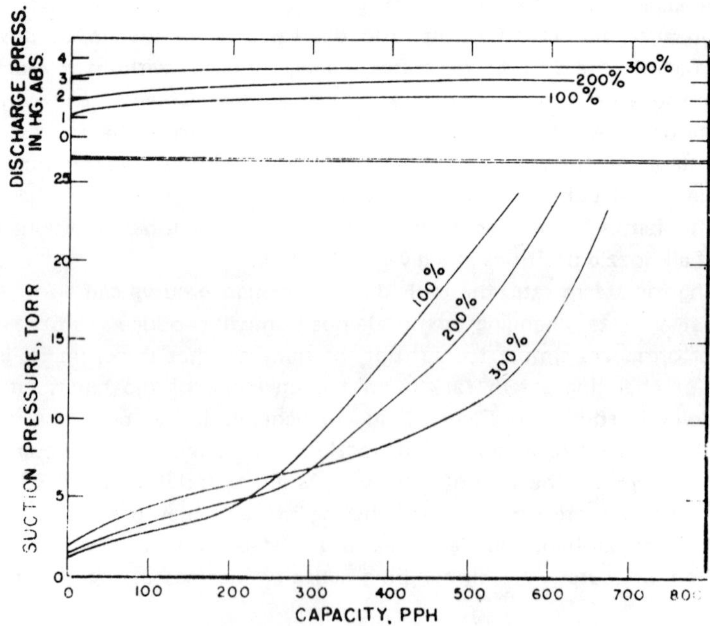

Figure 5.30. Typical effect of steam rates on stage performance.

The corresponding jet efficiency curves, Figure 5.31, for the various nozzle position curves shown in Figure 5.29, illustraté that maximum efficiency is achieved at only one point, near the knee in the curve. In addition, it also reveals that the jet efficiency is best at some nozzle position midway between the extremities of the operation. The information in Figures 5.29, 5.30 and 5.31 point out that although the air-capacity, suction, and discharge-pressure characteristics can be obtained by numerous combinations of nozzle position and steam rate, there will be only one combination that will produce the desired characteristic for the minimum amount of steam.

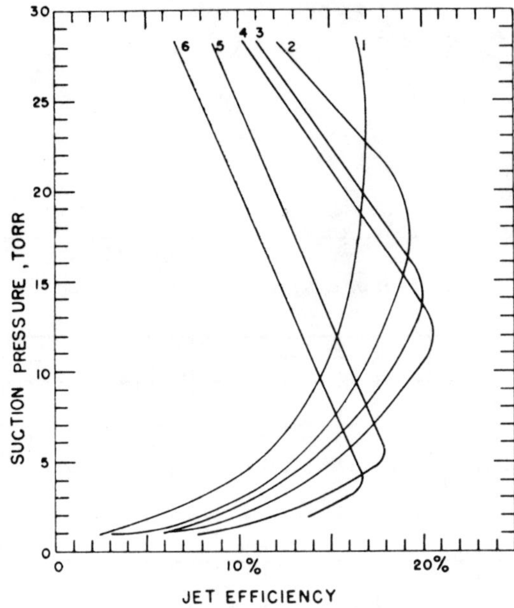

Figure 5.31. Typical effect of nozzle position upon jet efficiency of one stage.

From this unit selection it is now possible to establish a set of performance curves based on various motive pressures similar to that illustrated in Figure 5.27. Figure 5.32 shows jet ejector performance for varying motive pressures between 6.3 kg/cm^2 and 14.0 kg/cm^2. Suction load capacities increase with increasing motive pressure in a curve which is typical for all jet-ejectors. Selection of a particular unit to meet specific operating criteria is made between points 0 and 1 in Figure 5.32. Movement of this flat portion of curve to either the right or left is possible through variations in expansion in both the nozzle and inlet taper to the diffuser. This movement corresponds to increased suction capacity handling capabilities at a given suction pressure. Figure 5.33 illustrates a single stage jet-ejector with eight

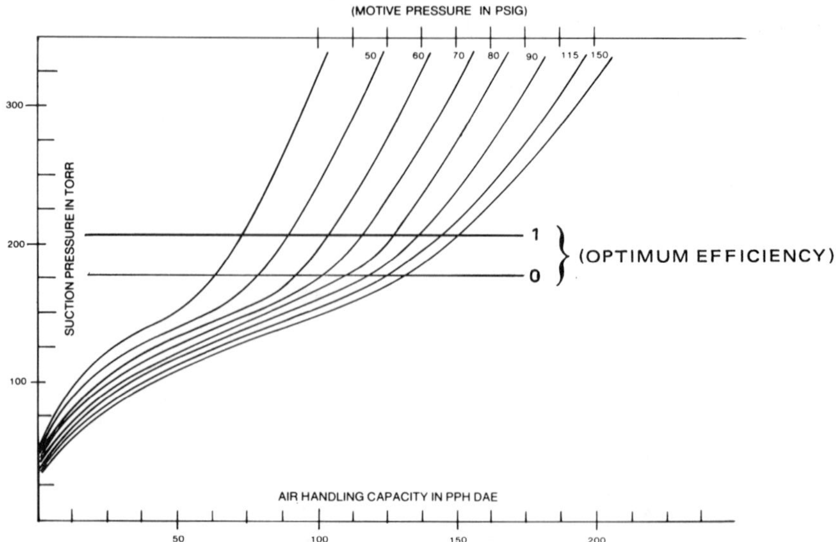

Figure 5.32. Typical effect of varying motive pressure upon a jet ejector.

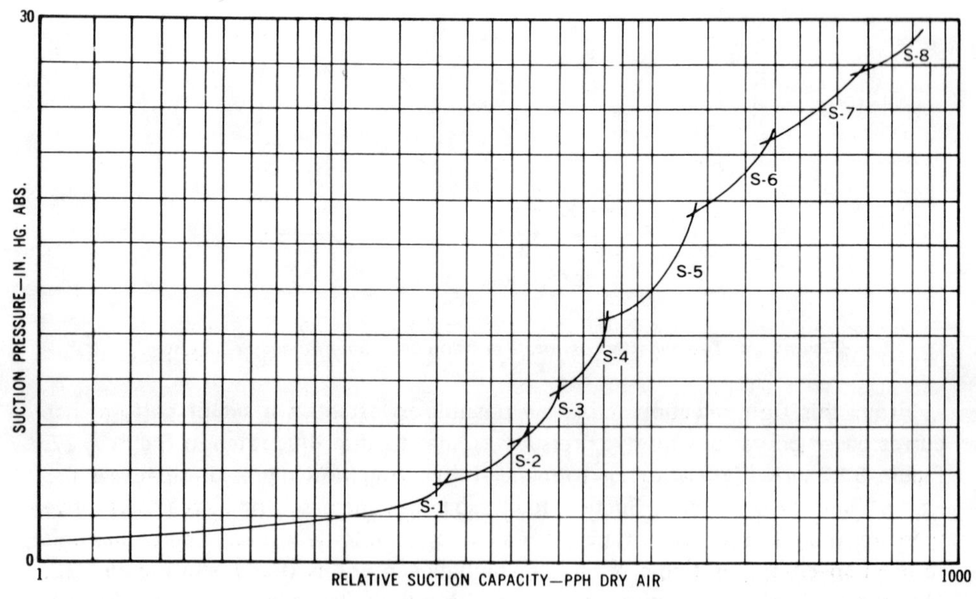

Figure 5.33. Suction pressure and capacity ranges for typical single stage ejector.

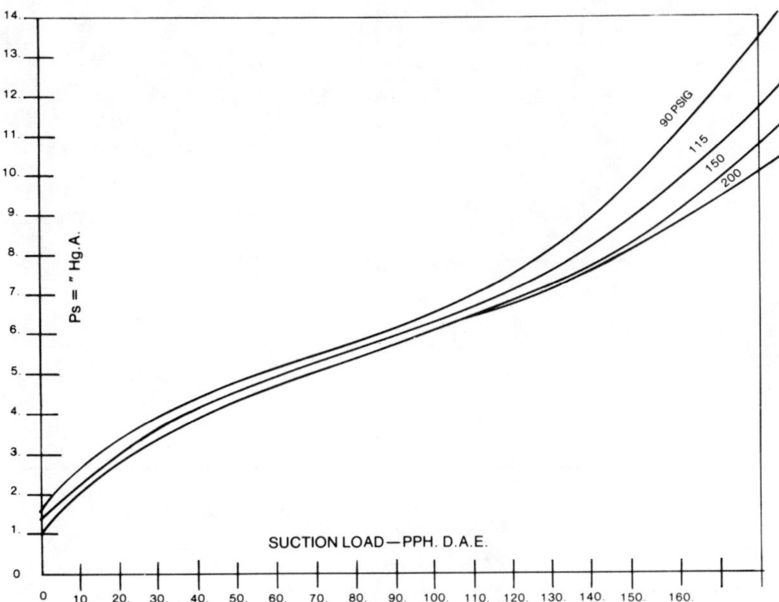

Figure 5.34. Typical performance for S-3 unit from Fig. 7.7 with varying steam motive pressures.

expansion changes in the nozzle and diffuser. As is shown, this allows selection over a wide range of both suction pressure and suction capacities. Each one of the eight units would then have an individual performance curve similar to that illustrated in Figure 5.32. Figure 5.34 is one such curve for a single stage jet-ejector with steam motive.

The jet-ejector with air motive has the same operating and performance characteristics as the steam operated jet-ejector. In some instances, particularly where heating and diluting are undesirable, or steam and/or water availability is minimized, the air motive jet-ejector offers definite advantages. It must be remembered, however, that air motive increases the non-condensible load under which the jet-ejector is operating. In addition, it may also increase size requirements on successive jet-ejectors and condensers. In general the air motive jet-ejector requires a higher rate of motive flow than a steam operated unit to handle the same load. Figure 5.35 shows the performance of an air operated jet-ejector using 3.16 kg/cm^2 motive air, as compared to the performance of a steam operated jet-ejector using 3.16 kg/cm^2 motive. Note that the required motive rate for air is 261 kg/hr versus 190 kg/hr for steam. The percentage motive rate difference between air and steam motives increases as motive pressure increases in single stage units and require a considerable increase (in the range of 200—300%) for multi-stage units.

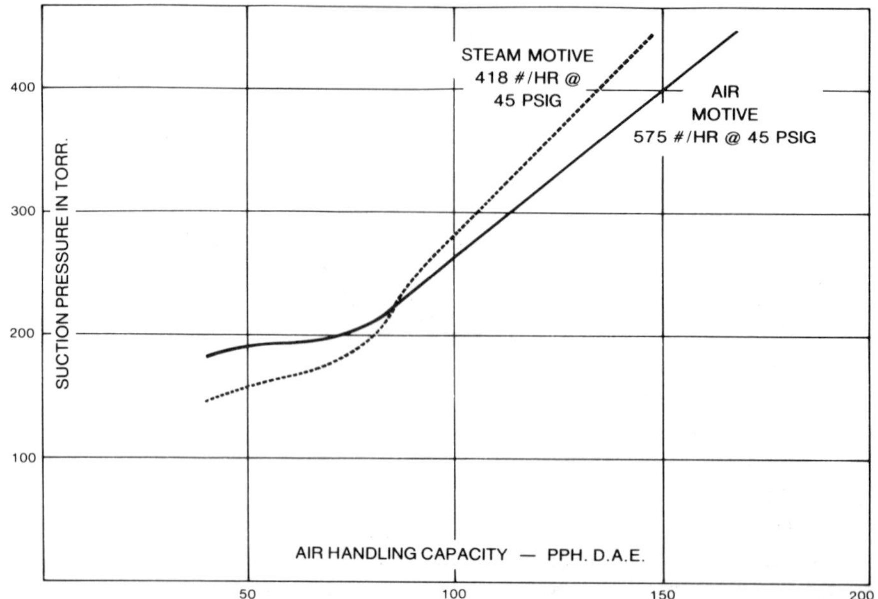

Figure 5.35. Performance of 1½" jet ejector with air versus steam motive.

MULTI-STAGE JET EJECTORS

Non-Condensing Type Multi-Stage Jet Ejectors

When combining stages the supporting units can be designed so that the system is completely stable over the entire primary stage performance curve as illustrated by lines B and D in Figure 5.37. The suction pressure-capacity characteristics, D, of the supporting stage are rarely identical to the discharge pressure characteristic B′ of the primary stage, B. As a result, at the design point a supporting stage for a completely supported and stable system will usually provide considerably more capacity than is necessary, as indicated by area between lines B′ and D in Figure 5.37. When it is desired to obtain minimum steam consumption and lower equipment cost, the steam consumption and size of the supporting stages can be reduced by supporting the primary stage in the region of the design point only, as illustrated by lines A and C. This produces a system which is stable at the design point, and has a steady and reproducible suction pressure at conditions other than the design point. Further reductions in system steam consumption are sometimes possible by using stages which have unstable shut-off points are previously illustrated by Curve 6 in Figure 5.29. Such systems will be stable at the design point, but unstable and unsteady at the shut-off point.

Assuming no rigorous limitations on design variables — steam consumption, unit size, number of stages required, equipment cost — systems can be designed to meet

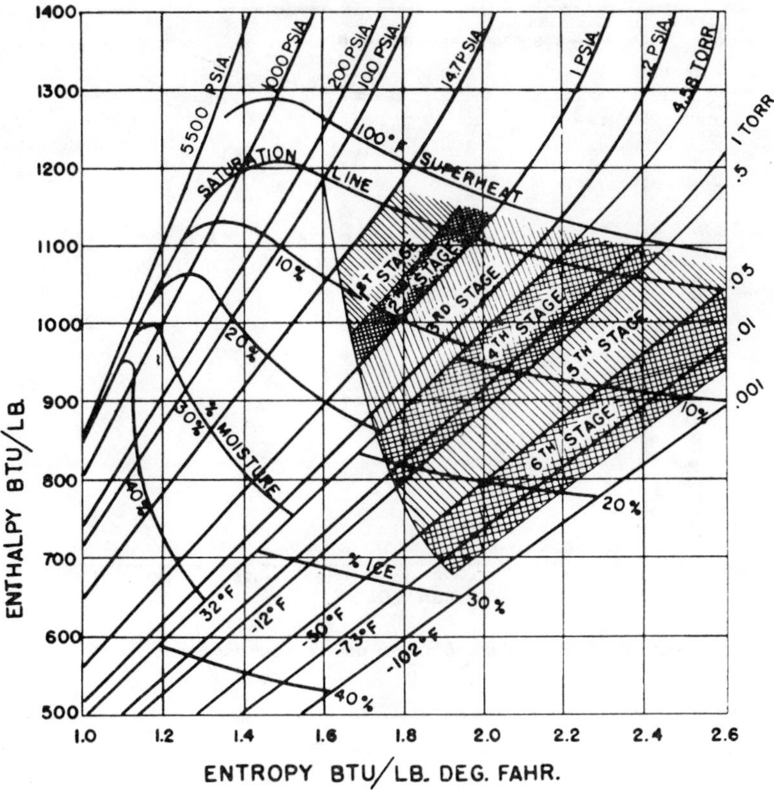

Figure 5.36. Mollier diagram to 10^{-3} torr showing operating regions of steam jet-vacuum pump stages.

special requirements. It could be advantageous, for example, to design a system with (a) two design load points at different suction pressures; or (b) vacuum capacity at 150 percent or greater than the design point; or (c) the capability of evacuating vessels to a given pressure within a specified time limitation. All of the aforementioned systems can be optimum designed systems with the individual jet ejectors sized and arranged to operate at or near the maximum efficiency points.

The general trend in this country is to use compression ratios up to approximately 10 to 1 for the primary and supporting non-condensing stages. As indicated in Figure 5.28 higher stage efficiencies and consequently lower system steam consumption can be obtained by using stage compression ratios in the range of 4:1 to 6:1. In sections of the world where water is scarce, users often find it more economical to use the lower compression ratios. This requires using one or two more stages than is general practice in this country.

Figures 5.38 through 5.40 illustrate typical vacuum producing capacities and

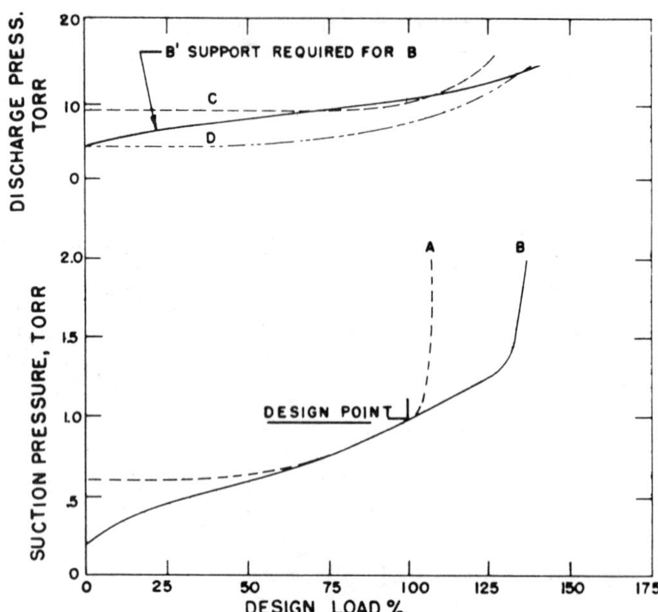

Figure 5.37. Typical combination of two noncondensing stages.

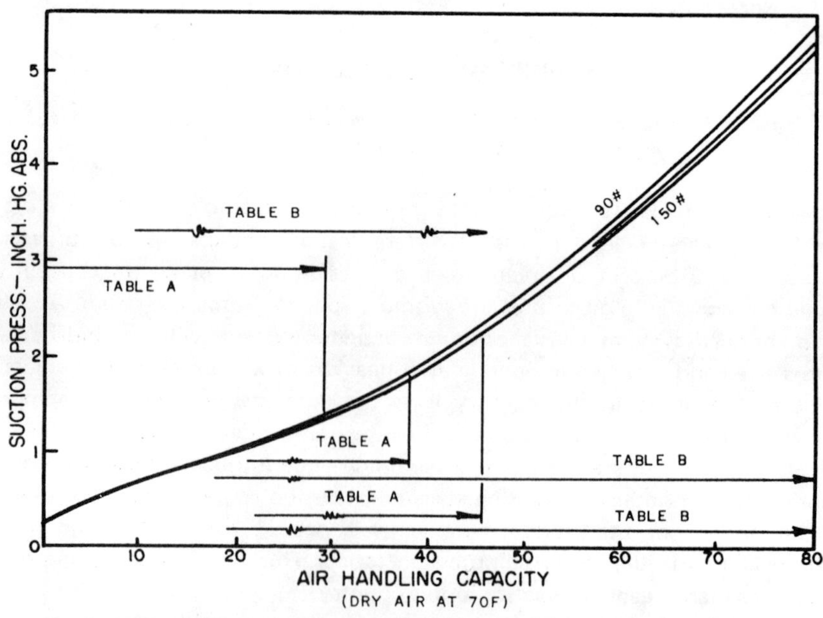

Figure 5.38. Performance curve of two stage non-condensing multi-stage jet ejector system with type T2 high vacuum unit and S-3 low vacuum unit with 90, 115, and 150 PSIG motives.

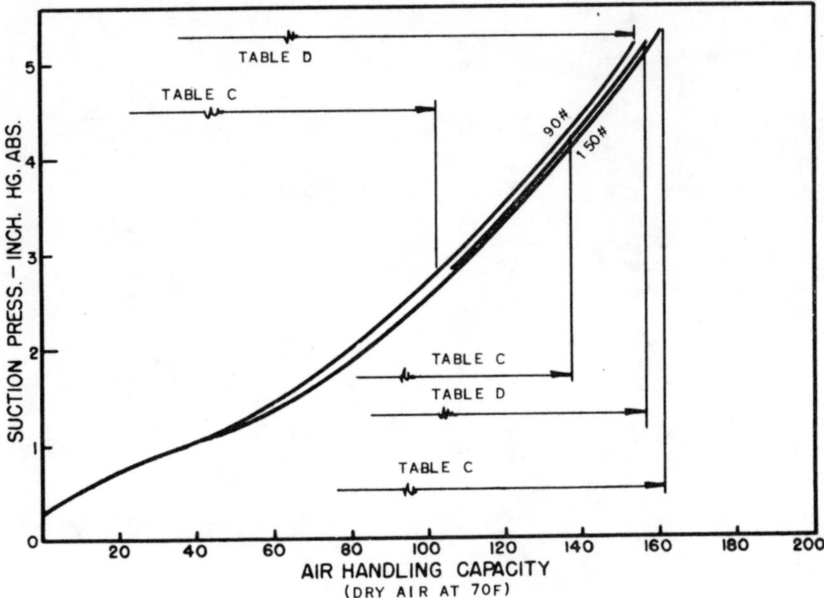

Figure 5.39. Performance curve for two stage non-condensing system as as Fig.8.1.2 except type TC2 high vacuum unit used.

corresponding steam consumptions of various multi-stage non-condensing steam jet-ejector systems. The curves shown are for systems that are stable in the general region of the design point and steady down to the shut-off point. Because of differences in the shapes and characteristics of the individual stage-performance curves, and the manufacturing necessity to construct units to standard sizes, there are some sets of conditions of capacity and suction pressure which cannot be accomplished at the rates shown by the generalized curves. At the same time there are also sets of conditions where steam-consumption rates bettter than those shown are possible.

Direct Contact Condensing Type Multi-Stage Ejector Systems

Because there are many various methods of arranging multi-stage ejectors, particularly when used with inter-stage condensers, variations in descriptive nomenclature may also be encountered. Some users and manufacturers count stages from the process side to atmosphere, others from the atmospheric side to the process. Since nomenclature is critical in discussing jet-ejector systems, the Heat Exchanger Institute developed a standardized indentification system which is illustrated in Figure 5.41. The stage discharging to atmosphere is considered the "Z" stage with preceeding stages denoted alphabetically backwards to the process. Intercondensers are

Table	Size	Steam Pressure-PSIG				Cap. Factor
		90	115	150	200	
A	2T2	51	49	45		1
	2-1/2S3	330	316	292		
	TOTAL	381	365	337	337	
B	2T2	51	49	45		1
	3S3	450	427	392		
	TOTAL	501	476	437	437	
	1-1/2T2	28	28	30		5
	2S3	224	214	200		
	TOTAL	252	242	230	227	
C	3T2	102	98	90		2
	4S3	779	745	700		
	TOTAL	881	843	790	787	
D	3T2	102	98	90		2
	5S3	1161	1133	1018		
	TOTAL	1263	1231	1108	1108	

Figure 5.40. Steam motive requirements for two stage non-condensing system shown in Fig. 8.1.2. (Note: Number preceding T2 or S3 is size in inches).

then labeled by their position between jet-ejectors. In a four stage condensing system the jet-ejector closest to the process would be the "W" stage, followed by the "X," "Y," and the "Z" stage jet-ejectors. Intercondensers would then be denoted accordingly, i.e. the "X-Y" condenser if it is located between the "X" and "Y" jet stage or "Y-Z" condenser if it is located between the "Y" and "Z" jet stage.

When using multi-stages, the supporting stages have to be large enough in capacity to handle the initial suction load plus the motive steam from the preceeding stages. In a multi-stage arrangement, the steam consumption and size of the last two supporting jet ejectors can be reduced substantially by using condensers between the stages. The condensers decrease the suction load to the supporting jet-ejectors by removing most of the condensible vapors. Usually they are interstaged between the last three jet-ejectors near atmosphere, at various interstage pressures from 30 Torr up to atmosphere. However, it is not uncommon to have a condenser located before the jet-ejector stages, called a pre-condenser, when the vapor load consists of a higher percentage of condensibles. Nor is it uncommon to utilize a condenser after the multi-stage jet-ejector system to eliminate exhausting steam to the atmosphere. This final condenser, called an after-condenser, can be a direct contact type unit or a surface type unit.

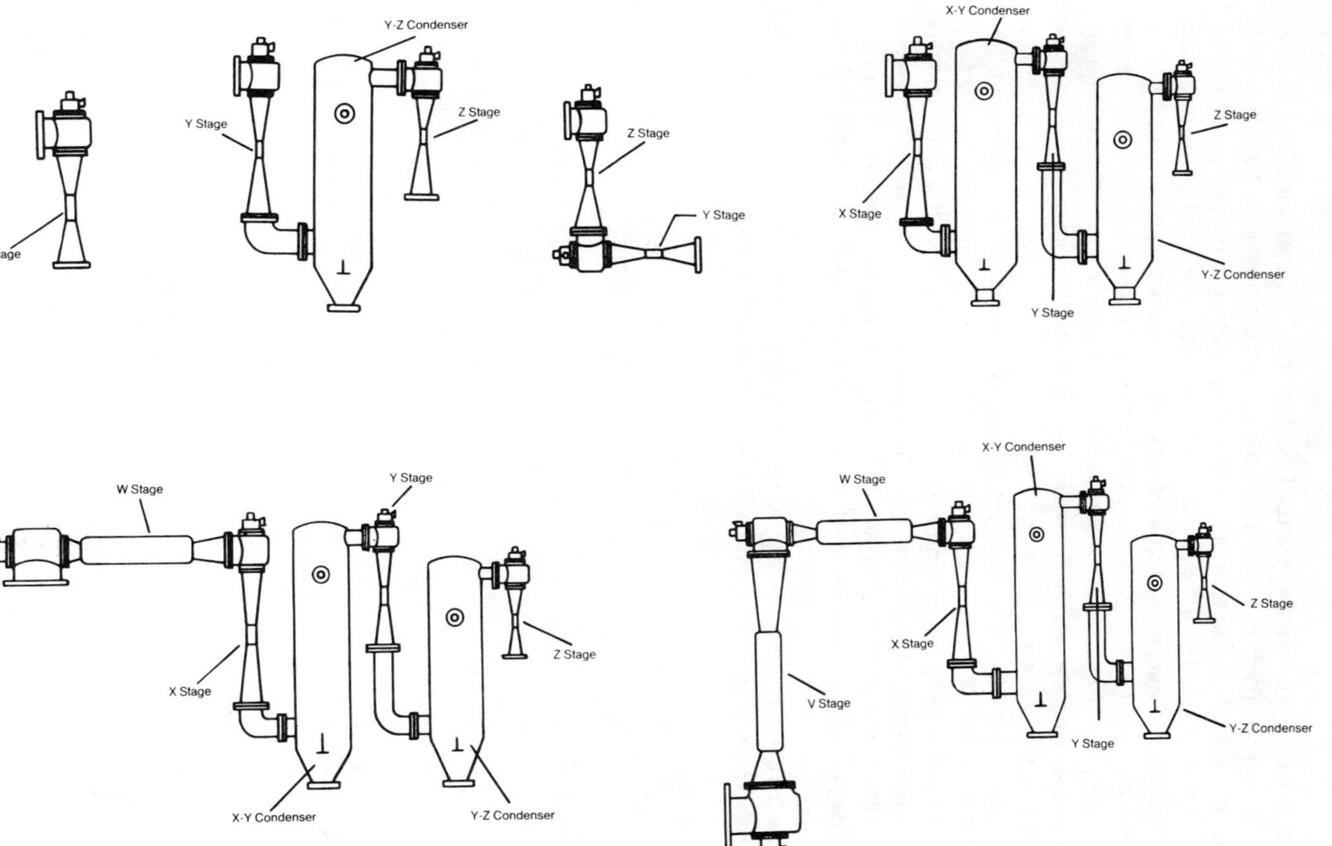

Figure 5.41. Heat Exchanger Institute standard idenfication of stages.

Direct contact type condensers used in jet-ejector systems are of the counter-current, or counter-flow type. Injection water enters the condenser through a water nozzle at the top of the unit. A distribution tray, weir, or disk and donut arrangement in the shell provide a "water curtain" through which the vapor must pass. Most of the vapor entering through a side inlet is condensed in the lower part of the shell, and the non-condensibles are then required to travel upward through the water curtain. A baffle arrangement is normally provided at the air suction connection to reduce the carry-over of water that may be entrained as the non-condensibles pass through the condenser.

The direct contact, counter-current condenser operates with a terminal difference of 1.7 to 2.8°C between tail water and vapor dew point temperature. Large condensers will cool the non-condensibles to within 2.8°C of the water temperature. In cases where large percentages of non-condensibles are present, however, terminal differences as high as 16 to 28°C may be expected since the condenser is actually performing as a gas cooler. This is particularly true in multi-stage jet-ejector condensing systems.

Figure 5.42 illustrates a counter-current condenser with a distribution tray and Figure 5.43 shows a condenser with a disk and donut arrangement.

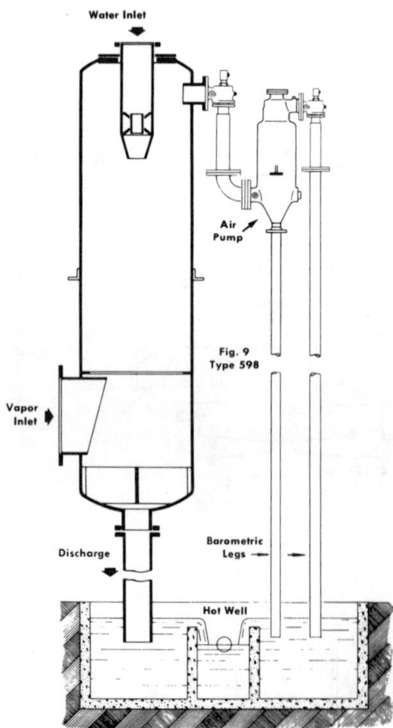

Figure 5.42. Typical counter-current, direct contact condenser.

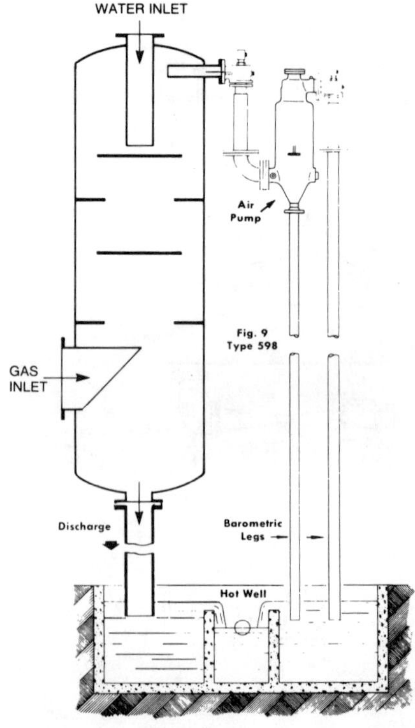

Figure 5.43. Typical disc and donut counter-current, direct contact condenser.

The performance characteristics of a two stage condensing system utilizing the same jet-ejectors shown in Figure 5.38 and 5.40 are illustrated in Figures 5.44 through 5.46. As indicated in the curves, a two stage non-condensing system handling 29 kg/hr dry air (at 21°C) and 89 Torr requires a 2″ by 3″ jet system using 228 kg/hr steam at 6.3 kg/cm^2 or 199 kg/hr steam at 10.55 kg/cm^2. At the same operating condition, and using a two stage ejector system with an intercondenser, it can be noted that only a 2″ by 2″ jet system is required. Also note that the motive steam required is lowered to 125 kg/hr at 6.3 kg/cm^2 or 111 kg/hr at 10.55 kg/cm^2. The intercondenser requires .88 1/sec at 24°C.

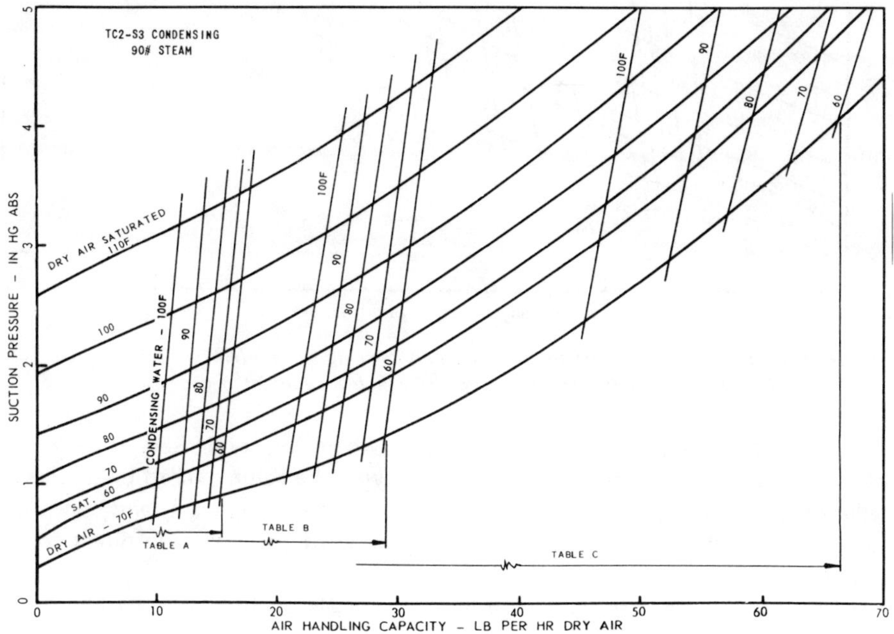

Figure 5.44. Performance curve of two stage condensing multi-stage jet ejector system with T2 high vacuum unit and S3 low vacuum unit with direct contact intercondenser.

It then becomes obvious that the choice of system to be utilized is dependent on many variables. Duration of operation, availability of motive steam, availability of water, water temperature, initial cost, cost of steam and water, etc., are all required to properly define the system best suited for any application.

MULTI-STAGE SYSTEMS WITH SURFACE CONDENSERS

Under some conditions, the use of surface condensers provides certain, definite advantages over the use of direct contact condensers.

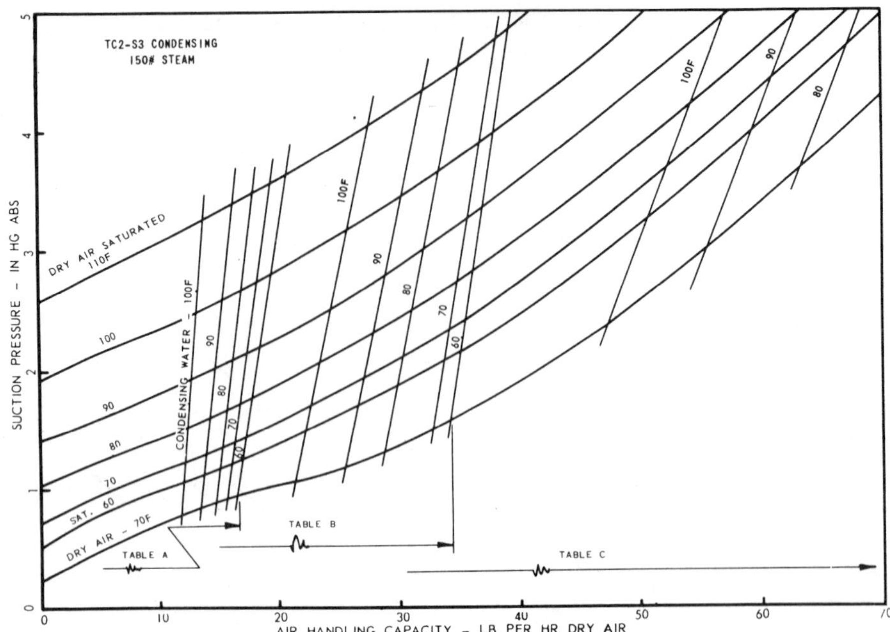

Figure 5.45. Performance curve for two stage condensing system (Note: System is identical to system in Fig. 8.2.3 except Type TC2 high vacuum jet ejector used.

In the case of objectionable contaminants, where the condensate effluent must be treated before reuse or discharged to the sewer, the volume of effluent is greatly reduced by the use of surface condensers since the cooling water and condensate streams are held separate as compared to mixing of the two as in direct contact condensers.

In the case of the recovery of process chemicals, such as solvents, the surface condenser yields less total volume of condensate to separate or refine.

Separation of the cooling and condensate streams by surface condensers provides not only clean condensate but ancillary conservation opportunities. In power plants, for example, it is common to use the turbine condensate as cooling water in the air ejector condensers before it is returned to the feed water system. This conserves both the heat from the steam used in the air ejector as well as pure water fit for boiler use. In marine systems which use sea water for cooling, surface condensers keep the condensate separate from the sea water and allow the reuse of the condensate, which would not be possible if direct contact condensers were used.

Surface type condensers used in conjunction with multi-stage jet-ejector systems are identified in the same manner as direct contact type units. That is, their location in the system will identify them as the "XY," "YZ," etc. condenser. The Heat Exchange Institute and Tubular Exchanger Manufacturers Association have both

Table	Size	Steam Pressure-PSIG				Cap. Factor	Condenser +		Noz.* Size In.
		90	115	150	200		Size	GPM	
A	2TC-2	51	49	45		1.0	#1	9	3/4
	1S3	62	60	59					
	TOTAL	113	109	104					
	3TC-2	102	98	90		2.0	#1	18	1-1/4
	1-1/2S3	111	105	105					
	TOTAL	213	203	195	193				
	4TC-2	206	198	184		4.0	#3	36	1-1/2
	2S3	224	214	200					
	TOTAL	430	412	384	384				
	4TC-2	206	198	184		4.0	#3	38	1-1/2
	2-1/2S3	330	316	293					
	TOTAL	536	514	477	477				
B	1-1/2TC-2	28	28	30		0.5	#1	7	3/4
	1S3	62	60	59					
	TOTAL	90	88	89					
	2TC-2	51	49	45		1.0	#1	10	1
	1-1/2S3	111	105	105					
	TOTAL	162	154	150	150				
	3TC-2	102	98	90		2.0	#2	20	1-1/4
	2S3	224	214	200					
	TOTAL	326	312	290	290				
	4TC-2	206	198	184		4.0	#3	41	2
	3S3	450	427	392					
	TOTAL	656	625	576	576				
	3TC-2	102	98	90		2.0	#3	30	1-1/2
	2-1/2S3	330	326	293					
	TOTAL	432	424	383	383				
C[t]	1-1/2TC-2	28	28	30		0.5	#1	7	3/4
	1-1/2S3	111	105	105					
	TOTAL	139	133	135	129				
	2TC-2	51	49	45		1.0	#2	14	1
	2S3	224	214	200					
	TOTAL	275	263	245	245				
	3TC-2	102	98	90		2.0	#3	35	1-1/2
	3S3	450	427	392					
	TOTAL	552	525	482	482				
	4TC-2	206	198	184		4.0	#4	75	2-1/2
	4S3	777	745	699					
	TOTAL	983	943	883	883				

Figure 5.46. Steam motive requirements and water requirements for two stage condensing system shown in Fig. 8.2.4. (Note: Number preceding TC-2 and S-3 is size of the ejector in inches.

further described this type condenser with alphabetical head and shell designations. That designation plus its shell diameter, tube diameter, and length will completely define the condenser.

Surface type condenser size is based on required surface area to achieve proper heat transfer. First the heat load is determined by the amount of desuperheating, condensing, and condensate subcooling required. This is most accurately calculated by determining the total enthalpy of the gases/vapors at the inlet of the condenser and subtracting the sum of the enthalpy of both vent gases and the condensate drains.

Surface type condensers are in reality, and are therefore correctly named, partial condensers – for two reasons. First, some amount of non-condensibles are always present in vacuum work. Secondly, a portion of the condensibles escape condensation and are flushed through the unit; the quantity of escaping condensibles depends on the amount of non-condensibles present, the vent temperature, and the operating pressure of the condenser at discharge. An additional influence on operation characteristics is that of temperature. Condensing temperature is the dew point of the gas/vapor mixture at any given point within the unit. Condensation is not an isothermal process in a partial condenser as would be the case in "pure" condensation (condensation of a "pure" vapor). Rather, condensing temperature decreases as the gas/vapor mixture flows down along the tube bundles. Concurrently, as the condensibles are condensed out, the percentage of non-condensibles increases. This increases the partial pressure of the non-condensibles and decreases the partial pressure of the condensibles. Because the dew point of the condensibles is dependent on the partial pressure of the condensibles, the dew point (condensing temperature) will decrease as the condensibles are condensed out of the mixture.

In "pure" condensers, where the condensing temperature is isothermal throughout the condenser and the cooling water temperature is reasonably constant, it is proper to use a log mean temperature difference (LMTD) (the derivation of which can be found in most books on heat transfer). But in the case of partial condensers where the condensing temperature varies, the LMTD is not the proper temperature difference to use – a weighted mean temperature difference is required.

As the condensibles condense, the mass (flow) velocity decreases and a film of non-condensibles builds up on the surface of the tubes. The condensing vapors must diffuse through this film. Each one of these phenomena causes the rate of heat/mass transfer to decrease, not as a straight line function, but as an irregular curve. The required surface calculations are, therefore, done in small sections and the total surface required becomes the sum of that required by the smaller, individual sections. Pressure drop calculations are performed in the same manner. In partial condensers for vacuum service, the pressure drop calculations are just as important as the thermal sizing calculations since the miscalculation of either can cause failure of the total vacuum system.

The classical equation for heat transfer, $Q = UA\Delta t$ is also applied to partial condensers even though the condensing is really a heat/mass transfer process. The work of Colburn and Hougen (6) has furnished equations for converting the mass transfer and diffusion to strictly heat transfer units. The overall heat transfer coefficient is the reciprocal of the sum of all the thermal resistances between the bulk of the gas/vapor condensing mixture and the bulk of the cooling water. The total of all thermal resistances would include the resistance of the water film, water side fouling, tube wall and vapor side fouling (all of which are constant or relatively constant) as well as non-condensible film and the sensible heat transfer (which do vary considerably from the inlet to the discharge of the condenser). Accurate determination of these latter resistances necessitates the division of heat load into sections for calculation of total thermal load.

The values of the overall U and Δt are determined at beginning and end of each section and each is averaged for that section. The surface area required for each section then is:

$$A\ (\text{section}) = Q\ (\text{section}) / U_{\text{average}}^{\text{section}} \times \Delta t_{\text{average}}^{\text{section}}$$

The total surface is then determined by summing the surfaces required for the individual sections.

The single most important parameter in evaluating the performance of condensers is the temperature of the cooling water. Water temperature must be low enough to effect condensation at the pressure involved with a direct proportion between the two: the lower the pressure, the lower the water temperature required. Conversely, for a given water temperature, the dew point of the gas/vapor mixture must be high enough (above the water temperature) for condensation to occur at all. For a given gas/vapor mix, if the dew point is not high enough to bring about a sufficient amount of condensing to make the condenser effective, a steam jet booster can be used to raise the pressure (and the dew point) high enough to allow condensing to take place. On the other hand, if the dew point is high enough and/or water temperature is low enough, and the suction load contains a reasonable amount of condensibles, a "precondenser" can be used. The "precondenser" condenses as much of the suction load as possible and thereby reduces the size of the following steam jet and condenser, therefore the utilities required.

When the vacuum load consists of a great quantity of condensibles at a pressure (and dew point) too low to condense with plant cooling water, but high enough to be condensed with "chilled water," a refrigeration system should be considered, and the economics weighed.

As should be noted once more, judicious design choices require care: the evaluation of various types of multi-stage jet-ejector condensing systems, the location, type, and number of both jet-ejectors and condensers required, and the overall economics of each, must all be fully analyzed to make a proper selection.

INSTALLATION AND TROUBLE SHOOTING JET VACUUM EQUIPMENT

The installation of ejectors is normally quite simple due to the inherent advantages of equipment with no moving parts, and relatively light weight. It is necessary, however, to adhere to certain requirements in order to obtain satisfactory operation. An ejector is essentially a fixed-flow device with nozzle and throat dimensions configured at manufacture for operation at a fixed set of conditions. Thus, it is essential to adhere as closely as possible to the design conditions for optimum operation of the unit. For high velocity units, such as steam jet-ejectors, the motive fluid pressure can be quite critical. To obtain optimum performance in these cases, it is always essential that the steam pressure be not less than the design pressure, with as little over-pressure as possible. It is also essential that the proper cooling water be supplied to any condenser involved in multi-stage ejectors and, to achieve low maintenance requirements, the motive fluid should be of as high quality as possible. A strainer and separator should always be installed — directly at the inlet to the ejector.

The installation position of ejectors is not critical. Some water jet-ejectors and high vacuum units (such as jet condensers) should be installed with the motive fluid flowing vertically downward.

The majority of steam jet-ejectors are installed at barometric height, especially when direct-contact condensers are used. It is essential that a full barometric leg be used to provide water removal from the condenser; allowance must be made for the mixtures of air and water discharged from jet condensers in calculating the required height. It is also important that the barometric leg (tail pipe) be as straight as possible and free from any horizontal runs where air entrained in the water can separate to cause discontinuities in the flow. In general, any ejector installation should allow for free flow of the fluid without odd arrangements which will cause turbulence and excess pressure drop in the piping. The supply pipes to liquid jets must be such as not to cause any tendency toward vortexing in the entrance chamber.

The following seven steps present an orderly approach to diagnosing system failure which will save appreciable time and eliminate errors.

Step 1: Determine whether the fault lies with the vacuum pump or the system. This is accomplished by blanking off the suction to the ejector and operating the entire unit. If the no-load or shut-off condition approaches the minimum suction pressure of the ejector, then the fault lies in the system. This shut-off pressure would be approximately 50 Torr for a single stage unit, 6.3 to 12.7 Torr for a two stage, 1 to 3 Torr for a three stage, and 100 to 200 microns for a four stage. If the test shows the ejector operating satisfactorily at shut-off, it can usually be expected to operate satisfactorily at the load.

Step 2: Next check the system for leaks and correct. To determine the amount of air leak, approximate the total volume of the system, operate the ejector to secure a pressure somewhat less than 381 Torr, and then isolate the ejector from the system. Measure the time required for a rise of 50 Torr in pressure in the vessel. It is essential that the absolute pressure does not rise above 381 Torr during this time. The following formula will then give the leakage:

$$W_L = \frac{0.095\, V\, \Delta p}{t}$$

where

W_L = Leakage, kilogram per hour
V = System volume, cubic meters

Δp = Pressure rise, inches Hg
t = Time, seconds

If the volume of the system is not known, the leakage can still be determined, but two tests will be required. First, the test described above must be run. Then a known air leak must be introduced to the system. This can be done by means of a calibrated air orifice. A second test is then made to obtain a new pressure rise and time. The unknown leak is then given by:

$$W_L = \frac{\dfrac{W'}{(\Delta p')\ (t)}}{(\Delta\ p)\ (t')} - 1$$

where
W' = Known leak, kilograms per hour
$\Delta p'$ = Second pressure rise, inches Hg
t' = Second time, seconds

If the air leakage is greater than the load for which the jet was designed, then the alternatives are to correct the leaks or to use a larger ejector. The ejector must be large enough to handle not only the leaks but the normal load from the process.

Step 3: Assuming that the test indicates the ejector is not operating satisfactorily, the blank should be left in and the ejector utilities checked. First install calibrated steam gages (with pig tail syphon) on the steam chest or immediately adjacent to the inlet to each stage of the ejector. The operating pressure is normally stamped on the nameplate or parts of the ejector, and the pressure determined by the gage must not be lower than this rated steam pressure. If steam pressure checks satisfactorily, the steam should be checked for excess moisture. The drainage lines from steam separators should be checked to make certain that they are draining properly through a steam trap or bleeding through a valve. Open the discharge from the separator by means of a valve and let it bleed to see whether this makes any change in the operation. A bucket trap discharging in cycles may cause fluctuating steam pressures.

Step 4: Assuming that the steam pressures have been found satisfactory, all of the ejectors except the atmospheric stage should be shut off and the resulting vacuum read. If this pulls down to approximately 50 Torr, then this stage is operating satisfactorily. If it does not pull down to this vacuum, then either the atmospheric stage is at fault or there is a leak in the ejector unit. The atmospheric stage should be isolated by blanking off its suction and checking shut-off pressure as before. If this stage does not operate it should be checked for:

(a) Clogged steam nozzle, or steam strainer reducing flow through the nozzle. If nozzle inlet is red or black, look for a scale deposit which can be removed by careful scraping and subsequent polishing. If nozzles are graphite or plastic, measure the orifice to check for any shrinkage or wear.

(b) Excess pressure at discharge of ejector. If the low stage is discharging directly to atmosphere, check the discharge line to make certain that it is clear. Check for any places where condensate may accumulate to cause excessive back pressure or pulsating back pressure. If the discharge line is connected to a hotwell or to an after-condenser, disconnect the well or after-condenser from the ejector and allow the ejector to discharge to atmosphere.

If the above points do not provide satisfactory operation, then the ejector must be disassembled. The nozzle and throat dimensions should be checked against the original manufactured dimensions. The diffuser should be examined for corrosion or wear. Pock-like irregular indentations indicate corrosion. Longitudinal wear marks in the steam orifice and diffuser are normally evidence of wet steam if corrosion is not present. A roughening of the inlet tapers through the diffuser, without any material change in diameter may cause malfunctioning of the ejector. The diffuser should be smoothed out or replaced.

Step 5: Assuming that the atmospheric stage ejector has been found to operate satisfactorily, it will be necessary to look for difficulties elsewhere. Inter-condensers operating under vacuum should be carefully checked for satisfactory operation. The inlet water temperature must be equal to or less than the design temperature. The tail pipe temperature should be approximately 11°C above inlet temperature for two stage units, and 5.6°C for three stage. Note that these figures are approximate, and will vary depending upon equipment design and the amount of non-condensibles present. The ejector manufacturer can furnish design temperatures. Malfunctioning of the condenser can occur with either too little or too much water. Too little water will be indicated by a high tail pipe temperature, and a high temperature in the head of the condenser before the suction of the atmospheric stage. Too much water may cause flooding of the condenser and would be indicated by pulsations in the atmospheric stage and discharge of water from that stage. If the atmospheric stage is removed from the condenser, this carryover of water would be indicated by scale deposits in the suction connection — which will reduce the area and cause excess pressure drop. Water carry-over can also be caused by a leaking tail pipe. Air rising in the pipe will prevent flow from the condenser and cause surging of condensing water. Look for leaks at the condenser outlet flange and at the surface of the water in the hotwell.

Assuming that the temperature and quantity of the utility supply to the condenser are correct the condenser itself should be checked for malfunctioning. In the usual cylindrical condenser, water-steam contact is obtained by a series of trays which give curtains of water. At points where these curtains strike the wall of the condenser there should be a uniform low temperature without any breaks. At points where the water is free from the wall of the condenser there should be a uniform high temperature. The steam temperature should decrease toward the top of the condenser, and the water temperature increase toward the bottom of the condenser. The head of the condenser should be relatively cool, approaching the inlet water temperature by 2.8 or 5.6 degrees, so that the load to the atmospheric stage is as low as possible. High temperatures at this point, with adequate water flow, indicates by-passing of the curtains. This may be caused by clogging of the inlet distribution piece, either nozzle or trays, or failure or clogging of the subsequent trays. This would then indicate that the condenser should be disassembled and examined.

Step 6: Assuming that the condenser check indicates satisfactory performance, proceed to the stage immediately ahead of the condenser and examine it in the same fashion as specified for the atmospheric stage.

Step 7: The same procedure should be followed step by step from the atmospheric end of the multi-stage unit up to the highest stage in order to eliminate the difficulties.

CODE STANDARDS (ASME/HEI/TEMA)

A steam jet-ejector, a direct contact condenser or a surface condenser can each be considered a pressure vessel, and as such would come under the rules for such vessels as set up by the American Society of Mechanical Engineers (ASME). In 1911 the ASME created a committee (now known as the Boiler and Pressure Vessel Committee) for the purpose of establishing rules of safety in the design, fabrication, and inspection of boilers and pressure vessels, and for the interpretation of these rules in case of question.

Three publications of the Society are pertinent. Section VIII, Division 1, furnishes formulas for calculating wall thickness for shells, heads, and nozzles, grades of materials and allowable stresses for each. In addition, it delineates criteria for the fabrication of ferrous, non-ferrous, alloy, and cast iron units, forgings and the heat treatment after fabrication, flange, gasket, and bolt loading calculation, hydrostatic and pnuematic testing and magnetic particle and liquid penetrant examination. Section II, "Materials," specifies chemical and physical requirements – Ferrous Materials in Volume A and in Volume B, Non-Ferrous Code Materials. Section IX gives the Welding Qualifications for Code Welders.

The Heat Exchange Institute (HEI), is an association of manufacturers of such products. Its objective is to promote and further (in every lawful manner) the interests of the manufacturers of heat exchange and steam jet vacuum apparatus, and the interests of the public in manfuacturing, engineering, safety, transportation, and other problems of the industry.

HEI's publication, "Standards for Steam Jet Ejectors," contains nomenclature, operating principle types, design specifications, materials of construction, HEI flanges for vacuum service, pressure protection and hydrostatic testing methods. A large part is apportioned to "Test Standards" and the last part to "Vacuum Engineering Data."

HEI's "Standards for Direct Contact and Low Level Condensers" gives nomenclature, terminology, performance, construction, and installation of both "barometric" and "low level units."

In their "Standards for Steam Surface Condensers," HEI provides nomenclature and terminology for surface condensers, a detailed section on performance which includes materials and thermal sizing calculations, required venting capacities for both fossil and nuclear power plants, and data for the sizing of hogging ejectors and of atmospheric relief valves.

There may appear to be a duplication of effort in pubications issued by the ASME and the HEI. In reality, the ASME is concerned with the physical strength (safety) of *all* types of pressure vessels and thus established standards for materials, design, and welding. The HEI has furnished standards for the manufacture of *specific* types of pressure vessels, all operating under vacuum – the steam jet ejector, the direct contact condenser, and the steam surface condensers, the latter especially for large steam engines and turbines.

The Tubular Exchanger Manufacturers Association (TEMA), has produced a manual specific to shell and tube heat exchangers for all pressures and materials entitled "Standards of Tubular Exchanger Manufacturer Association." The standards are complimentary to ASME VIII, Division 1, and invoke the Code Stamp unless otherwise specified by the purchaser. The standards furnish nomenclature for the various types of fixed heads, return heads and shell side flows and are divided into three classes: "R" for refineries, "C" for commercial and "B" for chemical process, the last usually includes alloyed materials. Minimums are specified for wall thickness and corrosion allowances for vents, drains, temperature and pressure connections, and formula for calculation of tube sheet thickness are provided. Typical fouling factors, physical properties of fluids and some useful general engineering information are also included.

EXAMPLE PROBLEMS

Problem No. 1 – Dry Air Equivalent (DAE) Conversion

Convert 50 PPH of dry air saturated with water vapor at 8″ Hg Abs and 140°F to DAE at 70°F.

Calculations: At 8″ Hg Abs and 140°F, one pound of air holds 1.8 PPH of water vapor at saturation, so that 50 PPH air would hold 50 × 1.8 = 90 PPH of water vapor. The total load then is 50 PPH air plus 90 PPH of water vapor.

To convert to DAE, refer to Figure 5.5.

Air	=	50 PPH/1	= 50	DAE @ 140°F
Water Vapor	=	90 PPH/.81	= 111.1	DAE @ 140°F
			161.1	DAE @ 140°F

To convert DAE @ 140°F to DAE @ 70°F, refer to Figure 5.5.

Air	50 DAE/.982	=	50.9 DAE @ 70°F
Water Vapor	111 DAE/.975	=	115.0 DAE @ 70°F
			164.9 DAE @ 70°F

Answer: 164.9 DAE @ 70°F

(Note: Unless otherwise specified, the capacity curves for steam jet exhausters (ejectors) use PPH of DAE @ 70°F as the suction load units, SCFM of air at 70°F, if more appropriate is sometimes used instead of DAE).

The DAE of the above suction load can also be determined from Figure 5.6 after the quantity of water vapor has been calculated. 50/(50+90) = 35.7% air in mixture. From 3.2 the entrainment ratio is 0.85. Dividing 140 by .85 gives 164.7 DAE @ 70°F.

Problem No. 2 — Steam Jet Syphon

Calculate the size and steam consumption of a steam jet syphon to evacuate 200 cu. ft. of air from atmospheric pressure to 20″ Hg vacuum (10″ Hg Abs) in five minutes or less using the #60 nozzle and 30 PSIG steam.

Refer to Figure 5.10. Note that the sizing chart has a capacity factor of 1.00 for the 1½″ syphon, which means that the accompanying curves are for the 1½″ size unit. At 20″ Hg vacuum, the #60 nozzle gives a rate of 2.4 seconds per cu. ft. of volume to be evacuated.

200 cu. ft. would require 200 × 2.4 seconds or 480 seconds. But the time limit is 5 minutes or 300 seconds, so that a unit 480/300 or 1.6 times the capacity of the 1½″ unit is required.

Referring again to the sizing chart, the smallest unit that has at least 1.6 times the capacity of the 1½″ size is the 2″ size. The steam consumption from the same chart and #60 nozzle is 523 PPH.

The above calculations can be condensed to:

$$\frac{\text{volume} \times \text{time (seconds) per cu. ft.}}{\text{time allowed in seconds}} = \text{capacity factor}$$

If 60 PSIG steam is available, the time rate is 1.7 seconds/cu ft. and

$$\frac{200 \times 1.7}{5 \times 60} = 1.17 \text{ capacity factor required.}$$

From the sizing chart under #115 nozzle (60 PSIG steam) the smallest size with at least 1.17 capacity factor would be the 2″ syphon with a capacity factor of 1.38. The corresponding steam consumption would be 375 PPH @ 60 PSIG.

Problem No. 3 – Water Jet Exhauster

What size water jet exhauster would be required to handle 15 SCFM of air at 10″ Hg. Abs when using 30 PSIG motive water at 80°F? What would be the water consumption?

Refer to chart Figure 5.12. Enter the chart at (1) on left hand margin at the 80°F water temperature. Move horizontally to the right to (2) at 10″ Hg Abs suction pressure. Proceed vertically upward to the 30 PSIG water pressure (3), and then horizontally to the right to the 15 CFM line (4). Read the size corresponding to this line at point (5). The size indicated is a No. 6 with a capacity factor of 7.5 (from "Capacity Table").

The quantity of water required for a 2″ unit (with a capacity factor of 1.0) using 30 PSIG motive pressure can be determined from Figure 5.13 which is 67 GPM. Since a No. 6 size exhauster has a capacity factor of 7.5 times that of a No. 2, the water required would also be 7.5 times that required for a No. 2 or 7.5 × 67 GPM or 503 GPM of motive water.

The answer then would be a #6 (size) water jet exhauster with 503 GPM of motive water at 30 PSIG and 80°F.

Problem No. 4 – Multijet and Multijet Spray Condenser

Calculate the size, cooling water required and air handling capacity of a multijet and a multijet spray condenser to handle 20,000 PPH steam at 4″ Hg Abs using 80°F cooling water.

Refer to Figure 5.19. The curves given for 4″ Hg Abs and 80°F cooling water gives a capacity of 16 PPH steam per GPM of cooling water. The total water required would be 20,000 PPH steam/16 PPH per GPM = 1,250 GPM water. From the sizing chart to the right of the curve, 1,250 GPM would require a 34″ condenser that has a maximum capacity of 1,300 GPM. Since the 33″ condenser has a maximum capacity of 1,100 GPM, the 34″ size would be required.

To determine the air handling capacity, it is necessary to determine the water discharge temperature from the condenser. Refer to Figure 5.17 A relatively close approximation of the condenser heat load can be had by using the latent heat of steam at the operating pressure of 4″ Hg Abs. Thus the total heat load would be 20,000 × 1,021 or 20,420,000 BTU/hr. The cooling water temperature rise would be 20,420,000/(1,250 × 500) = 32.6°F rise.

The water discharge temperature would be 80° + 32.6°F or 112.6°F and the corresponding vapor pressure is 2.8″ Hg Abs.

The "effective air pressure," the operating pressure minus the vapor pressure at the discharge temperature would be 4.0″ Hg Abs minus 2.8″ Hg Abs or 1.2″ Hg.

Using the 5 PSIG water pressure curve of Figure 5.17, for 1.2″ Hg gives the value of 2.8 PPH air per 100 GPM. Using this value, 1,200 GPM would handle 1,250 × 2.8/100 = 35 PPH air.

Problem No. 5 – Multispray Condenser

Calculate the size, cooling water required and air handling capacity of a multispray condenser for the same load and conditions given in Problem No. 4.

From Figure 5.20 the curve for 80°F water and 4″ Hg Abs gives a capacity of 20 PPH steam per GPM of cooling water. The total cooling water required would be 20,000/20 or 1,000 GPM. Referring to the sizing chart to the right (Figure 5.20), the minimum size condenser that can handle 1,000 GPM is 33″ with a maximum flow of 1,100 GPM. Therefore, a 33″ condenser using 1,000 GPM cooling water would be the answer.

The maximum air handling capacity of multispray condensers is not definite, since the precooler (direct contact, countercurrent condenser) would be sized to suit the air load and it in turn would be supported by a suitably sized air pump. Should the above 20,000 PPH steam contain 200 PPH air, the precooler size would be a #3 Fig. 597 and the water for it required would be 75 GPM. The vent temperature of the precooler would be 82°F so that the air pump load would be 200 PPH air saturated at 3.9″ Hg Abs and 82°F.

Problem No. 6 – Low Level Multijet Eductor Condenser

Calculate the size, cooling water required, and air handling capacity of a low-level multijet eductor condenser for the same load and conditions given in Problem No. 4. (Note, after start-up, the cooling water pressure is reduced from 25 PSIG to 9 PSIG).

Using Figure 5.24, the 80°F water curve at 4″ Hg Abs gives a capacity of 14.6 PPH of steam per GPM. The cooling water required would be 20,000/14.6 or 1,370 GPM. From the sizing chart to the right of the curve, the smallest condenser that will handle 1,370 GPM would be a 36″ size unit.

For the air handling capacity, refer to Figure 5.17. Note the statement that for these low-level units that the air capacity is only 1/5 the curve value. The heat load would be the same as Problem No. 4: 20,000 × 1,021 BTU = 20,420,000 BTU per hour. The water temperature rise would equal 20,420,000/(1,370 × 500) = 29.8°F. The water discharge temperature would be 80°F + 29.8°F = 109.8°F and the corresponding vapor pressure 2.58″. The effective air pressure = 4.0″ Hg – 2.58″= 1.42″ Hg. From the 9 PSIG curve 1.42″ Hg corresponds to 4 PPH of air per 100 GPM. The total air would be (1,370 × 4)/100 = 54.8 PPH.

Problem No. 7 – Steam Jet Exhauster (Ejector)

What size single stage jet and what would be the steam consumption required to handle the loading stated in Problem No. 1 using 90 PSIG steam? 200 PSIG steam?

The suction loading for 50 PPH air saturated with water vapor at 8″ Hg Abs and 140°F was calcualted in Problem No. 1 as 164.9 DAE @ 70°F.

1. Refer to Figure 5.34. From the 90# steam curve the capacity of a 1½" jet at 8" Hg Abs is 128 PPH DAE (at 70°F). If the suction load is 164.9 DAE and the capacity of the 1½"C jet is 128 DAE, then teh required capacity factor is 164.9/128 or 1.29 capacity factor.

From the sizing chart, the 2"A jet has a capacity factor of $1.24\overline{X}$ and requires 420 PPH of motive steam at 90 PSIG and the 2"B unit has a capacity factor of $1.57\overline{X}$ with a steam consumption of 538 PPH.

The decision as to which of the two above sizer should be made upon the accuracy of the 50 PPH air. If it is just an estimate and it is believed that 48 PPH of air would suffice then the smaller jet with less steam consumption could be chosen. If the 50 PPH of air is an absolute minimum than the larger jet with the greater steam consumption should be used.

2. Refer to the 200# steam curve on the same Figure 5.34. The suction capacity for the 1½"C jet at 8" Hg Abs is 149 DAE (70°F). The capacity factor required would be the same suction load divided by the 1½" jet suction capacity or 164.9 DAE/149 DAE or 1.197 $\overline{X}$ capacity factor.

The 1½"C jet with a capacity factor of $1.0\overline{X}$ is far too small so the minimum size suitable is the 2"A with a capacity factor of $1.24\overline{X}$, which requires 378 PPH motive steam at 200 PSIG.

Problem No. 8 – Two-Stage Condensing Steam Jet Vacuum Pump With Direct-Contact Condensers

What size two-stage condensing steam jet vacuum pump would be required to handle 20 PPH dry air saturated with water vapor at 2" Hg Abs and 90°F using 90°F cooling water and 90 PSIG motive steam? 150 PSIG steam?

1. Refer to Figure 5.44 for 90 PSIG steam. At 2" Hg Abs the intersection of 90°F saturation temperature and the 90°F cooling water gives an air handling capacity of 13 PPH of *Dry Air,* under Table A.

(Note: This set of curves, as well as those shown on Figure 5.45 for 150 PSIG steam, are set up for dry air and air saturated with water vapor at various temperatures, so that the suction capacity can be read directly in units of PPH of Dry Air. Thus, it is not necessary to determine the quantity of water vapor involved and convert the total air/water vapor load to DAE).

The capacity factor required would be 20/13 or $1.54\overline{X}$.

Refer to Figure 5.46. Under Table A, the tabulation gives a capacity factor of 2 as the minimum that could handle the calculated required capacity factor of $1.54\overline{X}$. Under "Size" corresponding to this capacity factor of 2, is listed a 3" TC-2 first-stage and 1½" S-3 second stage. The total steam consumption for the two jets is 213 PPH at 90 PSIG. For the inter condenser corresponding to this capacity would be a size #1 with 18 GPM at 90°F.

2. For 150 PSIG steam refer to Figure 5.45. At 2" Hg Abs under Table A, the intersection of the 90°F saturation temperature with the 90°F cooling water line

gives a suction capacity of 15 PPH of Dry Air.

The required capacity factor would be 20/15 = $1.33\overline{X}$.

Refer to Figure 5.46. Under Table A, the minimum capacity factor greater than the required $1.33\overline{X}$ is 2. The corresponding jets would be 3″ TC-2 first stage and a 1½″ S-3 second stage with a total steam consumption of 195 PPH @ 150 PSIG. The intercondenser would be a size #1 with 18 GPM at 90°F.

REFERENCES

1. Standards for Steam Jet Ejectors, Third Edition, Heat Exchange Institute, N.Y., N.Y., 1956.
2. Power Test Code 24-1976 "Ejectors" American Society of Mechanical Engineers.
3. "Thermodynamic Properties of Steam" J. H. Keenan and F. G. Keyes, First Edition, John Wiley and Sons, Inc., N.Y., N.Y. January 1967, page 76.
4. Standards for Steam Surface Condensers, Sixth Edition, Heat Exchanger Institute, N.Y., N.Y. 1970.
5. Standard of Tubular Exchanger Manufacturers Association, Fifth Edition, 1968 and 1970 Addenda.
6. Colburn, A. P. and O. A. Hougen, Ind. Eng. Chem. Volume 26 (1934).
7. "Characteristics of the Steam Jet Vacuum Pump" by L. S. Harris and A. S. Fischer, ASME Paper 63-WA-132.
8. "How to Get the Most From Ejectors," by C. G. Blatchley Petroleum Refiner, 12–58.
9. "Controlling Ejector Performance," by C. G. Blatchley, AMETEK/Schutte & Koerting Division.
10. "Selection of Air Ejectors," by C. G. Blatchley, Chemical Engineering Progress, 10–61 and 11–61.
11. "Jet Ejectors and Processing," by H. J. Stratton, Food Engineering, 11–68.
12. "Steam-Jet Air Ejectors," R. B. Power, Oil and Gas Equipment, 10–65 through 11–66.
13. "Steam-Jet Air Ejectors: Specification, Evaluation and Operation: by R. B. Power, ASME Paper 63-WA-143.

CHAPTER 6

POSITIVE DISPLACEMENT COMPRESSORS

GARY SZCZEPANKIEWICZ
Hooker Chemical Corp.
Corporate Engineering
Grand Island, N.Y.

THERMODYNAMICS OF GAS COMPRESSION

Pressure-Volume Relations

The basic cycle of a reciprocating compressor is illustrated by Figure 6-1.

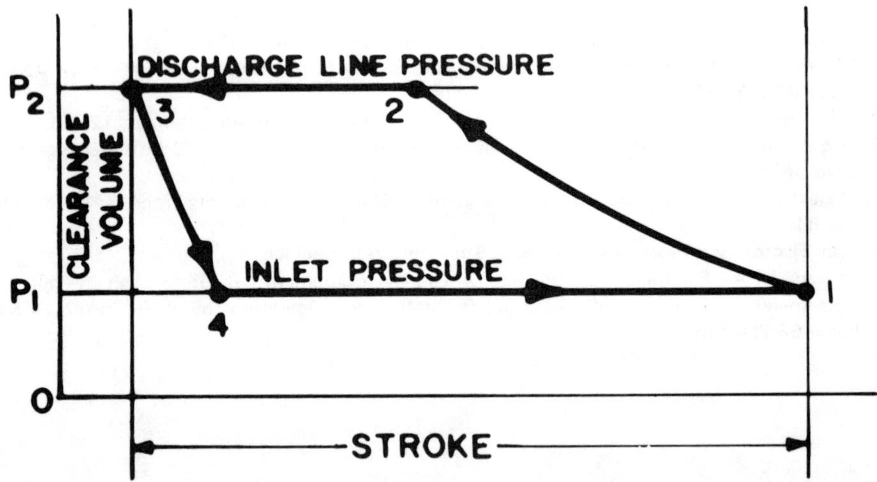

Figure 6-1.

Initially, the pressure within the cylinders is equal to the inlet pressure. As the piston moves to expand the volume of the chamber, gas enters the cylinder at low pressure. At the end of the suction stroke, the inlet valves close and the piston begins to travel back. As the piston reduces the volume within the cylinder chamber, the pressure of the gas trapped within the chamber rises (1–2). At point 2, the pressure within the cylinder equals the high pressure just outside the discharge valves. As the piston continues its stroke, it displaces compressed gas out through the discharge valves, until at point 3 it has displaced all but a small residual pocket of gas. The small volume left at the end of a compressor stroke is the 'clearance volume.'

The pocket of compressed gas trapped within the clearance volume begins to expand as the piston begins its return stroke (3–4) until at point 4 the pressure within the cylinder is once again at the inlet pressure.

The paths of compression (1–2) and expansion (3–4) can be described by one of three theoretical relations:

In the case of an adiabatic system, the governing relation is PV^k = a constant. (Eq. I-1)

In the case of isothermal compression, the relation is PV^1 = a constant. (Eq. I-2).

A more realistic relationship is described by the equation PV^n = a constant, (Eq. I-3) in which the exponent (n) is an experimentally determined figure. This is the polytropic relation.

Discharge Temperature

The increase in gas temperature during compression is dependent upon the initial temperature of the gas, the ratio of compression, and the path of compression.

In the adiabatic case:

$$T_2 = T_1 \times r_c^{\frac{k-1}{k}} \qquad \text{(Eq. I-4)}$$

In the case of polytropic compression:

$$T_2 = T_1 \times r_c^{\frac{n-1}{n}} \qquad \text{(Eq. I-5)}$$

When the polytropic efficiency is known, the exponent n – 1/n is determined by the formula:

$$\frac{n-1}{n} = \frac{k-1}{k} \times \frac{1}{\eta_{pt}} \qquad \text{(Eq. I-6)}$$

Compressibility

All real gasses deviate from the perfect gas law PV = NRT. In order to express the pressure-volume relationship correctly, a compressibility factor "Z" is introduced. The relation then becomes PV = ZNRT. (Eq. I-7)

Z is an emperical factor that varies with the pressure, the temperature and the nature of the gas being compressed. A number of compressibility charts for specific gasses are available (1) (2) (5). However, when it is necessary to predict the behavior of a gas for which no P-V-T data are available, a generalized coorelation may serve as an approximation. The correlation takes advantage of the observation that gasses diverge from the ideal gas law in similar ways, which suggests that a common scaling factor can be used to correlate values of Z for a wide variety of gasses. In actual practice, this scaling factor is determined by the critical point of the gas in question.

Generalized compressibility charts (3) (4) use the reduced parameters P_r and T_r to correlate Z, where $P_r = P/P_c$ and $T_r = T/T_c$.

Generalized charts are not entirely universal and should be used only when Z-P-T data for a specific gas are unavailable.

Equations of State

When a programmable computer is available, the engineer may make use of one of a number of equations of state to describe the behavior of a gas. At conditions ranging from low pressure up to approximately twice the critical density, the Benedict-Webb-Rubin (or Kellogg) (6) equation will give satisfactory results. Above this level, the Redlich-Kwong equation (7) may be used.

Gas Mixtures

PVT data for a number of gas mixtures can be found in the literature. However, when experimentally determined information is not available, a method of approximating this data must be used.

The pseudocritical point method is one of the simpler techniques. The method produces a pseudo T_c and a P_c that may be used as parameters to determine a reduced T_r and P_r. These reduced indeces may be used on a generalized compressibility chart to produce a value of Z.

The pseudocritical temperature equation is:

$$T_{pc} = Y_a \times T_{CA} + Y_B \times T_{cB} + Y_c \times T_{cc} + \cdots Y_n \times T_{cn} \qquad \text{(Eq. I-8)}$$

Similarly the pseudocritical pressure may be determined by:

$$P_{pc} = Y_A \times P_{CA} + Y_B \times P_{cB} + Y_c P_{cc} + \cdots\cdot Y_n P_{cn} \qquad \text{(Eq. I-9)}$$

The ratio of heat capacity (C_p/C_v (=k) can be approximated for a gas mixture by treating the mixture as a perfect gas at low pressure. In that ideal case, the relationship between heat capacity at constant volume and at constant pressure is

$$C_p - C_v = R. \qquad \text{(Eq. I-10)}$$

All that is needed then to determine the components of k, is a value of C_p. A mole weighted average may be used according to the equation:

$$C_p = Y_A \times C_{pA} + Y_B \times C_{pB} + Y_c \times C_{pc} + \cdots Y_n \times C_{pn} \quad \text{(Eq. I-11)}$$

Horsepower Requirements

The work required to compress a stream of gas adiabatically is approximately by the formula:

$$hp_{ad} = \frac{144}{33000} P_1 \times Q_1 \left(\frac{k}{k-1}\right) \left(r_c^{\left(\frac{k-1}{k}\right)} - 1\right) \left(\frac{Z_1 - Z_2}{2Z_1}\right) \quad \text{(Eq. I-12)}$$

where P_1 Q_1, and Z_1 are measured at intake conditions, and Z_2 is determined at discharge conditions.

Although this relation does not depend on the actual method of compression, the adiabatic formula approximates the performance of reciprocating compressors more accurately than dynamic units.

The brake horsepower of a reciprocating compressor is expressed as the adiabatic horsepower modified by empirical factors that account for fluid losses at the cylinder (CE) and mechanical losses throughout the compressor (ME). The relation is:

$$BHP_{recip} = \frac{hp_{ad}}{CE \times ME} \quad \text{(Eq. I-13)}$$

Centrifugal compressors are often rates in terms of adiabatic horsepower, but with a correction factor known as adiabatic efficiency (η_{ad}) applied to the denominator:

$$GHP = hp_{ad} / \eta_{ad} \quad \text{(Eq. I-14)}$$

Brake horsepower for centrifugal units may be approximated by the addition of a factor to account for bearing and seal losses. This factor will visually have a value between 7 and 50 hp (10).

In the case of air compression, the humidity of the gas often has a significant effect on the performance of the compressor. Charts are available for direct reading of the actual k value that should be used for any particular specific humidity (8) (9). This figure will vary from $k = 1.4$ to 1.32 at practical temperatures.

When adiabatic head is known, the horsepower requirement may be described by the equation:

$$GHP = \frac{W_1 \times H_{ad}}{\eta_{ad} \times 33000} \quad \text{(Eq. I-15)}$$

Centrifugal compressors may also be rated in terms of polytropic horsepower. The applicable equations are:

$$hp_{pt} = \frac{144}{33000} P_1 Q_1 \left(\frac{n}{n-1}\right)\left(r_c^{\left(\frac{n-1}{n}\right)} - 1\right)\left(\frac{Z_1 - Z_2}{2Z_1}\right) \quad \text{(Eq. I-16)}$$

$$GHP = hp_{pt}/\eta_{pt} \quad \text{(Eq. I-17)}$$

$$BHP = GHP + \text{mechanical losses} \quad \text{(Eq. I-18)}$$

$$GHP = \frac{W_1 \times H_{pt}}{\eta_{pt} \times 33000} \quad \text{(Eq I-19)}$$

The polytropic efficiency used in these formulas is related to the ratio of specific heats by the relation:

$$\frac{n}{n-1} = \left(\frac{k}{k-1}\right) \frac{1}{\eta_{pt}} \quad \text{(Eq. I-20)}$$

When a stated polytropic efficiency is not available, an approximation may be obtained from Figure 6.2.

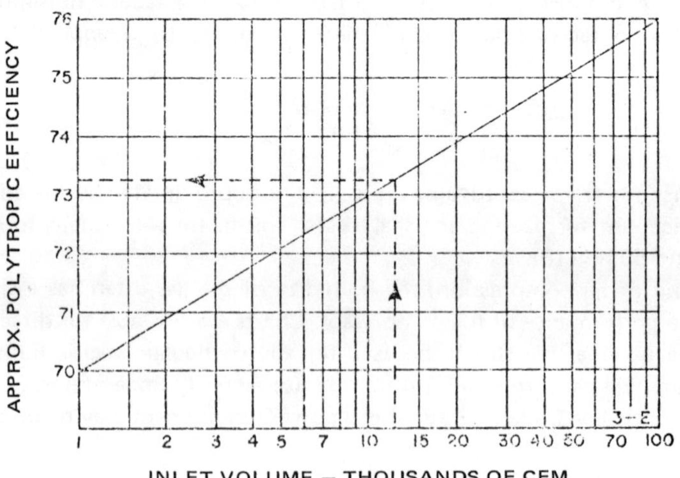

Figure 6-2.

Compressor Pressure Performance

Adiabatic head (H_{ad}) and polytropic head (H_{pt}) are concepts that are particularly useful to designers and users of dynamic type compressors. The head referred to is analogous to the head developed by a centrifugal pump. In the case of the pump, the head developed at a particular flow and at a particular impeller speed is the same regardless of the density of the fluid being handled. In the same way, the polytropic head developed by a specified impeller handling a particular volume rate of gas at a particular speed is constant regardless of gas density.

Polytropic head may be expressed in terms of enthalpy (BTU/lb) or units of ft-lbs/lb. It is determined by the formula:

$$H_{pt} = \left(\frac{Z_1 + Z_2}{2}\right) R\, T_1 \left(r_c^{\frac{n-1}{n}} - 1\right) \frac{n}{n-1} \qquad \text{(Eq. I-21)}$$

Adiabatic head is determined by the substitution of k for n in the above relation.

Adiabatic head may also be determined through the use of a Mollier diagram. Following a constant entropy line from inlet conditions to the required discharge pressure establishes two points on the chart from which an initial and a final enthalpy may be read. This enthalpy change is related to the adiabatic head by a simple unit-conversion factor:

$$H_{ad} \frac{\text{ft-lb}}{\text{lb}} = \Delta H \text{ BTU/lb} \times 778 \text{ ft-lb/BTU} \qquad \text{(Eq. I-22)}$$

RECIPROCATING COMPRESSORS

The basic components of a reciprocating compressor are the piston, cylinder, and valves.

Reciprocators are the most commonly used compressors in process services. They tend to be the most energy efficient of all alternative designs. This is particularly true of slower speed and large cylinder-bore units. The principle of compression is adaptable to a great number of services including those with variable suction or discharge pressures, variable volume load, low vacuum inlet, or high discharge pressures. Units range in horsepower from fractional to 12,000 HP.

A reciprocating compressor should be strongly considered for applications involving:

- High discharge pressure.
  Plunger-piston type units can compress gasses to over 25,000 psig the highest of any of the common mechanical compressor styles;
- Variable volume duty.
  Reciprocating compressors can be unloaded in discreet steps each changing

the inlet volume with a nearly proportional saving in power input;

– Simultaneous multiple services.

Larger reciprocating compressor frames can accommodate up to 10 crank throws some of which may drive one to three cylinders in tandem. The number of independent services possible with a single driver is obviously large.

Cylinder Arrangements

Smaller compressor units (less than 10 HP) are frequently single-acting in design. These units compress gas using only one side of a piston, the other end being open to the crankcase. The advantages of this design include a minimum of wearing parts, as well as the simplicity of splash lubrication of the cylinder in many cases. Single acting units are most often arranged with cylinders in a vertical or "Y" arrangement (Figure 6-3).

Figure 6-3. Single acting unit arrangement. (Courtesy Ingersol Rand Corp.).

When process gasses must be contained within the cylinder, the double acting cylinder, with a piston rod sealed at the stuffing box, is preferred

Vertical cylinder units range in power capability from fractional HP to about 250 HP. Vertical cylinders are preferred when non-lube service is required (see

Figure 6-4). Larger units, with cylinders measuring 5″ or greater in diameter are most frequently single cylinder in design. Vertical cylinders will not discharge liquids readily and are not recommended for wet or potentially wet services.

Figure 6-4. Multiple cylinder arrangement. (Courtesy Ingersol Rand Corp.).

Single-crank horizontal compressors provide flexibility in design for services ranging from 20 to approximately 200 HP. Cylinders are most commonly double-acting. Multiple cylinders may be driven in tandem from a single crank (Figure 6-5).

"V" or "Y" type compressors with two double-acting cylinders supported from a single frame range from 50 to approximately 500 HP. These units are among the most compact designs for their capacity when floor space is a consideration. A third cylinder mounted vertically may supplement the basic design for increased capacity (Figure 6-6).

Angle type compressors are often used in two stage service. The larger first-stage cylinder is placed vertically to minimize ring wear and to minimize space requirements. A smaller second stage cylinder is mounted horizontally from the same frame. These compressors are available for services ranging from 200 to 700 HP (Figure 6-7).

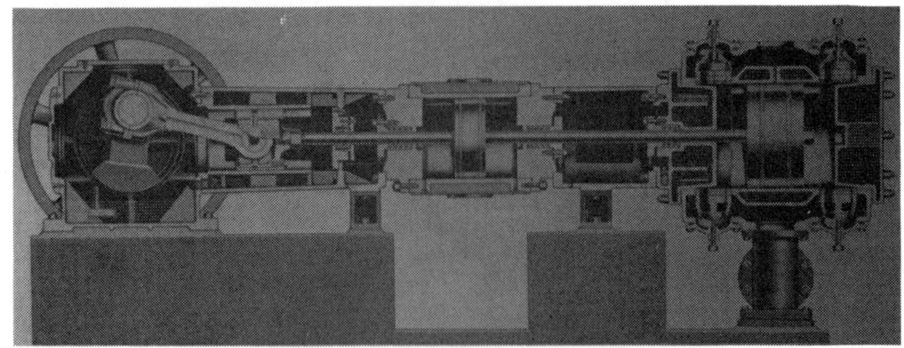

Figure 6-5. Multiple cylinders driven in tandem from a single crank. (Courtesy Ingersol-Rand Corp.).

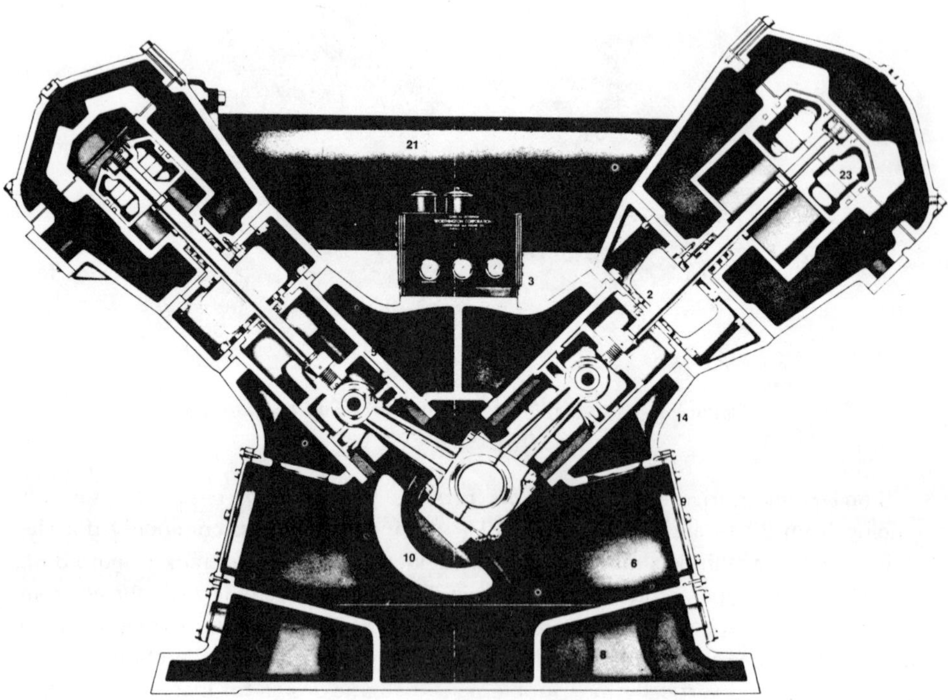

Figure 6-6. A third cylinder mounted vertically may supplement basic design. (Courtesy Ingersol-Rand Corp.).

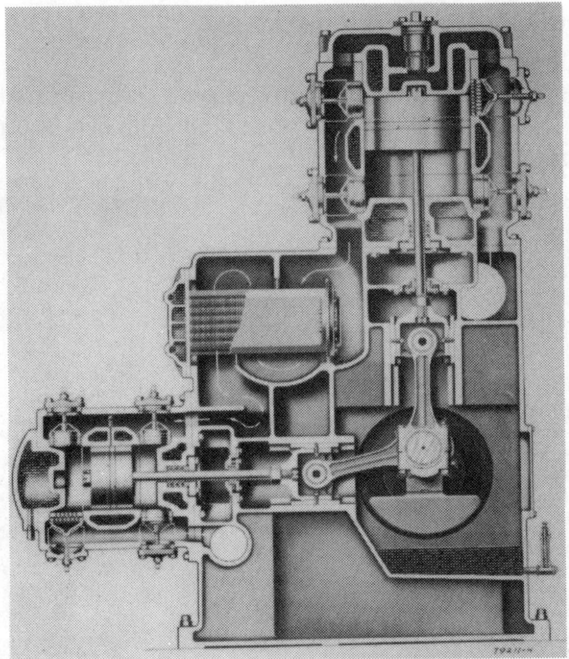

Figure 6-7. A smaller second stage cylinder mounted horizontally. (Courtesy Ingersol Rand Corp.).

Horizontal-opposed or balanced-opposed compressors are in common use for larger capacity services. These units range from 200 to several thousand horsepower (Figure 6-8).

Methods of Rating Compressors

Reciprocating compressors are categorized primarily by the frame. Each style of frame has an optimum horsepower range and speed range. Each frame style also has an array of cylinders, varying in bore diameter and pressure rating, that are designed for compatability with the frame.

The primary considerations that limit the choice of compressor style, presented somewhat in order of relative importance are:

number of services and/or stages;
inlet volume capacity;
discharge pressure;
inlet pressure;
physical characteristics of the gas including ratio of heat capacity, molecular weight, and compressibility;
inlet temperature;

anticipated allowable down-time;
type of motive power available.

Other factors such as chemical corrosiveness, chemical toxicity, available foundation space, or special oil-contamination limitations may also take on varying degrees of importance.

Figure 6-8. Horizontal opposed or balance opposed type unit. (Courtesy Ingersol Rand Corp.).

A preliminary calculation of the horsepower can be arrived at using equation I-12 which makes use of the first six parameters listed above. Allowances must be made for the mechanical efficiency and the compression efficiency as noted on page 282. At this point, the design engineer must rely on model availability data from manufacturers. The anticipated maximum horsepower will place the service in a range of frame styles. However, consideration of the seventh parameter, allowable downtime, will narrow the range considerably.

Higher speed units, although less expensive, will tend to have shorter run times between service periods than slower speed units. Again, the anticipated life of the process may be limited, or capital sensitive suggesting the application of high speed reciprocators (900 RPM and abofe).

The selection of a frame style from any individual manufacturer will fix the range of compatible cylinders that are available. Each cylinder has a specific internal diameter and a maximum working pressure. The total piston displacement (PD) of the cylinder being considered is determined by the internal area of the cylinder, the size of the piston rod, the stroke (piston travel distance allowed by the crankshaft), and the rotative speed of the crankshaft.

For single-acting compressors:

$$PD\ (CFM) = \frac{CA\ (in^2) \times stroke\ (in) \times S\ (rpm)}{1728} \qquad (Eq.\ II\text{-}1)$$

For double-acting compressors the piston displacement is twice the value predicted by the single-acting equation, less the displacement of the piston rod (and tail rod if applicable).

Consideration of piston displacements, which *must* begreater than the ACFM of the intended service, together with the cylinder pressure ratings will narrow the range of cylinders for the application.

Consideration must also be given to the "rod load" or "pin load" rating of the compressor frame. This must be compared to the maximum force imposed by the differential pressure on the cylinders, since this force is transmitted down the piston rod and/or connecting rod and onto the running gear and bearings within the frame.

For a typical double-acting cylinder the rod loading is determined by:

$$RL\ (lbs.) = PA\ (in^2) \times Pd\ (psia) - (PA\ (in^2) - rod\ area\ (in^2)) \times P_s\ (psia) \qquad (Eq.\ II\text{-}2)$$

When the range of cylinders has been narrowed to this point, a closer approximation of the gas handling capabilities of a cylinder may be obtained by an evaluation of the volumetric efficiency.

Volumetric efficiency, piston displacement, and cylinder capacity, are related as follows:

$$ACFM = PD \times VE \qquad (Eq.\ II\text{-}3)$$

Evaluation of the volumetric efficiency requires a knowledge of the cylinder clearance volume, which is commonly expressed as a percentage of the piston displacement. In most cases, the clearance designed into a cylinder will be between 3 and 16%.

Precise evaluation of the volumetric efficiency also requires information concerning the pressure losses at the inlet valves, as well as empirical data concerning the rate at which a particular gas slips past the piston rings, rod packing, and cylinder valves. Reliance on the manufacturer's method of calculation is, of course, necessary for guaranteed performance. However, for preliminary evaluations, the following approximation may be used:

$$VE = 0.97 - C \ \frac{r_c^{\frac{1}{k}}}{Z_{disch}/Z_{inlet}} - 1 \ - 0.01\ r_c \qquad (Eq.\ II\text{-}4)$$

Note that in most cases Z_{disch}/Z_{inlet} will be nearly equal to unity.

Horsepower Determinations

A number of factors cause the compression of a gas within a cylinder to require more horsepower input per unit volume than the theoretical isentropic equation will predict. Chief among the reasons for this extra power requirement is the pressure loss through the valves. Other factors are turbulence, valve leakage, heating of the incoming gas, and slippage past the packing rings. These losses collectively influence the compression efficiency of a cylinder design. Compression efficiency is expressed as the theoretical compression horsepower compared to the actual cylinder indicated horsepower. Present day compressors range from 85 to 93% in C.E.

In addition, there is a certain amount of horsepower expended in simply 'turning the machine over.' These losses include the work lost to friction in the bearings, packing, and piston rings. These losses influence the mechanical efficiency of an entire compressor. Mechanical efficiency is expressed as the ratio of cylinder indicated horsepower to the total brake horsepower of the unit. Preesent day compressors range from 88 to 93% in M.E.

The overall efficiency of a reciprocating compressor is the mathematical product of the mechanical and compression efficiencies. Most modern compressors range from 75 to 88% in overall efficiency.

There is unfortunately no exact and universal method of calculating brake horsepower from service data. A manufacturer's guaranteed rating is the final word in this matter short of startup and testing.

Staging

The selection of interstage pressures is frequently the responsibility of the process engineer. This may be the result of constraints on the process such as special intercooling, or the injection or removal of gasses at specific interstage pressures.

More frequently, the need for multistage operation in a reciprocating unit is determined by one or more of the following conditions:

1. Discharge temperature limitations.

Most compressor manufacturers set a nominal limit on the anticipated discharge temperature of a cylinder. The reasons include considerations of metal stress, machine tolerances, and thermal degradation of lubricants. In the case of oxidizing gasses, corrosion or lubricant flammability may have to be considered. In general, no cylinder should exceed 350°F in adiabatic discharge temperature. Adjusting the ratio of compression at each stage of compression, along with adequate interstage cooling, can insure this. (There are exceptions to this rule, as in the case of certain single stage air units which operate in the range of 500°F in adiabatic discharge temperature but which dissipate enough heat to reduce the actual temperature to between 375 and 425°F).

2. Power economy

A rough estimation of power savings can be made by assuming the following:

a. No stage will exceed a ratio of compression of four.

b. The ratios of compression per stage will be nearly equal and determined by the formula

$$\frac{r_c}{\text{stage}} \cong \sqrt[ns]{r_c \text{ overall}}$$

where ns = the number of stages anticipated. (This does not entirely apply if process conditions dictate interstage pressures).

c. Pressure loss between stages will be 5% of the upstream discharge pressure.

d. Where intercooling is possible, the temperature approach of the heat exchanger will be 20°F.

e. The discharge temperature of any stage will be 325°F or less.

These assumed limitations will lead to an optimum number of stages. Each stage can then be treated as a separate compressor when making approximations of horsepower.

The resulting total power requirement may be used as a basis of comparison when determining whether increasing the number of stages still further is likely to be cost effective.

For precise cost figures, a manufacturer's empirical knowledge of compressor performance, as well as vendor's pricing information, is needed.

3. Rod Loading.

A high ratio of compression can result in a compressive or tensile load on the piston rod that exceeds safe limits. Increasing the number of stages reduces the differential pressure across the piston of the intake cylinder which reduces the rod load. Frame bearings, as well as valves, and piston rings are all subject to less wear as the operating differential pressure per cylinder decreases. Lower maintenance costs can be expected.

Compressor Components

Figures 6-9 and 6-10 illustrate the components of typical double acting compressors.

1. Frame
2. Cylinder
3. Piston and piston rings
4. Crankshaft
5. Valves
6. Connecting rod
7. Oil reservoir
8. Main bearings

9. Crosshead
10. Piston rod
11. Piston rod packing
12. Cylinder cooking jackets
13. Distance piece
14. Oil scraper rings
15. Oil deflection collar
16. Wrist pin

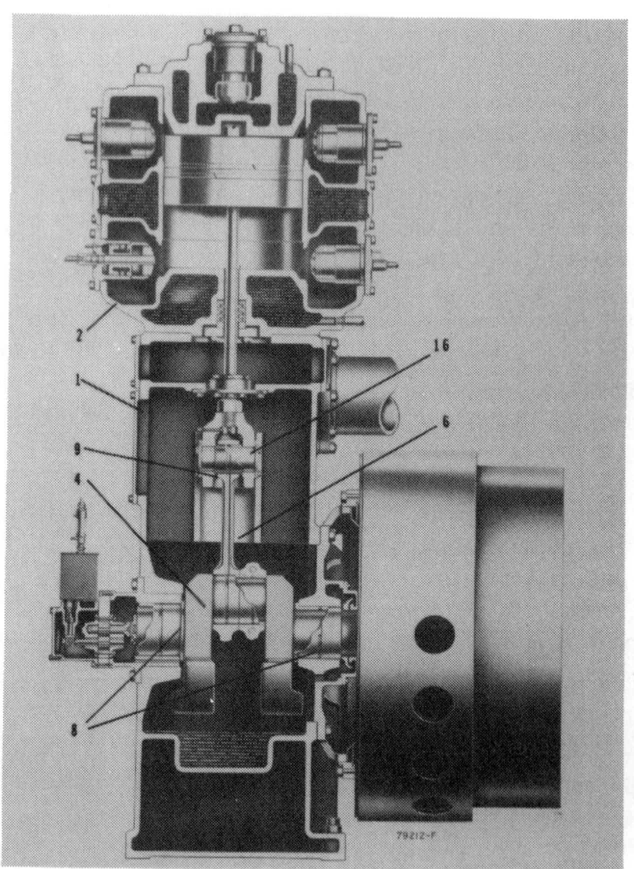

Figure 6-9. (Courtesy of Ingersol Rand Corp.).

Special Considerations

Liners – A cylinder liner is used where changing the diameter of the cylinder is anticipated at some time in the future. The bore diameter may be changed either to meet new capacity requirements or to produce a new refinished cylinder surface after the original surface has been damaged by wear.

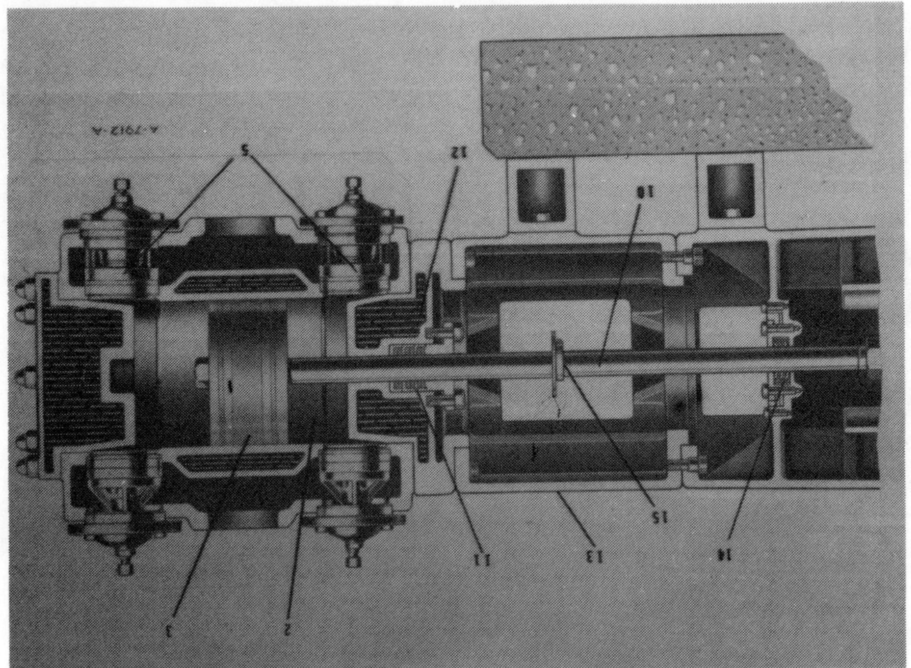

Figure 6-10. (Courtesy of Ingersol Rand Corp.).

Liners may also be required when the material used for the bulk of the cylinder will not provide proper wearing properties at the friction interface. This is the case in most steel cylinders.

Cylinder liners are commonly made of cast iron, although special wear resistant or chemically resistant materials may be ordered.

There are two types of liners in general use. The most common is the dry type, which is essentially a shrunk-fit or pressed sleeve within the original bore. These have the disadvantage of reducing the rate of heat rejection to the jacket water. The alternative wet type liners, are designed for jacket water circulation immediately behind the liner. These have a sealed seam between the compression chamber and the jacket which may leak under some extreme circumstances.

Piston Rods – Piston rods are subject to repeated compression-tension cycle loading. The surface finish of these rods should be as smooth as possible to avoid fatigue cracking. Heat treating, nitrating, or carborizing is sometimes used to harden the surface of this component.

Valves – Valves are the most frequently serviced of all compressor components. Valve designs must strike a compromise between maximizing operating cycles and minimizing valve losses. Maximum operating cycles are attained by valves with minimum lift, and few moving parts. Minimum valve losses (i.e. minimum pressure losses) are attained by valves with high lifts and/or several seating components.

An evaluation of compressor offerings should include a comparison of valve velocities. This parameter is found by dividing the total piston displacement of a cylinder by the total life area of all suction valves (11). In general, this figure should not exceed 7,500 ft/min. However, very low molecular weight gasses, hydrogen in particular, may not induce an adequate valve lift unless velocities are relatively high.

Suction and discharge valves are not interchangeable in function although they may sometimes be interchangeable in port seating dimensions. This is a particular hazard in order units. Care must be exercised by servicemen when replacing inlet and outlet valves, to avoid interchanging the two. These valves may be polarized if desired, by machining or inserting pins to prevent interchanges.

Packing — Piston rod packing in heavy duty compressors is almost invariably the full floating mechanical type. The most frequently used ring material is bronze, although micarta, phenolic resins, PTFE, and other materials or combinations or materials may be used for corrosion resistance (See Figure 6-11).

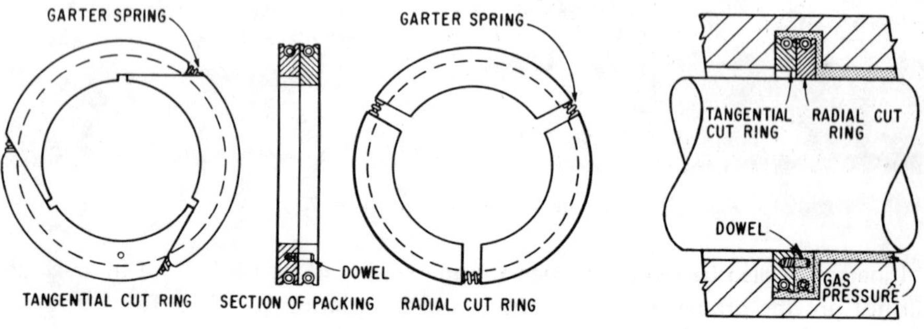

Figure 6-11. Single pair of full floating mechanical packing rings. (Courtesy Ingersol Rand Corp.).

Rod packings will wear out after long running periods even in the cleanest of services. Dirty, wet, or high pressure services will require packing ring replacement much more often.

Distance Piece — A distance piece may be installed between the cylinder and frame for one of three reasons.

1. When it is necessary to prevent carry over of frame lubricants into the cylinder, the distance piece assures that no portion of the rod can travel the distance from the frame oil wiper rings to the cylinder packing.
2. In other cases, the distance piece serves as a means of venting the process gas that leaks past the rod packing. Gasses may be simply vented to the atmosphere through large ports, or they may be purged from an enclosed distance piece with a stream of inert gas. The latter method is used when hazardous gasses are being processed.
3. The distance piece also serves as a service access to the piston rod packing and oil wiper rings.

Capacity Control

Reducing the volumetric capacity of a reciprocating compressor can be accomplished by one of the following means:

Start and Stop

This means may be used when the pressurized gas is being delivered to a storage receiver. Caution must be exercised when sizing a compressor for this type of service to assure that the compressor motor is not subjected to too many starts within a limited time period. Starting is most commonly done while the compressor is unloaded in order to limit the starting torque.

Constant speed controls.

These include cylinder inlet valve unloading mechanisms, special clearance pockets, and external bypassing.

Valve unloaders, the most commonly used capacity reduction device, hold the inlet valves open during the compression stroke of the cylinder, thereby preventing compression from occurring. In a double-acting cylinder, unloaders allow capacity reduction steps of 50% and 0% of full ACFM. Multi cylinder units allow still more combinations of loaded and unloaded cylinders (Figure 6.12).

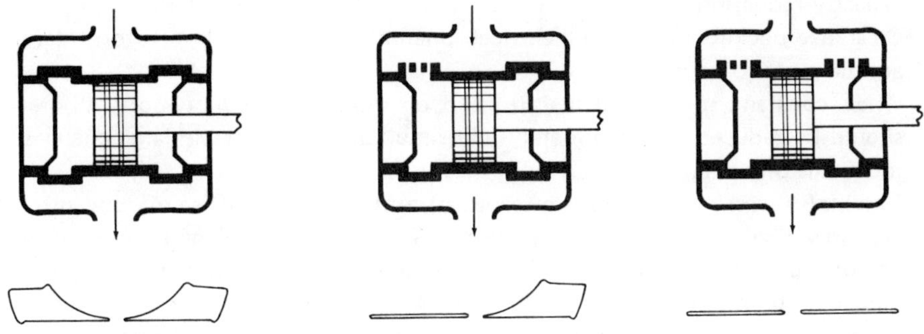

Figure 6-12. Capacity control using valve unloaders. The cylinder in the first figure is loaded on both the head and crank ends (as shown on the indicator cards). In the second figure the head end is unloaded for 50% capacity. In the third figure both ends are unloaded for 0% capacity. (Courtesy Ingersol Rand Corp.

Care must be exercised in the planning stage when a unit will have valve unloaders in a particular service. A check should be made to guarantee that rod load reversals (compression followed by tension loading) will occur when a cylinder is partially unloaded. In addition, multi-cylinder horizontal opposed units should be unloaded symetrically to avoid producing excessive unbalanced forces on the crankshaft and frame.

The power saving of this method of capacity control is not total since some work must be done to move the gas in and out through the valve ports. Generally, this work will be about 5% of the full load for the cylinder. The mechanical friction losses of the compressor also make a contribution to the power consumption. These factors result in a no-load horsepower demand of roughly 25% of full load.

Cylinder unloaders may have electrical, pneumatic, or manual operators.

Clearance pockets reduce the volumetric efficiency of a cylinder by effectively increasing the cylinder clearance by a fixed percentage, or by an incremental amount.

Although clearance pockets may be used at the crank end and/or the head end of a cylinder, they are more commonly located at the head end.

Clearance pockets, and fixed volume clearance bottles have an advantage over inlet valve unloaders in that they do not limit control to large increments of capacity reduction. They may be machined in order for small increments such as 10% reduction. More than one pocket may be used on a cylinder for several steps of reduction. Variable volume pockets are available for still finer control increments.

The power reduction effected by clearance pockets is approximately 85% of the capacity reduction.

Clearance pockets may have electrical, pneumatic, or manual operators. Manual are the most common.

When constant speed controls are used on multi-stage compressors, all stages should be reduced proportionately and simultaneously to avoid large changes in interstage pressure.

External bypassing as a means of capacity control allows the use of conventional gas flow regulating valves. This method is not often used with reciprocating compressors because of the lack of any saving in power, and because of the tendency to build up heat in the recycle loop which requires the use of a bypass-gas cooler. External bypassing of essentially all of a compressor's output is sometimes used as a means of unloading a compressor during startup. However, suction valve unloaders, when available, are the preferred means of unloading for startup.

Cylinder Cooling

The heat of compression and the heat developed by mechanical friction within the cylinders, can be detrimental to the operation of a compressor. High temperature within cylinder walls can cause lubricant failure and the formation of solid deposits. Excessive and uneven heating can change the dimensions of components resulting in faster wear of valves rings and packings. Cylinder castings may even fracture if subjected to uneven heating.

From a thermodynamic viewpoint, heated cylinder surfaces will contribute to a higher horsepower demand and fewer lb/hr of gas delivered.

There are two chief cylinder designs used to dissipate this heat:

Air cooled machines have external fins which act to extend the surface area of cylinders in order to transfer heat to the surrounding atmosphere. In most cases, the convention heat-dissipation rate is increased by incorporating an integral fan into the design.

Air-cooled cylinder compressors are nearly always small units (less than 100 total HP). They are very common in air package units for limited demand service.

Water cooled cylinders are the common choice for heavy duty applications. Cylinder castings include channels for circulating a cooling fluid to maintain a uniform working temperature within the metal walls of the cylinder.

Although the heat dissipation to the cylinder cooling fluid does have a measurable effect on the cylinder discharge temperature as well as the compression efficiency, most general methods of predicting this advantage are not precise. The conservative approach of neglecting the cooling effect of the cylinder on the gas is recommended for all but the most critical process calculations. In those cases, the manufacturer should be consulted for performance data.

In the case of the well-studied air compressor for utility service, the practice has been to assume that 15 to 20% of the accumulated heat that must be dissipated between stages and by aftercoolers, is actually taken by the jacket cooling system.

In order to determine the quantity of water needed for cylinder jackets, an approximate heat rejection rate of 500 BTU/BHP-hr with a 15°F rise in water temperature may be assumed for cast iron cylinders. The heat rejection rate will increase as the cylinder diameter is reduced. The use of dry type cylinder liners, or applications involving gasses with low k values (such as natural gas) will reduce the heat rejection rate by some 50%.

Cooling water should never be cold enough to cause condensation within a cylinder. Severe wear or sudden damage to the compressor can result. Incoming water temperature should be 10 to 15°F above the temperature of the incoming gas. An aftercooler or an intercooler may serve as a convenient supply of warm water.

Where only cool water is available, the rate of circulation through the jackets should be controlled to maintain the water outlet temperature at 15 to 20°F above the gas inlet temperature.

The discharge temperature of the cylinder jacket water should be less than 130°F except where condensation within the cylinder may result.

Some compressor applications will require no cooling. These include low temperature services such as refrigeration, and systems involving low ratios of compression (less than 3) with gasses having a low specific heat ratio as in the case of light hydrocarbons. In these cases, the cylinder jackets may be filled with a heat conducting fluid, such as oil or an antifreeze solution, to distribute temperatures evenly throughout the cylinder casing.

A simple thermosyphon system is often used to move the coolant slowly through the jackets. This involves the use of an external coolant reservoir tank and a small amount of piping to allow the coolant to circulate through a closed loop by heat-induced convention.

Materials of Construction

Compressor cylinder materials are selected primarily for strength.

Cast iron is widely used for services up to approximately 1,200 psig. Smaller bores may sometimes exceed this pressure.

Nodular iron is used in services up to 2,000 psig in smaller cylinders and 1,200 psig in larger bores.

Cast steel is reserved for higher pressure applications in the 1,000 to 2,500 psig range.

Forged steel is used for all pressures exceeding 2,500 psig.

Piston materials are selected for minimum weight to keep inertia forces as low as possible.

Cast iron is the material most commonly used. Larger pisons have internal voids to reduce weight.

Cast aluminum is frequently used in large diameter cylinders, but may be subject to corrosion in many applications.

Lubrication

Reciprocating compressor lubrication is divided into four main categories:

Frame lubrication;
Full cylinder lubrication;
Minimum cylinder lubrication;
Non-lube cylinder systems.

All compressors, including non-lube units, require frame lubrication.

Frame Lubrication

In many smaller units, splash lubrication is provided by the agitating action of the crankshaft rotating in an oil reservoir. In the single-acting compressors, both the frame bearings and cylinder are lubricated by the same atomized oil mist.

Single-acting spash lubricated compressors require frequent oil changes and occasional oil additions as part of a regular maintenance program.

Larger process compressors have force feed lubrication systems. The points of lubrication include the frame running gear, crossheads, the cylinders, and packing. In most cases, the running gear and crossheads will be pressure lubricated through a single oil system, while the cylinders are lubricated through an independent oil system. In most cases, the lubricating oil used in the cylinders will not be identical to the oil used in the running gear (Figure 6-13). This is particularly true in the case of services involving unusual process temperatures, or gasses, requiring special lubricants.

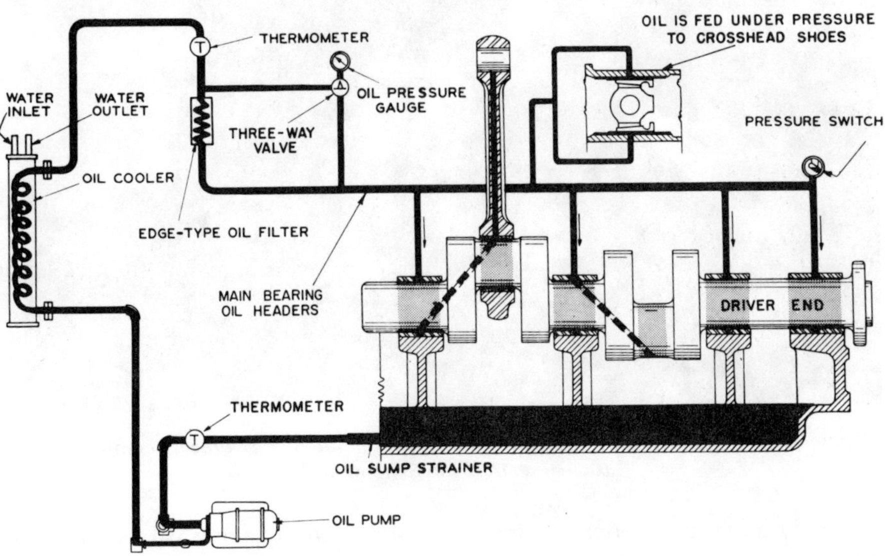

Figure 6-13. Force feel lubrication system. (Courtesy Ingersol Rand Corp.).

The frame lubrication system in larger compressors commonly includes a dust tight crankcase which serves as an oil reservoir, an oil strainer, a small gear-type or centrifugal pump which may be driven from a power take-off on the main drive and an oil filter. Larger units may use an oil cooler.

Some moderate size units (less than approximately 200 HP) may use a flood type system for lubricating the running gear. Frame oil is carried up by a crankshaft driven mechanism within the crankcase, and allowed to flow down to lubricate the jornals.

Full Cylinder Lubrication

The cylinder lubrication system in larger compressors usually includes a multipoint lubricator which is capable of individual and adjustable flow rates to each point. Lubricants are not recycled (see Figure 6-14).

The points of lubrication within a cylinder include the piston rings, the packing, and on occasion, the valves. Of these, the piston rings are the most critical and are invariably fed through ports in the cylinder bore.

Figure 6-14. Multiple feed motor driven lubrication for a large process compressor. (Courtesy Ingersol Rand Corp.).

In many cases, the lubricant being fed to the cylinder bore is carried over onto the piston rod and onto the packing in sufficient quantity to eliminate the need for a separate feed to the packing.

Valves usually do not require a separate lubricant feed. Oil is carried to the valve seats with the gas stream. In some installations, oil may be injected into the gas stream before the compressor inlet to guarantee lubrication of the inlet valves.

Care must be exercised in the selection of cylinder lubricating oils. Lubricants are highly individualized from process to process. Halocarbon-compressor oils cannot be interchanged with lubricants for ammonia compressors, utility air compressor oils cannot be interchanged with lubricants for high pressure cylinders, etc.

Thermal decomposition of an incorrectly prescribed lubricating oil in an air compressor can result in accumulations of soot in the discharge piping. These deposits are combustible.

Compressor manufacturers are able to specify the optimum characteristics of a lubricant for a particular machine in a specified service. However, the final recommendation on the exact lubricant to be used should come from a reputable oil supplier who is willing to guarantee the performance of his product.

In general, the break-in period of a new reciprocating compressor will require relatively large amounts of lubricant. Most often, this oil will not be identical to the lubricant that will be used during the process run.

The rate of cylinder feed is not completely predictable. Proper oil feed rates are determined by observation at startup, and every six months or so thereafter.

Minimum cylinder lubrication is a compromise between full lubrication and non-lube design. Full lubrication offers mechanical reliability to minimize down time, but invariably results in a gas stream contaminated by the lubricant. Non-lubricated cylinders will not contaminate the gas stream to any measurable degree, but are more subject to mechanical wear.

Minimum lube systems incorporate the materials of construction of non-lube cylinders. The mechanical wear of these components is minimized by supplying them with a fine coating of a liquid lubricant.

In the most common design, the rate of lubricant addition is so small that the packing is the only lubricated component. In this case, oil is carried into the cylinder by the rod and slowly migrates to the cylinder walls and valves. The rate of oil feed to the packing itself may be minimized, or eliminated altogether if sufficient crankcase lubricant reaches the packing during normal operation (see Figure 6-15).

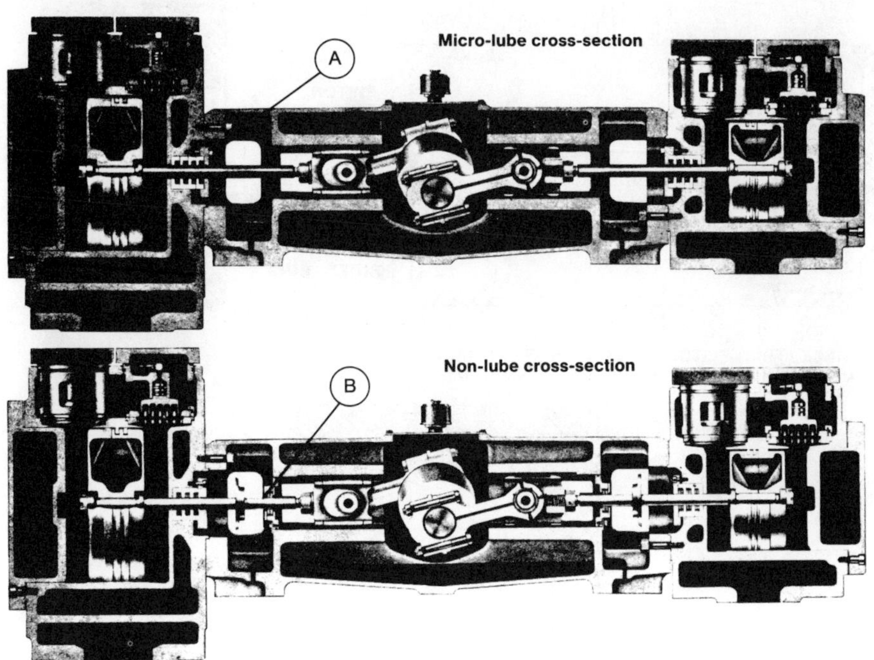

Figure 6-15. (Courtesy Ingersol Rand Corp.).

Non-lubricated Cylinders

Non-lubricated cylinder compressors incorporate the following construction features:

1. Packing rings and compression rings made of a self-lubricating solid such as carbon, reinforced PTFE, combinations of carbon and PTFE, or molybdenum disulphide;
2. A distance piece between the frame and cylinder to guarantee that frame oil cannot enter the cylinder through the packing. An oil-deflection collar may be used to stop the migration of frame oil along the metal surface of the rod;
3. Special valve inserts at the points of wear made of materials similar to those used in the piston rings and packing;
4. An extra ring or set of rings on the piston designed to support and guide the piston. These 'rider rings' are not designed to seal against gas leakage, but rather to take most of the mechanical load and wear that would otherwise be imposed on the piston rings; (see Figure 6-16).
5. A finely honed finish in the cylinder bore and on the piston rod. Cylinder surfaces are commonly honed with PTFE to fill casting pores etc. This process eases the wear on rings during break in.

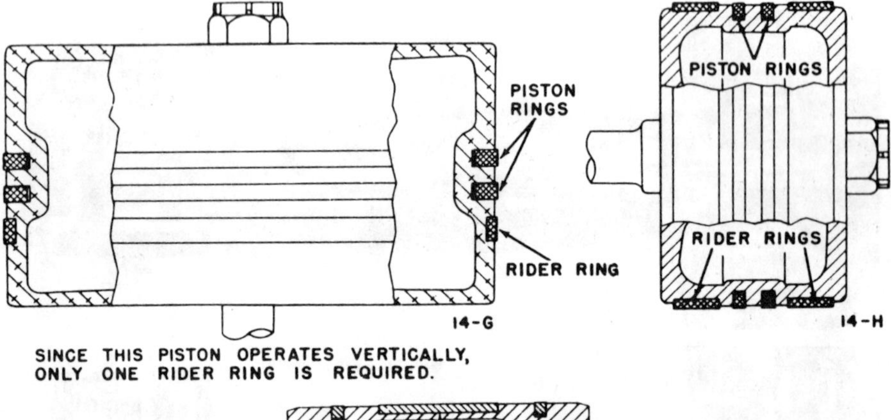

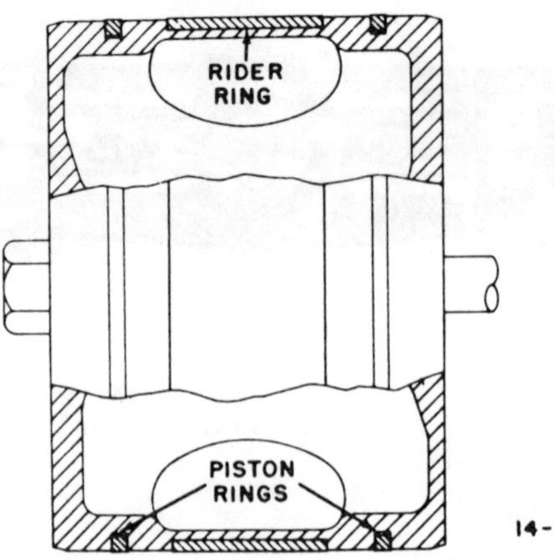

Figure 6-16.

In general, non-lube cylinders will have a lower compression efficiency than corresponding lubricated cylinders. This added CE loss is approximately 5%. This reduction in capacity does not result in a noticeable reduction in horsepower since work is done on virtually all the gas within the cylinder even though some of it slips back to the suction side during compression.

Special precautions must be observed when a non-lube compressor is to be used:

1. The cylinders are particularly sensitive to dirt. Thorough pre-startup cleaning of all process piping is necessary. A suction filter with a rating of 5 microns or finer is preferred to a suction screen. The rate of wear on cylinder walls, piston rings, rod packing, and on the piston rod can be expected to exceed that of a lubricated unit. Careful monitoring of the rate of wear is necessary;
2. Spare cylinder components will be used more frequently in a non-lube compressor than in a comparable lubricated unit;
3. Special prcautions must be taken to prevent the oxidation of metal surfaces during periods of downtime. Ferrous cylinder bores and piston rods are not protected by a lubricant coating and are subject to attack by small amounts of moisture and oxygen. The use of warm circulating jacket water or an inert gas purge may be necessary as part of a standard shut-down or equipment storage procedure.

Compressor Gas Piping

Compressor cylinder ports are most often sized larger than gas piping practices would predict. This is because gas velocities or nominal pressure losses based on steady state flow do not account for an all important criterion in reciprocating compressor line sizing, which is that pressure pulsations must be attenuated.

Gas piping to and from compressor cylinders should not run smaller in diameter than the corresponding port connection size.

Piping runs to the inlet and discharge side of compressor cylinders may have the unplanned for effect of resonating with pulses from the compressor. If unchecked, these oscillations can interfere with flow dynamics within the cylinders, causing a loss of capacity and an extra burden on the compressor driver. In high pressure installations, these pressure waves can do physical damage to the piping installation.

There are three main solutions to the problem of pressure pulsations. The surest of these is to rely on a firm specializing in the design and manufacture of pulsation dampeners. Commercial dampeners incorporate proprietary pulse attenuating devices within a vessel that have the effect of reducing downstream pulses to 1% of the lime pressure. They also may reduce upstream pressure pulses to a specified limit.

Pulsation dampeners tend to be expensive in first cost, but offer the best protection to compressors in services involving high horsepower and high pressure.

The next best alternative to the pulsation dampener is the surge bottle. These are enlarged chambers in the piping, located adjacent to the compressor ports, that act to detune the piping system. Some rules-of-thumb can be applied for evaluation purposes: In services up to 700 psig, they are at least seven times the single-acting displacement volume of the cylinders they serve. Higher pressures require larger bottles. At 2,000 psig, the suggested minimum is at least 15 times the swept cylinder volume (13).

Surge bottles should be located as close to the cylinder as possible. Pressure taps along the chamber may be useful in checking the performance of a volume bottle. Whenever possible, bottles should be equipped with drains.

The third method of averting pulsation resonance, is to design the inlet and discharge piping systems to avoid straight piping lengths, and equivalent piping lengths, that will oscillate in sonic resonance with the compressor. Compressor suppliers can usually assist in this planning. For large installations, an analog study may be needed.

Compressor piping must include an allowance for thermal expansion. It is an unwise practice to bolt a horizontal compressor cylinder down to a section of discharge piping that is rigidly braced against a floor. Other lines that may be subject to pipe thermal expansion such as those with tracing and those handling refrigerants, should not be positioned to create a strain on a cylinder casting or risk distorting cylinder-frame alignment.

Compressor discharge lines and manifolds must include a safety relief device with a pressure rating adequate to protect the cylinder, as well as a capacity rating selected in consideration of the volume between the compressor and the nearest point of isolation (12).

In installations where the inlet piping will not stand full discharge pressure, it is best to consider that the compressor valves may at some time leak high pressure gas back to the suction side. If the inlet piping can be isolated by an upstream valve, then a low-side relief valve is advised.

Basic instrumentation should include inlet and outlet pressure gage taps and/or installed gages with snubbers, a mounted dial type thermometer to monitor discharge temperature, and a means of measuring gas inlet temperatures. Where metallic rod packing is used, a means of measuring the packing temperature will prove useful.

Alarms

The need to monitor the likely trouble spots in a compressor increases in proportion to the importance of the compressor unit in process. It is often the responsibility of the process engineer to select the appropriate alarm points.

Process variables that may endanger the operation of a compressor include:

Suction pressure variation.

Reduction in inlet pressure will change the horsepower requirement of a compressor. If capacity control cannot accommodate this change, then the compressor driver may encounter a peak horsepower demand at some point. A well designed system will be able to operate in this peak range without overloading the driver or compressor frame rating. However, economic considerations may warrant the use of a low inlet pressure alarm.

A more significant effect of low inlet pressure is the probable increase in the ratio of compression. Piston rod loading may be exceeded, or discharge temperatures may run excessively high causing a risk of lubrication failure, cylinder damage, or valve damage.

Discharge pressure variation

Operating a loaded compressor against a dead head can be disasterous. Under no circumstances should a compressed gas system be designed without regard to the maximum allowable working pressure of the cylinders.

Variations in discharge pressure can also have horsepower, rod loading, and temperature effects similar to those caused by low inlet pressure.

Excessive discharge pressure also resents the danger of condensation within the gas being compressed. Lubrication failure, valve damage, or severe damage to the cylinder head and drive train may result.

Inlet temperatures.

At a constant ratio of compression, the discharge temperature of a compressor cylinder is very nearly a multiple of the absolute tempeature of the inlet gas. Good process designs should not depend on the cooling jackets of cylinders to relieve high adiabatic discharge temperatures.

Composition changes.

Failure to account for the addition of small amounts of corrosive or chemically active materials in the process stream can result in lubrication failure, accelerated corrosion of wearing parts, and the voiding of manufacturer's warranty.

Composition variables should be controlled to preclude condensation within cylinders.

Features within the compressor can also endanger its operation. These include:

High discharge temperature.

No single symptom is a more versatile diagnostic clue. High discharge temperature occurs when:

- ratios of compression are running high;
- valves are broken, worn, or seating improperly;
- cylinder cooling is inadequate;
- piston rings are worn or damaged, allowing excessive slip;
- cylinder walls are out of round, scored, or worn;
- cylinder lubrication is inadequate.

Low discharge pressure.

Loss of pressure may be attributed to valve failure, worn rings or worn packing. However, this symptom is more likely to be the result of conditions external to the compressor, such as a downstream increase in demand for gas or a plugged inlet filter.

Low frame oil pressure.
Probable causes include lube pump failure, auxiliary drive failure, contaminated oil, an oil leak, a plugged oil filter, or loss of oil.

Low frame oil level.

High frame oil temperature.
This condition is likely to occur when a bearing has failed. It might also be the result of a failure in the oil cooler.

Low frame oil flow.
This is a condition caused by plugged lubrication ports, a plugged filter, a mechanical failure of the pump, or a loss of oil.

Low cylinder lubricating oil level.

Cylinder lubricator rotation failure.

Excessive vibration.
This may be the result of a bearing failure, a valve failure, a valve unloader failure, liquid in a cylinder, a lubrication failure, a weakening in the compressor foundation, or a failure in the hold down bolts.

High interstage temperature.
Multi-stage compressors with integral watercooled intercoolers depend on a constant flow of cooling water to avoid a high inlet temperature on the stage downstream of the intercooler. A rise in the interstage pressure and an uncontrolled shifting in the ratios of compression will result from an intercooler failure.

Many compressor manufacturers are able to supply standard annunciator panels containing an audible alarm, labeled warning lights (usually arranged to identify the first point of failure), as well as timers and switch contacts to atuomatically shut the compressor down when necessary. The alarm points are often selected by the customer.

Drivers

Present day reciprocating compressors are driven by electric motors, steam engines, steam turbines, gas engines and turbines, diesel and gasoline engines.

The choice of driver type will be decided primarily by the available source of power and secondly by the relative added cost of the drive component. Once a driver preference is decided upon, it is usually best to allow the compressor manufacturer to assume the responsibility of engineering the drive system and obtaining the driver.

Electric motors rank among the most efficient of drivers and are generally considered first when in-plant process systems are being designed.

The squirrel cage induction motor often has a cost advantage over the alternative synchronous motor in services ranging to approximately 200 HP. (This nominal horsepower limit is by no means a hard rule. Applications of compressors operating at several thousand HP can be found).

A wide variety of motor designs, simplicity of application and general reliability make the induction type motor the most popular electrical driver in current use. On the other hand, they have the disadvantage of a lagging power factor, as well as a slightly higher power consumption to power output ratio when compared to the synchronous motor.

Induction motors may be direct coupled, belted, or flange mounted onto a reciprocating compressor.

Synchronous motors operate at a precise rotative speed. They tend to be slightly more power efficient than their induction motor counterparts, and have the often useful advantage of a leading power factor.

Synchronous motors are most often applied to larger compressors, usually without speed reduction. They may be direct coupled, flange mounted, or mounted on the compressor shaft in an engine type arrangement.

Induction and synchronous motors may be required to operate in corrosive or flammable atmospheres. Attention to the National Electric Code, the NFPA and local ordinances is advised. Special venting, a special enclosure, or special materials of construction may be required.

The choice of motor enclosure will also be influenced by the nature of the operating atmosphere. Standard casing designs include open, drip proof, splash proof, totally enclosed, and explosion proof, among others.

A *steam engine* may be used to drive a compressor in applications where surplus steam is available. Steam engine drive offers the advantage of relatively high efficiency in comparison to the alternative steam turbine. It also provides flexibility in capacity control since its rotative speed is infinitely variable through a wide range. Capacity reduction is achieved with a very nearly proportional saving in power consumption.

Units range from about 50 to 4,000 HP.

Compressors with integral steam engines are among the oldest and most reliable compressor-unit designs available. Some less common designs use a compressor-drive coupling arrangement.

Steam engines have three disadvantages:

The steam exhaust from most engines will contain small amounts of lubricating oil which can affect the performance of the boiler if condensate is to be reused. Non-lube adaptations are available, but only at a sacrifice of on-line availability.

Steam engine drives tend to be bulky in comparison to electric motor drives.

The efficiency advantage of the steam engine over the turbine tends to diminish when high pressure (over 300 psig) and high temperature steam is to be used.

Steam turbines operate at relatively high speeds and require a reducing gear to drive standard reciprocating compressors.

Steam turbines can match and exceed the range of horsepower covered by steam engines. They have the added advantage of oil-free discharge.

The chief mechanical disadvantage of the steam turbine is the necessity of the reducing gear.

Gas engines and gas turbines are the preferred type of driver where fuel gas is plentiful. These sites include gas transmission and well head stations, gas storage facilities, fuel gas processing plants, and gasoline refining facilities, among others.

Integral-engine compressors have been the preferred design in very large installations although direct coupled engine drives are not uncommon. Gas engines may be 2 or 4 stroke cycle in design, with supercharging added to some units to improve thermal efficiency. Adjustments in compressor capacity can be made easily by varying the speed of the engine driver.

Gas turbines require a speed reducing gear to drive reciprocating compressors.

Some applications have incorporated exhaust gas heat exchangers to improve the overall efficiency of plant processes.

Diesel oil fueled engines are used less frequently than other drivers. These may be integral compressor-driver units or direct coupled compressor-driver assemblies.

These engines may be designed for dual fuel application to take advantage of seasonal changes in fuel prices and fuel availability.

Installation

Reciprocating compressors vary in foundation requirements. Some compressors have multiple pistons and counterweights arranged to minimize net dynamic forces transmitted through the frame. These balanced units may be mounted on a structural steel skid. Smaller 'Y' type units may need no more than a strong concrete floor as an adequate foundation.

However, most reciprocating compressor designs require an independent poured concrete foundation to hold machine alignment, and to assure that vibration forces from the compressor are not transmitted to surrounding structures.

Basic and general recommendations on the type of foundation needed for the compressor in question are provided by the manufacturer. However, the compressor purchaser assumes the responsibility of investigating the bearing capacity of the supporting soil.

ROTARY COMPRESSORS

A number of positive displacement compressor styles provide alternatives to the reciprocating compressor. These include the helical or rotary screw type, the rotary lobe type, the sliding vane design and the liquid piston type among others. Compressors in this category offer the advantage of very low shaking forces, minimal foundation requirements, relatively few wearing components and relatively compact dimensions.

Helical Compressors

The basic components of a rotary screw compressor are the main rotor, the secondary rotor, and the housing or cylinder. Compression is accomplished without cyclical valve operation.

The principle of compression involved may be envisioned by examining either rotor separately. Each rotor, encased in a closely fitted cylinder housing, forms hollow spiral-shaped cavities. Gas enters these hollow chambers as they rotate past an inlet port in the housing. The intermeshing between rotors effectively forms a moveable seal within each of the cavities. Compression occurs during rotation as the cavity travels down the housing toward the discharge port. Cavity volume is reduced until the chamber encounters an opening discharge port.

Like many other compressors the most frequent application of the rotary screw machine is in compressing air. Helical compressors in this service are nearly always supplied as packaged units.

Apart from air service, the rotary screw compressor has been used for handling HN_3, coke oven gasses, fluorocarbon refrigerants, wet H_2S, various organic vapors, saturated steam, and a variety of other applications.

The oil-flooded design and the dry type are the most popular units in current use.

Dry Type Rotary Screw Compressors (See Figure 6-17)

The mating rotors of a dry type rotary screw compressor do not make contact with each other within the compression chamber. Timing gears are used to maintain a small but finite clearance between the mating surfaces as the rotors turn. Typically, a dry type units operate at rotative speeds that are three times faster than oil flooded models. Speeds exceeding 10,000 rpm are possible depending on the size of the unit. Alternatively, some models are designed for direct coupling to 3,600 to 1,800 rpm motors.

Dry type screw compressors may be considered for applications ranging up to 9,000 h.p. Capacities ranging to 26,000 CFM (of air) are available.

The differential pressure across the compressor is limited by the allowable deflection of the rotors. In some standard units this differential limit is approximately 80 psi, in others, it extends to 170 psi.

The highest discharge pressure attainable is dependent on the strength of the casing. Special designs are available for single-stage services exceeding 450 psig.

Discharge temperatures are limited to those values that do not cause a significant rotor growth through thermal expansion. Designs for temperatures up to approximately 450°F are available.

These compressors should be strongly considered for services involving:

1. Gasses with entrainments.

Non-contact operation of the rotors often permits the handling of droplets and fine solid particles that would cause unacceptable wear in a reciprocating or centrifugal unit of comparable size. Screw compressors can provide reliable service in 'dirty' applications as found in pollution abatement systems, in compressing vertical pyrolysis furnace off gases, or in wet applications such as the mechanical recompression of vapors in evaporator systems.

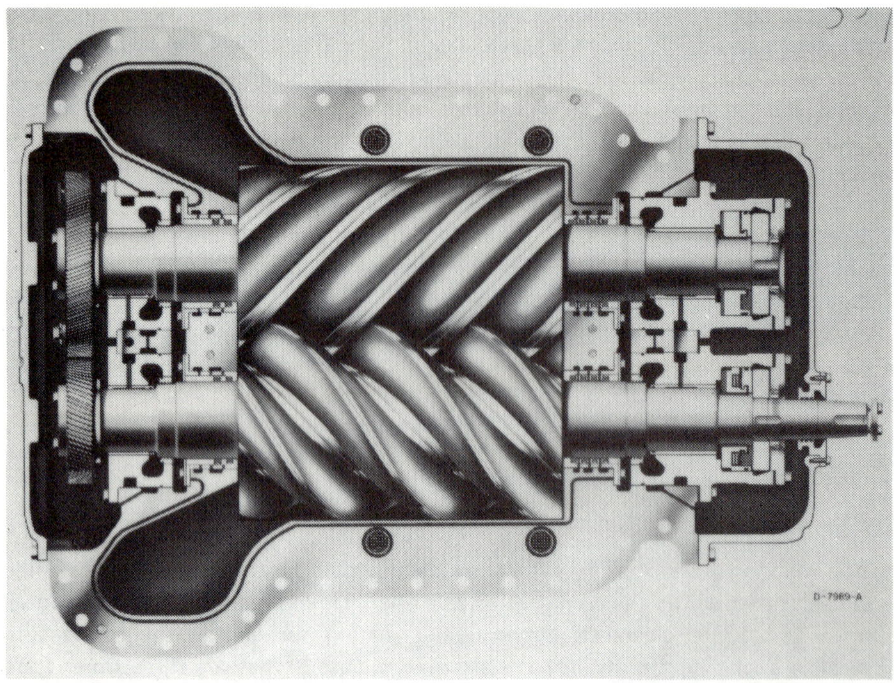

Figure 6-17. Dry type rotary screw compressor. (Courtesy Ingersol Rand Corp.).

Although a properly applied screw compressor may be the best choice for a specific difficult service, some wear and eventual loss of efficiency must be expected.

2. Gasses with entrainments that contain fouling materials.

Some rotary screw compressor designs will accept the continuous injection of a liquid into the compressor cavity. The liquid acts as both a gas coolant and a rotor cleaning solvent.

Examples include the injection of water into a unit compressing vapors from an evaporator concentrating an inorganic solution, where entrained droplets might otherwise result in encrustation. In this case, the injection of water prevents solids accumulation on the rotors. Another example is the compression of spray tower discharge gasses which may have entrainments with dissolved hard-water minerals or scrubber-liquor compounds.

The recommendation of a particular solvent is usually the responsibility of the design engineer. Water is the most common medium except where strongly reactive gasses are encountered. Alternatives include orthodichlorobenzene, gas-oil, kerosene, liquid NH_3 and others. The injection of a coolant or solvent may increase the compression efficiency of a helical unit by partially sealing the rotor gap and reducing the gas slip loss. On the other hand the injection of a liquid will induce some hydraulic power losses. In addition, any vapors generated by an evaporative cooling effect must also be compressed with the main stream flow. The net horsepower advantage or disadvantage of coolant injection is often negligible when compared to dry adiabatic performance.

3. Gasses that polymerize at elevated temperatures.

In some applications, the cooling effect of liquid injection can avert the formation of sludges, films or resins that would otherwise form at higher temperatures and pressures.

Wet hydrogen cyanide and acrylic acid vapors are examples of materials that have been successfully compressed in rotary screw machines in spite of their tendency to polymerize under conditions developed by adiabatic compression.

4. Gas streams of variable molecular weight.

The screw compressor, being a positive displacement unit, will move gas at a fixed volume rate despite variances in gas density.

5. Gasses that dissociate or react at high temperatures.

The injection of a liquid coolant can reduce the heat of compression in gas streams down to acceptable level.

Phosgene is an example of a temperature sensitive compound that has been successfully handled.

Construction

Casings are most often cast iron for services to 200 psig. Cast steel can be supplied for higher pressure services.

Rotor shafts are commonly made of steel. The rotors themselves may be independently machined components made of cast iron, or they may be steel components fabricated as one integral piece with the shaft for still greater mechanical strength.

Special materials of construction such as Ni resist, Hastalloy® coatings, or stainless steels can be supplied for particularly corrosive services.

Shaft seals are commonly carbon ring labyrinth design, but may be modified by standard options such as purge, vacuum vent, or liquid end fittings. A number of other more specialized designs, including mechanical seals, are also available.

Bearings may be splash lubricated in light duty compresssors. Force feed lubrication of the bearings and timing gear is the preferred method in large units. Lube oil pumps may be shaft driven or external and motorized. Lubrication systems may be of a manufacturers standard design. to suit the installation site and operating conditions or built specially to meet particular industry or user specifications. Oil pumping, cooling, filtering, and pressure control for single or multiple services are common to all.

Operating Characteristics

The theoretical indicator card for the rotary screw compressor resembles that of a reciprocating compressor with zero cylinder clearance. However unlike the reciprocating compressor, the opening of the discharge port depends on axial port placement and not on differential pressure between the compression cavity and the discharge line. A rotary screw compressor will compress gas within its lobes through a specific volume reduction ratio despite the line pressure outside of the machine. If, when the compressor discharge port opens, the pressure within the chamber exceeds the pressure in the line, then the gas compressed within the machine will expand into the discharge line. On the other hand, if the line pressure exceeds the pressure within the compression chamber when the discharge port opens, then some of the gas from the line will flow back into the compression chamber. Consequently, any mismatch between the design ratio of compression and the actual service ratio will result in a small loss of power efficiency.

The optimum ratio of compression (p_2/p_1) for any rotary screw compressor is determined by the thermodynamic properties of the gas and by the volume reduction ratio built into the machine in question. Typical ratios of compression, based on atmospheric air inlet, are 1.5:1 and 2:1 for units capable of 35 psig discharge pressure, and 4.2:1 for high pressure machines.

Multiple-stage compression is achieved through the use of multiple casing compressor units with a common drive shaft. The overall ratio of compression within two-stage compressor arangements is typically 10.

External intercooling between stages is often more power efficient than direct injection of a coolant into the compression chamber.

The horsepower requirements of rotary screw compressors are fundamentally dependent on features that vary from design to design. These features include lobe clearances, rotor cross-section designs, and liquid injection capabilities. For this reason, the relative efficiencies of these compressors varies widely between styles.

However, as a first approximation it may be assumed that a dry rotary screw compressor discharging above 20 psig will have a power efficiency comparable to that of a centrifugal compressor operating at its design point.

In general, as the ratio of volume flow rate to clearance area increases, the efficiency of a rotary screw machine rises. Thus these compressors tend to be more efficient at higher volumetric flows.

Cooling Water

Except for low temperature or very low ratio services, dry type compressors will require cooling water to maintain uniform temperatures throughout the casing for dimensional stability. The bearing and gear lubrication systems may also have heat exchangers that require cooling water.

The effect of jacket water cooling on the thermodynamic performance of a rotary compressor is usually negligible. However, the use of cool water in the jackets, and consequential condensation within the compressor, is less of a hazard in these machines than it is in reciprocating cylinders. The compressor manufacturer should be consulted for temperature limitations.

In general, cooling water flow rate requirements range from approximately 1 to 9 gpm. A typical cooling water discharge temperature is 160°F.

Performance Quotations

Since the volumetric efficiencies of rotary screw compressors are not easily calculated by the process engineer, the relative volumetric displacement is of little value as a basis of comparison between competitive units. Vendor performance quotations are essential in the early stages of design.

Performance guarantees for domestic designs are most often limited to ± 4% of capacity and horsepower.

Noise

Rotary screw compressors discharge gas in pulses. The number of pulses per second and the energy dissipated to the surroundings with each pulse often produces objectionable noise within the audible range. Most machines can exceed 103 dB in normal use. Some generate much less sound, and some installations have required little noise control equipment.

If noise is seen as a problem but the gas being compressed contains materials that can foul the complex internal surfaces of commercial silencers, then alternative control methods must be used. These alternatives may include the use of an acoustic enclosure surrounding the installation, or personnel access restrictions.

The proper selection of compressor silencers and noise abatement enclosures is part of the skill of most application engineers who are concerned with rotary screw compressors as a product line. It is usually wise to allow the vendor to specify and supply the necessary silencing equipment.

Special Precautions

Rotary screw compressors can operate as expanders. A discharge check valve is recommended to prevent reverse rotation after shutdown. If two or more compressors are installed in a parallel arrangement, a means of isolating the discharge of each machine is necessary to prevent possible damage to any compressor, and to prevent the back flow of pressurized process gas.

Unrestrined reverse rotation poses the danger of exceeding the rotative tip speed limit of the driver, or reversing the loading direction on the thrust bearings restraining the compressor shafts.

When simple check valves cannot be used because of fouling problems, then valves with powered operators and suitable controls may substitute.

Capacity Control

Dry rotary screw compressors are best applied as base load units. The common constant-speed methods of capacity control tend to be relatively inefficient from the standpoint of power savings.

Variable speed control is possible with adjustable speed drivers such as engines and turbines. However, the rate of gas slip becomes more significant as the inlet capacity decreases. The lower limit of this method of control is reached when the volume of gas slipping back to the inlet is equal to the forward rate, or when the discharge temperature limit is reached. In general, these compressors rarely operate at less than 50% of design speed. Nevertheless, variable speed control is easily the most efficient means of capacity reduction.

Throttling the inlet of a dry type compressor usually does not result in a substantial power saving. Moreover, throttling the inlet may result in excessively high discharge temperatures or unacceptable differential-pressure loading. This method of capacity control is less preferable than gas bypass or speed reduction.

Although returning part of the discharge stream back to the compressor inlet offers no power saving as a method of capacity control, the practice does permit better control of heat buildup. If necessary, a bypass-gas heat exchanger may be installed to remove the heat of compression.

Oil Flooded Rotary Screw Compressors

This style of compressor incorporates a means of injecting a lubricating oil directly into the compression chamber. This is done for three principal reasons: (1) The oil also acts as a cooling agent allowing the compressor to operate with low discharge temperatures. (2) In many units, the presence of a lubricant also allows the primary rotor to drive the secondary rotor directly within the compression housing. (3) The oil acts to seal the clearance space between rotors and casing thereby reducing gas slip loss. These particular designs do not require timing gears to keep the rotors from making contact with each other.

Capacities of oil flooded designs vary from 7½ to 700 horsepower, with discharge pressures up to 150 psig for air applications, and approximately 300 psig in some process units. The largest standard compressors are capable of handling more than 3,000 CFM of air.

Rotative speeds vary but are always slower than comparably sized dry type rotary screw designs. In general, larger units operate at 1,800 RPM and smaller designs run at 3,600 RPM. Approximately 5,000 RPM is a typical upper limit.

Optimal tip speeds are in the range of 65 to 115 ft/sec. Below this speed range, the rate of gas slip past the rotors, which tends to be constant, becomes significant in comparison to the volume of gas being compressed. At speeds above the optional range, dynamic losses adversely affect the operating efficiency. These include hydraulic losses introduced as a result of oil injection.

The majority of oil injected rotary screw compressors are supplied as part of air package units. However, these compressors may also be considered for various other services where oil-gas contact is not objectionable. These include the compression of ammonia and fluorocarbons in refrigeration systems, and the compression of inert gasses, such as nitrogen, for process or utility service.

Oil flooded compressors tend to be most competitive in the horsepower ranges between the very small (less than 10 hp) units where air cooled reciprocating units dominate, and the low ranges of the centrifugal compressors (approximately 500 hp) which dominate the large-capacity field.

Operating Characteristics

The oil injected into the compression chamber absorbs much of the heat of compression. This results in a compression cycle that is closer to isothermal than adiabatic. The heat transfer process is so complete that for all practical purposes oil and gas are discharged at a single temperature. Usually, this temperature is well below 200°F.

Oil is cooled to approximately 140° F before it is reinjected into the compressor. Lower temperatures would present a risk of moisture condensation and consequent oil contamination. Therefore, automatic oil temperature controls are most often supplied as standard package equipment.

The relatively low discharge temperatures generated by these compressors permits them to operate at high ratios of compression. Most standard oil flooded machines are capable of discharging to 125 psig from atmospheric air inlet, in a single stage. Some are capable of still higher ratios.

Variances in design prevent any precise and universal correlations for predicting horsepower from gas service data. However, single stage oil flooded rotary screw compressors for air service require approximately 20 hp for each 100 SCFM of air compressed to 110 psig. In general, an oil flooded screw compressor will require slightly less than 110% of the power supplied to a two-stage water cooled reciprocating compressor in the same service.

Two-stage oil flooded compressors may offer some improved power efficiency over their single-stage counterparts. However, they are applied much less often than the simpler and less costly single stage machines.

Oil flooded rotary screw comprssor units incorporate oil separation devices which reduce the carry-over of oil within the discharge stream down to 8—10 ppm by weight, in most cases. Dual demisters are able to reduce this concentration to 2—5 ppm, and specialized units claim removal down to less than 1 ppm.

Special Considerations

Rotary screw compressors will act as expanders. In order to prevent back flow through these machines after shutdown, a check valve is commonly installed in the discharge piping. This check valve is usually included in the compressor equipment package.

Noise

Rotary screw compressors discharge pressurized gas in rapid pulses. This can result in objectionable noise levels unless acoustic control devices are used. These may be built into the compressor equipment package.

In general, oil flooded rotary screw compressors generate less nosie than their oil free counterparts because of the dampening effect of the injected oil.

Capacity Control

Capacity reduction is frequently accomplished by throttling the inlet stream. Automatic controls, designed to maintain a relatively constant receiver pressure, are used to modulate the inlet throttling valve to reduce capacity down to approximately 30% to 100% of full load. Some modest power savings may be realized, but most of the sources of power inefficiency are not reduced. Thus at 0% load, these compressors still require approximately 75% of full load power.

Inlet throttling is sometimes supplemented with a means of relieving pressure at the discharge side of the compressor whenever the modulating controller persistently signals for low flow. This method can reduce the idling power draw to less than 25% of full load.

Long no-demand periods may warant the use of start-stop control in addition to constant speed control in a dual control system.

Internal control devices are available on some units. These act to change the inlet or discharge ports in the compressor cylinder, which reduces the effective length of the rotor within the compressor. These displacement-control systems present an opportunity for significant savings over inlet throttling in installations where large changes in loaded operation are anticipated.

Liquid Ring Type Rotary Compressors

The essential components of the liquid ring compressor are the rotor, the eccentric or oval-shaped body, and a quantity of liquid formed into a ring within the body by the turning rotor.

The principle of compression is illustrated in Figure 6-18. As the rotor turns, a series of forward curved rotor vanes induce the liquid within the body to follow the rotor in a cyclical path. The moving liquid forms a ring that aligns with the contours of the body (Figure 6-18).

Each set of blades forms a chamber that changes in volume as the surface of the liquid within the chamber follows the contour of the adjacent body wall. Inlet and outlet ports are located on a central cone surrounding the rotor shaft.

Liquid ring compressors are available in capacities ranging from 2 to 16,000

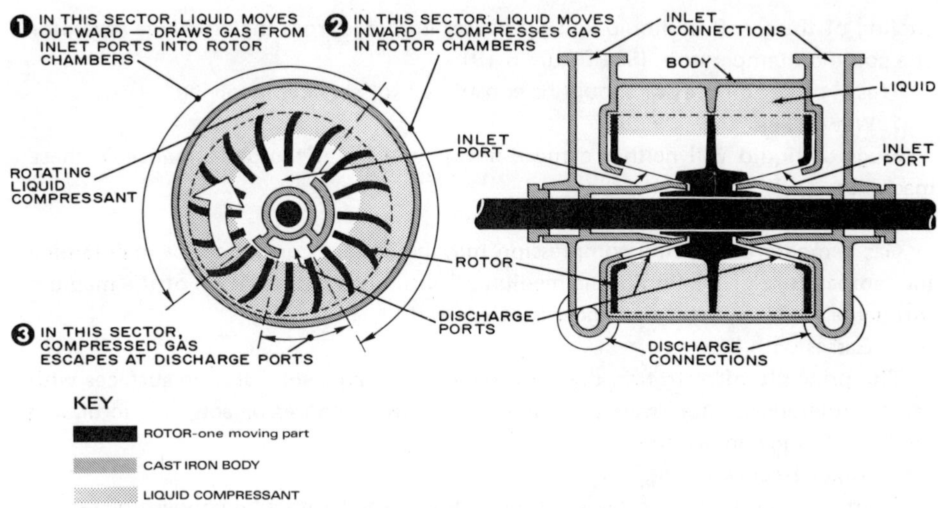

Figure 6-18. Schematic of a type CL compressor. (Courtesy Nash Engineering Co.)

Figure 6-19. Components of a typical eccentric lube liquid ring compressor (Courtesy Nash Engineering Co.).

SCFM. Discharge pressures range up to 125 psig. Vacuum service is one of the most frequent applications.

Intimate contact with a liquid "piston" during compression assures thorough cooling of the gas. Both liquid and gas phases immerge from the discharge separator at a common temperature (See Figure 6-19).

This type of compressor should be considered for services involving:

1. Wet Gasses.

Slugs of liquid will neither damage nor greatly upset the performance of these machines.

2. Temperature Limitations.

Gas temperature during compression may be closely controlled by maintaining the temperature of the liquid ring medium. Continuous recirculation of the medium through an external heat exchanger is often used.

3. Gasses with Dust.

The principle of operation does not depend on large solid sealing surfaces with careful tolerances. Moreover, the liquid within the compressor acts as a scrubbing medium during compression.

4. Vapor Recovery.

These compressors can serve the function of gas separators by condensing saturated components out of a gas stream while passing the non-condensibles.

5. Corrosive Gasses.

The liquid medium coats most of the internal surfaces of the compressor. For example, flooding a compressor built with standard iron construction with an alkaline liquor allows the machine to handle gasses with acid vapors.

6. Limited Allowable Down Time.

The simplicity of these condensers is relfected in minimal servicing and down time.

Operating Characteristics

Liquid ring compressors require a constant supply of liquid. The liquid compressant may be supplied in a once-through piping arrangement, or it may be recirculated from the discharge separator, through a heat exchanger, and back to the inlet. In the recirculation arrangement, the compressor itself may provide the pumping action needed to move the liquid. (See Figure 6-20).

In virtually every case, a discharge liquid-gas separator is necessary. This is often provided as a standard compressor accessory.

Liquid ring compressors tend to be less power efficient than other designs. Their actual performance characteristics vary with the conditions of service. Standard performance data is published for air service using water as a liquid compressant. The compressor capacity and horsepower for most services can be approximated from the standard tables by using an equivalent CFM with a calculated allowance for the vaporization of the compressant. More precise predictions of performance for gasses other than air or a liquid compressant other than water will usually require consideration of the following:

1) Liquid compressant specific gravity, viscosity, specific heat and vapor pressure.

Figure 6-20. Instrument air compressor package. (Courtesy Nash Engineering Co.).

2) Solubility of the gas in the liquid compressant.

3) Effect of condensibles in the gas stream on liquid compressant characteristics. Manufacturers' emperical data describing the effect of these factors on compressor performance is useful when available.

The method of compression and the porting arrangments designed into these compressors permit compression with a near absence of pulsation.

Multiple stage, single casing liquid ring compressors are available for increased ratios of compression.

Construction

Cast iron is the most common material used in the bodies of liquid ring compressors. Bronze may be used for internal fittings, and in some units, for rotors and cones. All iron construction is also available.

Special corrosion resistant alloy construciton may also be supplied.

Two styles predominate:

The *eccentric lobe* provides one compression cycle per chamber per revolution. This design imposes lateral forces on the rotor shaft as compression occurs. It is therefore limited to the lower pressure ranges (approximately 20 psig and under). The eccentric lobe design tends to be more power efficient than the alternative double lobe design, and is therefore the preferred style in the higher volume ranges.

The *double lobe* design has two cycles of compression per chamber per revolution. Forces on the shaft are balanced, permitting higher ratios of compression than the eccentric lobe design. (See Figures 6-21 and 6-22).

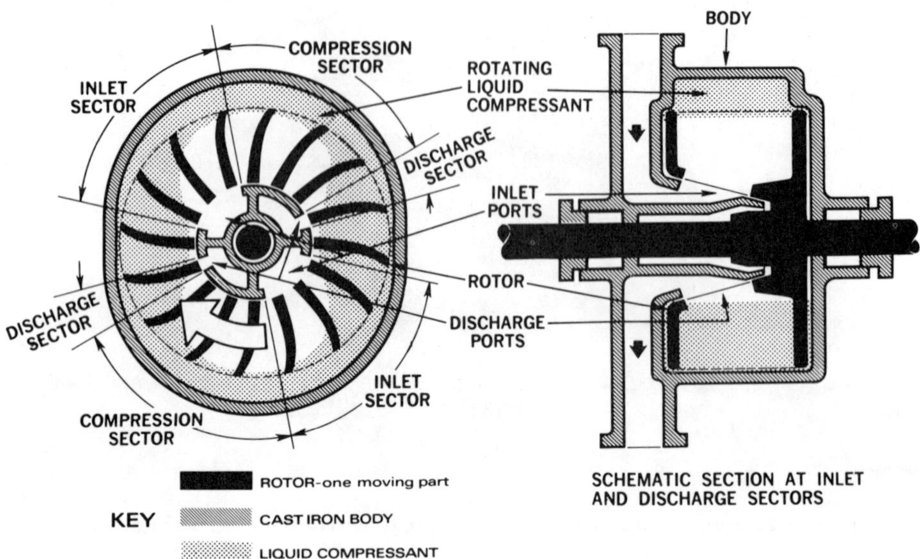

Figure 6-21. Functional schematic of double lobe pumps. (Courtesy Nash Engineering Co.).

Capacity Control

Discharge presssure control and capacity control is most often accomplished through an external bypass line. An alternative method of constant speed control used with some units consists of automatically draining part of the liquid "compressant" which allows the units to operate unloaded for a limited period while conserving power.

Conventional start-stop control may also be used.

Alternative Designs

Several other compressor styles contribute their particular advantages to the field of rotary compressors.

The *sliding vane* is among the oldest and most reliable of rotary models.

Figure 6-22. Components of double lobe design liquid ring compressor. (Courtesy Nash Engineering Co.).

Auxiliary package components are similar to those used in rotary screw units.

Single-stage jacket water-cooled vane type compressors range in capacity from 33 to 3,250 CFM and to 50 psig in discharge pressure. (See Figure 6-23).

Two-stage water-cooled units are available for capacities ranging from 115 to approximately 3,000 CFM and pressures to 125 psig.

Special three stage units may be supplied for pressures to 250 psig.

All jacket water cooled designs are force feed lubricated.

Smaller oil flooded vane type compressors are availble in capacities ranging from 60 to 600 CFM with discharge pressures in the range of 80 to 150 psig. The oil flooded design reduces the temperatures of the discharge stream to under 200°F.

Straight lobe type compressors are available in capacities ranging from 5 to 30,000 CFM and pressures to 12 psig in single-stage operation. In some services, two-stage units may provide up to 20 psig. (See Figure 6-24).

These compressors are more properly termed positive displacement blowers because they move quantities of gas from one level of pressure to another without an internal volume reduction mechanism.

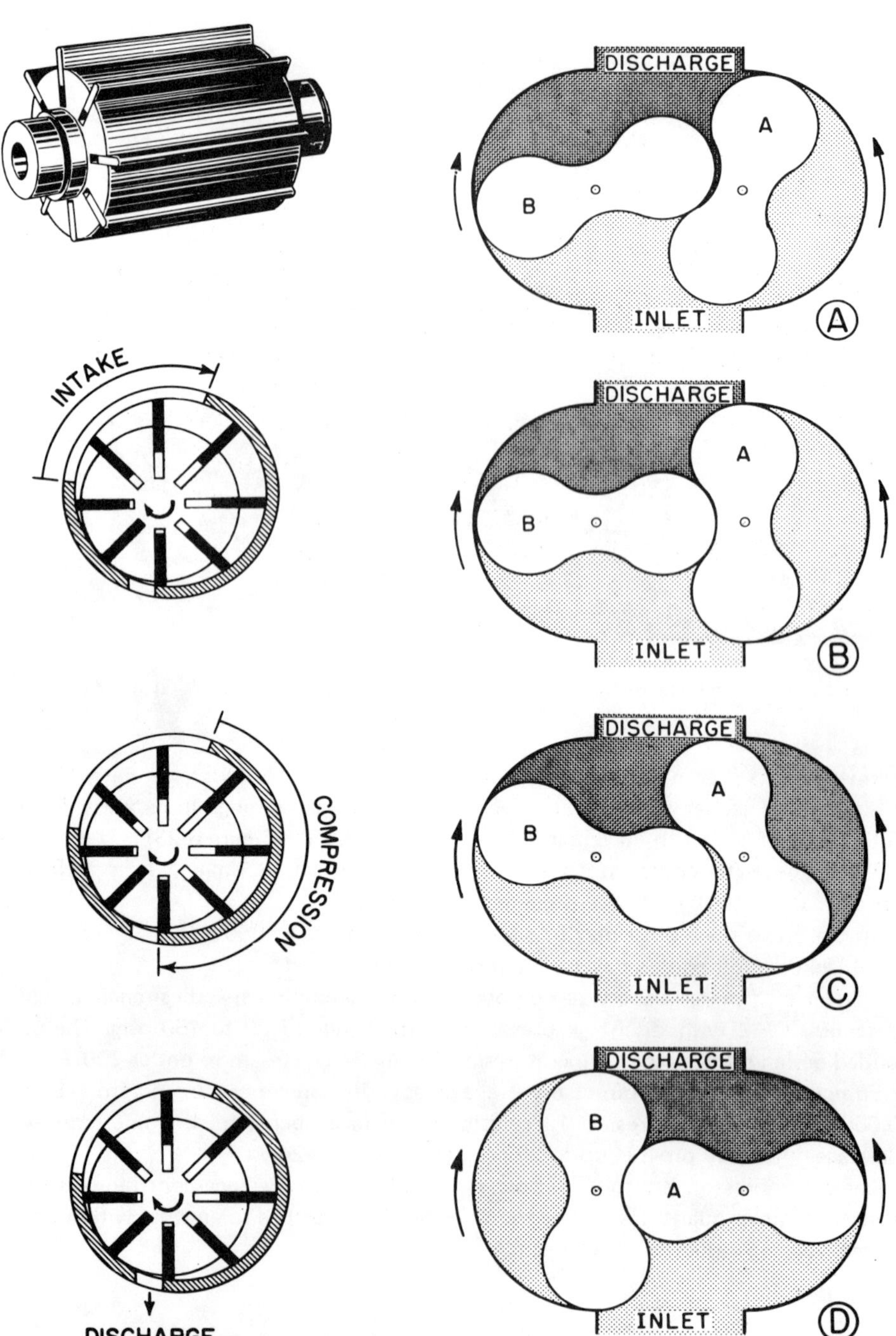

Figure 6-23. Operational principles of the sliding vane compressor. (Courtesy Ingersol Rand Corp.).

Figure 6-24. In the rotary lobe compressor gas is moved from inlet to outlet without a significant volume change.

NOMENCLATURE

ACFM Actual, or measured cubic feet per minute, $ft^3 m$
BHP Brake horsepower, hp
C Cylinder clearance, in^3/in^3
CA Cylinder cross-sectional area, in^2
CE Compression efficiency, dimensionless
C_p Heat capacity at constant pressure, Btu/lb-mole
C_y Heat capacity at constant volume, Btu/lb-mole °k
GHP Gas horsepower, hp
H Enthalpy, Btu/lb
H_{ad} Adiabatic head, ft-lbs/lb
H_{pt} Polytropic head, ft-lbs/lb
hp theoretical power, horsepower
k ratio of heat capacities, c_p/c_v, dimensionless
ME Mechanical efficiency, dimensionless
MW Molecular weight, dimensionless
N Number of moles, lb-mole
n Polytropic exponent, dimensionless
ns Number of stages, dimensionless
P Absolute pressure, psia, lbs/in^2
PA Piston area, in^2
P_c Critical pressure, lbs/in^2
PP Piston displacement, ft^3/min
PN Discharge pressure, lbs/in^2
P_r Reduced pressure, P/P_c, lbs/in^2
P_i Inlet pressure, lbs/in^2
Q Cubic feet of gas per minute, ft^3/min
R Ideal gas constant, 1595 ft lb/lb mole °R or 1.986 Btu/lb mole °R
r_c Ratio of compression, $p_{final}/p_{initial}$, dimensionless
SCFM Standard cubic feet per minute at 14.7 psia and 60°F, ft^3/min
T Temperature, °R
T_c Critical temperature, °R
T_r Reduced temperature, °R
V Volume, ft^3
VE Volumetric efficiency, dimensionless
w Mass rate, lb/min
Y_A Mole fraction of component A in a gas mixture, dimensionless
Z Compressibility factor, dimensionless
η_{ad} Adiabatic efficiency, dimensionless
η_{pt} Polytropic efficiency, dimensionless

REFERENCES

1. Chemical Engineer's Handbook, Fifth Ed., R. H. Perry and C. H. Chilton, editors, McGraw-Hill, 1973, pp. 3–105, 109.
2. Compressed Air and Gas Data, C. W. Gibbs, editor, Ingersoll Rand Co., 1969, p. 34–8.
3. Introduction to Chemical Engineering Thermodynamics, J. M. Smith and H. C. Van Ness, McGraw-Hill, 1959, pp. 95 to 98.
4. Chemical Engineer's Handbook, Fifth Ed., R. H. Perry and C. H. Chilton, editors, McGraw-Hill, 1973, pp. 3-232 and 3-233.
5. Engineering Data Book, Gas Processors Suppliers Association, Ninth ed., 1977, pp. 16-11, and 16-14.
6. M. Benedict, G. B. Webb, and L. C. Rubin, J. Chem. Phys., 8, 334 (1940).
7. O. Redlich and J. N. S. Kwong, Chem. Rev., 44, 233 (1949).
8. Compressed Air and Gas Handbook, 4th ed., J. P. Rollins, ed., Compressed Air and Gas Institute, 1973, p. 3–77.
9. Compresed Air and Gas Data, C. W. Gibbs, ed., Ingersoll Rand Co., 1969, p. 34–117.
10. Compressed Air and Gas Data, C. W. Gibbs, ed., Ingersoll Rand Co., 1969, p. 4–30.
11. American Petroleum Institute Standard 618, Second Ed., July 1974, Sec. 2.7.
12. Applicable standards may be found in Section VIII of the ASME Boiler and Pressure Vessel Code, in the American Petroleum Institute Standard 618, and in ANSI 31.8.
13. Gas Processors Suppliers Assn. Engineering Data Book, Ninth Ed., Third rev. 1977, pp. 4–17.

PROCESS EQUIPMENT SERIES

Volume 3 Index

D

E

T

U